Design, Measurement and Management of Large-Scale IP Networks

Bridging the Gap between Theory and Practice

Designing efficient IP networks and maintaining them effectively poses a range of challenges, but in this highly competitive industry it is crucial that these are overcome. Weaving together theory and practice, this text sets out the design and management principles of large-scale IP networks and the need for these tasks to be underpinned by actual measurements. Discussions of the types of measurements available in IP networks are included, along with the ways in which they can assist both in the design phase as well as in the monitoring and management of IP applications. Other topics covered include IP network design, traffic engineering, network and service management and security. A valuable resource for graduate students and researchers in electrical and computer engineering and computer science, this is also an excellent reference for network designers and operators in the communication industry.

Antonio Nucci is the Chief Technology Officer at Narus Inc., and he obtained his Ph.D. in Telecommunication from Politecnico di Torino, Italy, in 2003. He was awarded the prestigious Infoworld CTO Top 25 in 2007 for his vision and leadership within Narus and the IT community. His research interests include network design, measurement and management, traffic analysis, security and surveillance.

Konstantina Papagiannaki is Senior Research Scientist at Intel Research, Pittsburgh, Pennsylvania, and received her Ph.D. in Computer Science from University College London, UK, in 2003. Her dissertation work on "Provisioning IP backbone networks based on measurements" was awarded the Distinguished Dissertations Award of the British Computer Society. Her research interests include network design and planning for wired and wireless networks, network measurement and security.

Design, Measurement and Management of Large-Scale IP Networks

Bridging the Gap between Theory and Practice

ANTONIO NUCCI

Narus Inc.

KONSTANTINA PAPAGIANNAKI

Intel Research, Pittsburgh, Pennsylvania

CAMBRIDGE
UNIVERSITY PRESS

Shaftesbury Road, Cambridge CB2 8EA, United Kingdom

One Liberty Plaza, 20th Floor, New York, NY 10006, USA

477 Williamstown Road, Port Melbourne, VIC 3207, Australia

314–321, 3rd Floor, Plot 3, Splendor Forum, Jasola District Centre, New Delhi – 110025, India

103 Penang Road, #05–06/07, Visioncrest Commercial, Singapore 238467

Cambridge University Press is part of Cambridge University Press & Assessment,
a department of the University of Cambridge.

We share the University's mission to contribute to society through the pursuit of
education, learning and research at the highest international levels of excellence.

www.cambridge.org
Information on this title: www.cambridge.org/9780521880695

© A. Nucci and K. Papagiannaki 2009

First published 2009

A catalogue record for this publication is available from the British Library

ISBN   978-0-521-88069-5   Hardback

For Konstantina: To my wonderful husband, Richard. For his love and unconditional support, and for always making me try a little harder. To my little Daphne, who came into my world to provide me with a new perspective to life.

For Antonio: To my beautiful wife and best friend, Maria Rosa. For always be there for me, filling up my life with true love and pure joy and striving me toward making my dreams a reality. I love you. To little son, Mattia, who has recently arrived new to our world, now the only world we shall ever dream about. To my parents, Assunta and Vittorio, for all the shelter, care, love and support they have given me over all these years.

Cover graphic

In theory, there is no difference between theory and practice; In practice, there is.
                                              – *"Yogi" Berra, American baseball player and manager*

The figure that appears in the center of the cover describes, in a very cohesive way, properties of an empirical process whose internals are completely unknown but can be studied through the analysis of its behavior under varying probing schemes. The "strip" between the two orange lines represents an admissible region, i.e. support of the process, where samples of the process were observed while changing its inputs. Thus, the strip defines the space where the unknown process resides. The energy of the process is concentrated on horizontal lines departing from the y-axis and moving toward the center of the strip. The black area is associated to high energy levels of the process, while the bright yellow area describes low energy levels of the process. The superimposed contour, represented by a set of dashed lines, annotated with different numbers, gives an idea of the probabilities corresponding to the "pixels" of this shading, which are drawn at full screen resolution. Knowing where the density becomes negligible, and under which circumstances, identifies specific characteristics of the unknown process and therefore helps to achieve a better definition and calibration of estimators.

Contents

Acknowledgments

Over the past ten years we have had the fortune and honor of working with so many talented people who have shaped us as scientists and helped us find our position in the applied research community. While it is our names that appear on the cover of this book, claiming the associated responsibility and credit, this book would not have been possible without the discussions and inspiration from a number of colleagues and friends. Although a complete list would be impossible to enumerate, we would like to extend our warmest thanks to our ex-collagues at Sprint, Dr. Christophe Diot, Dr. Nina Taft, Dr. Supratik Bhattacharryya, Dr. Gianluca Iannaccone, Dr. Sue Moon, Richard Gass and Ed Kress. Without Dr. Diot's enthusiasm and extreme energy, we would never have had the opportunity to study Sprint's IP network using fine granularity measurements, which are typically impossible to collect, and consequently make this book stand out. It is through our collaboration with Dr. Taft, Dr. Bhattacharryya, Dr. Iannaccone and Dr. Moon, that we managed to address the challenging problems presented throughout the book. Finally, Mr. Gass and Mr. Kress made the instrumentation and management of a vast amount of data possible. Although not part of the Sprint IP group, we would also like to sincerely thank Prof. Lixin Gao for the close feedback she provided during the preparation of the book, reviewing the entire structure and content; we are indebted to her for this time commitment and her comments that certainly improved the quality of this work.

Second, we would like to thank Sprint Corporation for the unfettered access to their network, data and operational personnel. Without their support the work presented in this book would not have been possible. This book comes four years after we decided to pursue a different career path. It is because of the support of our current employers that we managed to complete this project. We sincerely thank Narus Inc. and Intel Corporation for allowing us to spend the time needed to bring this project to completion and for exposing us to a wide range of problems that broadened our horizon with respect to the fundamentals covered in this book.

We also thank Cambridge University Press and the various reviewers that helped in the positioning, polishing and publishing of this book. It is through their encouragement that we managed to overcome any obstacles and succeed in completing this project.

Last, but foremost, we owe our families a huge debt of gratitude for being so patient, while we put their world in a whirl by injecting this lengthy writing activity into their already full lives. Antonio would like to extend a special acknowledgment to his wife Maria Rosa for her encouragement and support and for putting up with his late nights and long weekends. He would also like to thank his parents, Assunta and Vittorio, and his sister, Milena, who kept him excited and constantly charged for the entire duration of this project. Konstantina, better known as Dina, would like to extend a special acknowledgment to her husband, Richard Gass, for his constant support and for always being there to discuss anything and everything. It is because of him that she always tries to exceed herself, knowing that he will stand by her every decision.

1   Introduction

The Internet has become an integral part of our society. It not only supports financial transactions and access to knowledge, but also cultivates relationships and allows people to participate in communities that were hard to find before its existence. From reading the news, to shopping, to making phone calls and watching videos, the Internet has certainly surpassed the expectations of its creators.

This vital organ in today's society comprises a large number of networks, administered by different authorities and spanning the entire globe. In this book, we will take a deep look into what constitutes those entities and how they form the skeleton of the Internet. Going beyond the physical infrastructure, the wires and boxes that make up the building blocks of today's Internet, we will study the procedures that allow a network to make optimal use of those building blocks.

As in traditional telecommunication networks, the Internet constituents rely on a team of network designers and managers, for their design, operation and maintenance. While such functions are well studied within the traditional telecommunication theory literature, the Internet imposes constraints that necessitate a whole new array of methods and techniques for the efficient operation of a network. Departing from the traditional circuit-switched paradigm, every unit of information flows independently across the network, from the source to the destination, and thus is much harder to account for.

Since the late 1990s, the research community has been devising techniques that allow the operators of this new kind of network to allocate and manage their networks' resources optimally. However, most such solutions typically remain on paper and hardly make a transition to actual deployment. There are a few reasons behind such a state of affairs. First, there is a clear disconnect between the research and operations community; second, some of the proposed techniques fail to take into account the degree of complexity that is present in today's Internet. Any proposed solution needs to scale to thousands of network elements and millions, or even billions, of simultaneous "conversations" across the network. Solutions need to deal with an unprecedented scale.

In this book, we discuss the fundamental processes in the design, management and operation of a network that forms part of the Internet. We will outline ways in which one can capture information on the actual scale of such networks and attempt to bridge the gap between theory and practice. Using measurements collected from operational networks, we demonstrate how previous theoretical techniques can be used to address topics of interest to the network operators. We clearly describe the constraints that need to be taken into account and the ways traditional methods need to be adjusted to cope with the scale of the Internet.

Our discussion will revolve around the design of large-scale IP networks, focusing in particular on the worldwide Sprint IP backbone network. We have both spent four years researching the properties of that network, while employed at the Sprint Advanced Technology Labs, in Burlingame, CA. Throughout those four years, we have engaged in multiple discussions with network operators to identify critical operations and provide the tools to assist in the design and management of one of the largest IP networks in today's Internet. What makes the solutions unique is the fact that they are based on actual measurements collected from the operational Sprint IP backbone network, measurements that were obtained after an extensive effort to instrument the appropriate network elements and collect a variety of data that would allow the complete understanding of the phenomena guiding the behaviors observed. Note that all discussions in this book deal with plain IP networks, which do not employ solutions such as Multi-Label Protocol Switching (MLPS) that may be used by other providers offering an equivalent of a circuit within an IP context.

The use of operational measurements in the design of the different tools has two important benefits. First, one can ensure that the tool can cope with the amount of input information provided, i.e. that it scales well with the size of the network. Second, one can quantify the magnitude of performance gain that would be observed under realistic conditions. Given the fact that some of the tools described in the book have been adopted by the Sprint network operations, such benefits are indeed appreciated within an operational context.

1.1   Skeleton of the book

The rest of the book is organized as follows. First we provide a background section that lays the foundation of the terms and concepts that will be used in the rest of the book. We describe the Internet architecture, and the elements and protocols guiding its behavior. We then outline issues associated with the design, management, optimization and security of such a complex infrastructure, topics that will be the focal points in the following chapters.

The rest of the book is then organized in three thematic entities. Part 1 describes the state of the art in the measurement of IP backbone networks, and ways in which one can manipulate such measurements to define new metrics that characterize behaviors that are elusive to capture with today's technology. In particular, after discussing the tools used for the monitoring of network elements and protocols in operation in today's IP networks, we also present techniques to measure quantities, such as the instantaneous queue size in routers (in a scalable manner) and the entity of an IP traffic matrix, which forms a fundamental input in most network design and planning tasks.

Part 2 builds on Part 1 to present techniques making use of the collected information in order to build and provision a highly performing IP network. Part 2 outlines the entire IP network design process and its associated constraints. In particular, it separates the stage of network design from its optimization, given that the former should be a one-time task, while the latter is a continuous one. Throughout Part 2 the reader will understand the basic constraints driving the design and optimization of today's IP networks and ways to perform those tasks in a way that is theoretically sound but also based on realistic assumptions, as obtained through operational IP network measurements. Part 2 addresses the mapping of logical IP links on top of a fiber optic physical infrastructure, appropriate design techniques that allow robustness to physical layer failures, and the optimization of the IP routing layer in order to offer applications the illusion of error-free communication links. Part 2 concludes with a technique that allows a network operator to evolve the highly performing IP network by applying appropriate capacity upgrades to those parts of the infrastructure that experience growth rates, which would lead to excessively high utilization and could compromise quality of service.

Upon the design and optimization of an IP network, a network operator needs to deal with a number of tasks that ensure optimal service delivery to the end users. Such tasks typically involve the management of the network itself to avoid cases of overload, while offering a diversity of services in a secure fashion. This is the focus of Part 3. Given the connectionless nature of IP traffic, we present techniques that enable network operators to identify the types of traffic utilizing a network's resources. Such a task is very hard in today's networks. Nonetheless, the identification of applications such as VoIP, multimedia, and peer-to-peer, are becoming requirements for optimal provisioning of an IP network. Part 3 describes two different techniques that can achieve such a goal. Once we manage to gain visibility into the nature of the traffic flowing on top of an IP network, we gain the additional advantage of visibility into types of traffic that may signal the onset of security threats that could compromise the quality of services across the network. This is the focus of the latter part of Part 3. We should emphasize that the area of Internet security is rather vast, and therefore our coverage of security primarily deals with the detection of oncoming threats, at the control and data layers, rather than countermeasures.

At this point, we should note that, while Internet Service Provider (ISP) networks will continue to evolve in the way they are designed, operated and managed, we believe that the concepts, models, processes and techniques presented in this book will still be very helpful to operators. The same holds true for the classification of emerging services that ISPs will deliver to their customers in the years to come, and the detection of hidden threats that might use them as vehicles to deliver the malicious code to their targets. The general concepts that constitute the foundation of the techniques presented are still applicable.

1.2 How to read this book

Throughout the book we assume that the reader has some basic knowledge of networking, Transfer Central Protocol/Internet Protocol (TCP/IP), and is relatively well versed with statistics and optimization theory. While we have made every attempt to include the basic knowledge required for the reader to follow the exposition of the different techniques, given the limited space of this book we have also used a number of references to textbooks, whose primary focus is the coverage of their fundamentals.

We would greatly encourage any readers who wish to learn more about the underlying theory to read these textbooks.

Finally, we note that, even though each individual chapter could stand on its own, and in fact some of the chapters are adapted from previous publications, Chapter 3, Chapter 6 and Chapter 12 lay the foundation of terms and concepts that will be explored in the remainder of the respective theme. Consequently, we would encourage the reader to go through these chapters before diving into the detailed exposition of particular problems in later chapters. Lastly, Chapter 2 serves as an introductory chapter to network measurements, and as such could provide a basic overview of the types of input data required for different tasks, as well as appropriate techniques for their collection.

2   Background and context

Before embarking on the exploration of techniques to assist operators in the management and design of IP networks, this chapter lays the foundation of the terms and concepts that will be used in the rest of the book. We describe the Internet architecture and the elements and protocols guiding its behavior. We then outline issues associated with the design, management, optimization and security of such a complex infrastructure, topics that will be the focal points in the following chapters.

2.1   What is the Internet?

The Internet is a diverse collection of independent networks, interlinked to provide its users with the appearance of a single, uniform network. Two factors shield the user from the complex realities that lie behind the illusion of seamlessness: (i) the use of a standard set of protocols to communicate across networks and (ii) the efforts of the companies and organizations that operate the Internet's different networks to keep its elements interconnected.

The networks that comprise the Internet share a common architecture (how the components of the networks inter-relate) and software protocols (standards governing the exchange of data), which enable communication within and among the constituent networks. The nature of these two abstract elements – architecture and protocols – is driven by the set of fundamental design principles adopted by the early builders of the Internet. It is important to distinguish between the public Internet and the Internet's core technology (standard protocols and routers), which are frequently called "IP technology." The public Internet is distinguished by global addressability (any device connected to the Internet can be given an address and each address will be unique) and routing (any device can communicate with any other). IP technologies are also employed in private networks that have full, limited, or even no connectivity to the public Internet; the distinction between public and private blurs when private networks acquire connections to the Internet, and thus by definition become part of it.

If one looks at the elements that physically make up the Internet, one sees two categories of objects: (i) communication links, which carry data from one point to another, and (ii) routers, which direct the communication flow between links and, ultimately, from senders to receivers. Communication links may use different kinds of media, from telephone lines to cables, originally deployed or used in cable television systems, to satellite and other wireless circuits. Internal to networks, especially larger networks in more developed parts of the world, are links that carry relatively large amounts of traffic, typically via optical fiber cables. The largest of these links are commonly said to make up the Internet's "backbone." Links closer to users, especially homes and small businesses, typically feature considerably less capacity. Large organizations, on the other hand, tend to have high-capacity links. Over time, the effective capacity of links within the network has been growing. Links to homes and small businesses – the so called "last mile" – have, until recently, with the emergence of cable modem, Digital Subscriber Line (DSL) and other technologies, been largely constrained to the relatively low speeds obtainable using analog modems running over conventional phone lines.

Routers are computer devices located throughout the Internet that transfer information across the Internet from a source to a destination. Routing software performs several functions. It determines the best routing paths, based on a set of criteria that quantitatively capture the quality of an Internet path, and directs the flow of groups of packets (data) through the network. Path determination at each step along the way depends on information that each router has about paths from its location to neighboring routers as well as to the destination; routers communicate with one another to exchange relevant path information. A number of routing algorithms that determine how routers forward packets through the network are in use; routing protocols mediate the exchange of path information needed to carry out these algorithms.

The Internet can be divided into a *core*, made up of the communications links and routers operated by ISPs, and an *edge*, made up of the networks and equipment operated by Internet users. The line between core and edge is not a sharp one. Users who connect via dial-up modems attached to their computers clearly sit at the very edge. In most business settings, as well as in an increasing number of homes, Local Area Networks (LANs) sit between the ISP and the devices that the Internet connects. These LANs, and the routers, switches and firewalls contained within them, sit near the edge, generally beyond the control of the ISP, but not at the very edge of the network.

Software applications running on these computer devices – today typically PCs – use Internet protocols to establish and manage information flow in support of applications carried over the Internet. Much as a common set of standard protocols lies at the core of the Internet, common standards and a common body of software are features of many applications at the edge, the most common being those that make up the World Wide Web (WWW or Web). More specialized software, which also makes use of the Internet's basic protocols and is frequently closely linked to Web software, supports applications such as real-time audio or video streaming, voice telephony, text messaging, on-line gaming and a whole host of other applications. In light of the prominence of the Web today, Web-based applications and the content and services provided by them are sometimes viewed as synonymous with the Internet; the Internet, however, is a more general-purpose network over which the Web is layered.

Following usage from the telecommunications industry, the essential physical components – communications links and routers – of the network can be referred as "facilities." Internet Service Providers use these facilities to provide connectivity using the Internet protocols. What is done with the facilities and basic connectivity comes under the heading of "services." These services, which include access to content (e.g. viewing Web sites, downloading documents, listening to audio and watching video), electronic commerce (e.g. shopping, banking and bill paying), or video-telephony, are enabled by both devices and software in the hands of users and service providers. Some services are enabled merely by installing software on user computers, while others rely on functionality implemented in computers and software attached to the Internet by a third party. In either case, the general-purpose nature of the Internet has meant that there does not need to be any arrangement between the ISP and the provider of a particular service. While this statement generally holds true today, we are seeing the emergence of exceptions to it in the form of application-specific delivery networks (e.g. Akamai) that employ devices located throughout the network, generally near the edge.

A multitude of businesses are based on selling various combinations of these elements. For instance, many ISPs integrate connectivity with content or services for their customers. Some ISPs rely in part or in total on access facilities owned and operated by other providers, while others operate mostly on their own communication links. Also, ISPs may opt to own and operate their own communications links, such as fiber-optic cables, and networks owned and operated by other communications companies (just as companies have resold conventional voice telephony services for years).

The tale is well told about how, since the last decade of the twentieth century, the Internet evolved from a novel, but still developing, technology into a central force in society and commerce. Suffice to say that the transformation resulting from the Internet, along with the expectations for continued growth in its size, scope and influence, have given rise to widespread interest and concern on the part of government and society. The remainder of this chapter presents the Internet architecture, e.g. interconnection among ISPs, describes a specific ISP Tier-1 network architecture, and introduces a high-level description of protocols, services and issues associated with the design, management, optimization and security of such a complex infrastructure.

2.2 Maintaining end-to-end service through multiple ISPs: physical and logical interconnection agreements and the associated financial value

The Internet is designed to permit any end user ready access to any and all other connected devices and users. In the Internet, this design translates into a minimum requirement that there be a public address space to label all of the devices attached to all of the constituent networks and that data packets originating at devices located at each point throughout the network can be transmitted to a device located at any other point, e.g. "connectivity." Internet users expect that their ISPs will make the necessary arrangements that will enable them to access any desired user or service. At the same time, those providing services or content over the Internet expect that their ISP(s) will similarly allow any customer to reach them and in turn allow them to reach any potential customer. To support these customer expectations, an ISP must have access to the rest of the Internet. Because these independent networks are organized and administered separately, they have to enter interconnection agreements with one or more other ISPs. The number and type of arrangements are determined by many factors, including the scope and scale of the provider and the value it places on access for its customers. Without suitable interconnection, an ISP cannot claim to be such a provider – being part of the Internet is understood to mean having access to the full, global Internet.

Connections among ISPs are driven primarily by economics. Who may have access to whom, with what quality of access and at what price is something that is negotiated in private and translates to specific interconnection and routing policies implemented at the network layer. These business agreements, also called "Service Level Agreements" (SLAs), cover the technical form of interconnection, the means and methods for compensation based on the service provided, the grades and levels of service quality to be provided, and the processing and support of higher-level protocols.

A straightforward and useful way to categorize ISPs is in terms of the interconnection arrangements they have in place with other service providers. The backbone service providers, which include commercial companies as well as government-sponsored networks, use trunk capacities that are measured in gigabits, or billions of bits, per second (abbreviated to Gbps). Roughly one dozen ISP companies provide the backbone services that carry the majority of Internet traffic. These providers, termed "Tier-1," are defined as those providers that have full peering with at least the other Tier-1 backbone providers. Tier-1 backbone providers, by definition, must keep track of global routing information that allows them to route data to all possible destinations on the Internet. They must also ensure that their own routing information is distributed such that data from anywhere else in the Internet will be properly routed back to their network. Tier-1 status is a coveted position for any ISP, primarily because there are so few of them and because they enjoy low-cost interconnection agreements with other networks. They do not pay for exchanging traffic with other Tier-1 providers; the peering relationship is accompanied by an expectation that traffic flows – and any cost associated with accepting the other provider's traffic between Tier-1 networks – are balanced in terms of volume. Tier-1 status also means, by definition, that an ISP does not have to pay for transit service.

Below Tier-1 sit a number of so-called Tier-2 and Tier-3 ISPs, which connect corporate and individual clients to the Internet backbone and offer them varying types of service according to the needs of the target market place. This group spans a wide range of sizes and types of providers, including both a small set of very large providers aimed at individual/household customers (e.g. America On Line) and a large number of small providers. The latter include providers of national or regional scale, as well as many small providers offering service in limited geographical areas. The logical architecture established among all these different types of ISP is also known as the "Internet hierarchy" (see Figure 2.1).

Interconnection agreements between ISPs mainly cover three aspects: the physical means of interconnection, the agreed patterns of traffic that can be exchanged by the providers (transit and peer) and the financial arrangements that underlie and support the physical means and different traffic patterns.

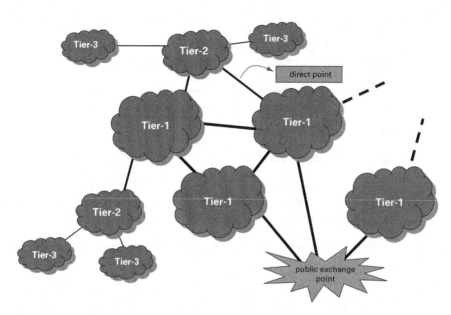

Fig. 2.1. The Internet hierarchy.

The introduction of *public exchange points* is a way of making the interconnection between a number of providers more cost-effective. If *n* providers were individually to establish pairwise interconnections, they would require $n(n-1)/2$ direct circuits. A public exchange point, where all *n* providers can connect at a common location, permits this to be done much more inexpensively, with *n* circuits and a single exchange point. A provider interconnects to an exchange point, either physically – by installing their own equipment and circuit into a specific location – or logically – by using a leased network connection to an interconnect provider through an ATM or Ethernet network. These interconnected networks are usually operated by large access providers, who hope to derive considerable revenue by selling access lines to ISPs wishing to attach to each other through the access provider's facilities. Examples of public exchange points are the London Internet Exchange (LINX) and MAE-WEST facilities.

Another option is to use a *private direct point*, e.g. a point-to-point connection. One motivation for point-to-point connections is to bypass the bottleneck posed by a public exchange point when traffic volumes are large. Between large providers, connections are usually based on high-performance private interconnections, for example point-to-point links at high speeds (DS3 or higher, please see Table 2.1). Direct connection can also allow for better management of traffic flow. The very large volume of traffic that would be associated with a major public access point can be disaggregated into smaller, more easily implemented, connections (e.g. a provider manages ten OC-3 connections to ten different peers in different locations rather than a single OC-48 connection to a single exchange point that then connects to multiple providers). Another reason for entering into private connections is the desire to provide support for the particular SLAs and quality-of-service provisions that the two networks agree to in their peering or transit agreements.

Table 2.1. Speed connection table

DS0 = one 64 Kbps unit of transmission bandwidth. A worldwide standard for digitizing one voice conversation and more recently for data transmission. Twenty-four DS0s = one DS1 (64 Kbps × 24) or a T1

PIPE	Traffic rate	Number of DS0s	Equivalent	European
DS0	64 Kbps	1	DS0	64 Kbps
DS1	1.544 Mbps	24	24 DS0	2.048 Mbps (E1)
DS3	44.736 Mbps	672	28 DS1	34.368 Mbps
OC-3	155.52 Mbps	2,016	3 DS3	STM-1
OC-12	622.08 Mbps	8,064	4 OC-3	STM-4
OC-48	2.488 Gbps	32,256	4 OC-12	STM-16
OC-192	9.953 Gbps	129,024	4 OC-48	STM-64
OC-768	39.813 Gbps	516,096	4 OC-192	STM-256

When two or more ISPs establish an interconnection, they exchange route advertisements to specify what traffic is to be exchanged between them. Route advertisements describe the destination Internet addresses for which each provider chooses to accept packets from the other. These advertised routes are loaded into each other's routing tables and are used to determine where (including which providers) packets should be routed based on their destination address. There are two common options for how providers accept each other's traffic: *transit* and *peer*. In the transit model, the transit provider agrees to accept and deliver all traffic destined for any part of the Internet from another provider that is the transit customer. It is possible that two providers in a transit agreement will exchange explicit routing information, but more typically the transit provider provides the transit customer with a default route to the transit network, while the transit customer provides the transit provider with an explicit set of routes to the customer's network. The transit customer then simply delivers to the transit provider all packets destined to IP addresses outside its own network. The transit provider will then distribute routing information from the transit customer to other backbone and network providers and will guarantee that full connectivity is provided. Usually, the transit provider tries to deliver the transit customer packets as fast as possible through its routing infrastructure, also known as "hot-potato routing."

Today, the preferred way for large providers to interconnect is through *peering agreements*. In contrast to transit arrangements, where one provider agrees to accept from the other traffic destined to any part of the Internet, in a peering relationship, each provider only accepts traffic destined to that part of the Internet covered by the agreement. Peers exchange explicit routing information about all of their own addresses along with all of the addresses of their customers. Based on that routing information, each peer only receives traffic destined to itself and its transit clients. This exchange of routing information takes the form of automated exchanges among routers. Because the propagation of incorrect routing information can adversely affect network operations, each provider needs to validate the routing information that is exchanged.

The issue of compensation for interconnection is a complex one [98]. The essence of interconnection is in the explicit agreement to forward packets towards their final destination, according to the routing information exchanged. Compensation reflects the costs associated with provisioning and operating sufficient network capacity between and within ISP networks for the relaying of customer traffic at the appropriate quality of service level (without excessive packet loss or delay). As a basic unit of inter-connection, packets are somewhat akin to call minutes in voice telecommunications. However, architectural differences between the Internet and the Public Switched Tele-phone Network (PSTN) make accounting in terms of packets much more complicated than call-minute-based accounting. Even if an infrastructure were to be put in place to count and charge on a packet-by-packet basis, the characteristics of packet routing would make it difficult to know the cost associated with transmitting a given packet, since the amount of resources consumed will depend on the time of day and the actual network conditions. As a result, interconnection schemes that are used in other contexts, such as the bilateral settlements employed in international telephony, are not used in the Internet, and interconnection has generally been established on the basis of more aggre-gate information about the traffic exchanged between providers. Some of these issues have to do with the cost of the physical link that allows for the interconnection, traffic imbalances (e.g. one provider originates more traffic than it terminates), relative size (one provider offers greater access to users, services and locations than the other) and the value attached to particular content exchanged (e.g. the value attached to either the transit or peer relationships is no longer based only on the number of bits exchanged, nor is it based solely on the origin, destination or distance – it also reflects the type of application and service being provided). New types of applications and services will continue to emerge on a regular basis. Related to this is the variability in the value attached to Internet data packets.

2.3   Typical Tier-1 ISP network architecture: the Sprint IP backbone

In order to formalize all concepts mentioned so far, this section presents the Sprint backbone as a typical example of a network architecture for a large Tier-1 ISP.

The topology of a Tier-1 ISP backbone typically consists of a set of nodes known as Points-of-Presence (PoPs) connected through multiple high-capacity links. The Sprint IP backbone consists of approximately 40 PoPs worldwide (as of 2004). Out of those 40 PoPs, approximately 17 are located in the USA (excluding Pearl Harbor). Figure 2.2 presents the Sprint IP backbone topology for the continental USA.[1] Each neighboring PoP pair has a minimum of two parallel logical links, a maximum of six and an average of three, for a total of 35 bidirectional links.

Each Sprint PoP is a collection of routers following a two-level hierarchy, featuring (i) *access routers* and (ii) *backbone routers* (Figure 2.3). Such an architecture is typical

---

[1] Visual representations of a large number of commercial, as well as research, networks can be found at http://www.caida.org/tools/visualization/mapnet/Backbones

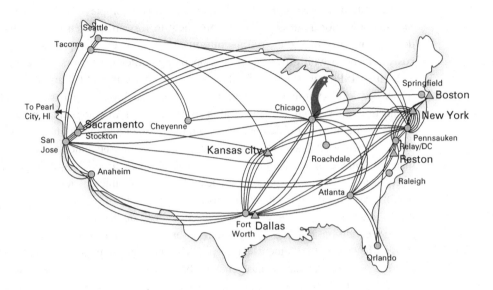

Fig. 2.2.   North American Sprint IP backbone network topology (third quarter, 2001): 17 PoPs with 35 bidirectional inter-PoP links.

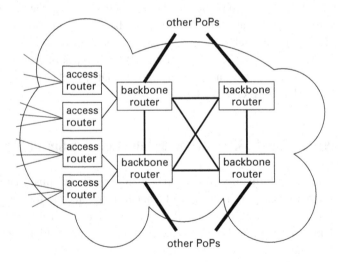

Fig. 2.3.   Typical configuration for a PoP.

among large Tier-1 ISP providers [60]. Access routers are normally lower-end routers with high port density, where customers become attached to the network. These routers aggregate the customer traffic and forward it toward the PoP's backbone routers. Note that typically the sum of incoming client traffic rarely matches the provisioned capacity to the core of the network. This practice is termed as *oversubscription*, employed by ISPs to recover some of the infrastructural costs. The backbone routers receive the aggregate customer traffic and forward it to other PoPs or the appropriate access routers inside the same PoP (in case the destination networks can be reached through the same PoP). Public and private peering points, where one ISP exchanges traffic with other ISPs, are usually accommodated by selected backbone routers inside a PoP.

The capacity of links interconnecting routers inside the PoP depends on their level in the hierarchy, i.e. the level of traffic aggregation they correspond to. For instance, customer links are usually 45 Mbps (T3 or DS3) or greater. The links connecting access routers to backbone routers are an order of magnitude larger, reaching 622 Mbps (OC-12). The backbone routers inside a PoP are densely meshed (not necessarily fully meshed) and interconnected through higher-speed links, i.e. OC-12 to OC-48 (2.5 Gpbs). The inter-PoP links are long-haul optical fibers with bandwidth of 2.5 Gbps (OC-48), 10 Gbps (OC-192) or 40 Gbps (OC-768).

IP backbones are engineered for high availability, and resilience to multiple link and router failures, while meeting the contracted service level agreements. Traffic is guaranteed to transit the network experiencing bounded edge-to-edge delays. In addition, even in cases when certain IP paths through the network become unavailable, traffic is guaranteed to be delivered to its destination through alternate paths. Consequently, each PoP is designed to connect to multiple other PoPs through multiple high-capacity links (Figure 2.3). A similar approach may be followed by edge networks, termed in that case as *multi-homing*; a customer leases more than one line towards the same or different providers for increased reliability and potentially load balancing.

Dense connectivity between PoPs guarantees that traffic will go through the network transiting a bounded number of PoPs. In the presence of link failures, there should exist other links available to take on the affected load without introducing additional delay or loss. Given that the inter-PoP links are covering large distances, thus accounting for the largest part in the edge-to-edge delay, re-routing traffic across PoPs should be employed only as a last resort. For that reason, adjacent PoPs are interconnected through multiple links. These links, when feasible, follow at least two physically disjoint paths, thus offering resilience against fiber cuts, which may affect more than one physical link sharing the same fiber path.

Underlying the IP network layer resides the physical infrastructure that is in charge of carrying bits of information from one location to another. The physical network includes both the communication links (copper wire, optical fiber, wireless links and so on) and the communication switches (packet routers, circuit switches and the like). In the case of IP-over-DWDM (IP-over-Dense Wavelength Division Multiplexing), each communication link represents an optical fiber, while each communication switch represents a high-data-processing device, e.g. Optical Cross Connect (OXC) switch. A simplified version of the Sprint optical fiber infrastructure, composed of 51 OXC and 77 WDM fibers that have between 16 and 48 channels, is shown in Figure 2.4.

Historically, a multi-layer stack was used to maintain a division of labor. The Asynchronous Transfer Mode (ATM) layer was used as an access technology, while the IP layer was used as an edge/core technology. In order to maintain a homogeneous edge/core, ISPs converted ATM packets into IP packets at the very access. From the access to the edge to the core, all packets were then IP packets. Below the IP layer

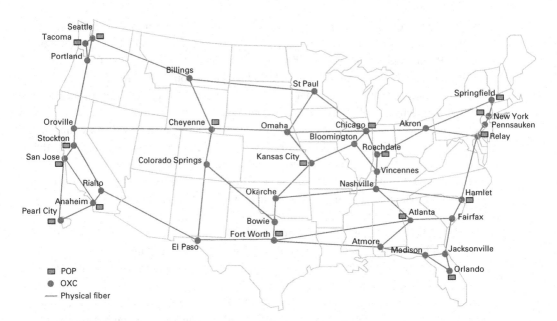

Fig. 2.4.   North American Sprint fiber-optic backbone topology (third quarter, 2001, simplified version): 51 OXC and 77 WDM fibers.

(e.g. looking at the network from the edge to the core and avoiding considering the access), ISPs used a Synchronous Optical NETwork (SONET) layer that sits on top of a Dense Wavelength Division Multiplexing (DWDM) system for increasing bandwidth demand. SONET is a standard for connecting fiber-optic transmission systems and plays the middle layer between the optical and IP layers. It takes care of scheduling packets to be transported by way of Time Division Multiplexing (TDM). It also handles rate multiplexing, traffic grooming, error monitoring and restoration.

More recently, the SONET layer has been removed because ISPs realized that SONET introduced more problems than benefits. Functional overlap is one such problem. Both IP and SONET layers try to perform restoration in the event of a failure, thereby creating more havoc in the system. The SONET interface is advantageous for constant bit rate traffic, but not for highly bursty traffic like Internet traffic. The mere presence of high capacity in the system obviates the need for TDM and traffic grooming. Thus, a multi-layer stack introduces undesired latency. Moreover, SONET provides fast provisioning only in the form of redundancy. All these problems have made it necessary to redesign the system without having too many layers. As a consequence, ISPs have decided to remove the SONET layer and transmit IP packets reliably over a fiber-optic network adopting an IP-over-DWDM system. In this architecture, the transport layer is all-optical and is enriched with more functionality, previously found in the higher layers. This creates a vision for an all-optical network where all management is carried out in the photonic layer. The optical network is proposed to provide end-to-end services completely in the optical domain, without having to convert the signal to the electrical domain during transit. Transmitting IP directly over DWDM has become a reality and is able to support bit rates of 10 Gbps (OC-192) and above. As has been clearly seen, it holds the key to the bandwidth glut and opens up the frontier of a terabit Internet.

2.4   Overview of TCP/IP

The existence of an abstract bit-level network service as a separate layer in a multi-layer suite of protocols provides a critical separation between the actual network technology and the higher-level services through which users interact with the Internet.

Imposing a narrow point in the protocol stack removes from the application builder the need to worry about implementation details and the potential for change in the underlying network facilities. In parallel, it removes from the ISP the need to make changes in response to whatever standards are in use at the higher layers. This separation of IP from the higher-layer conventions is one of the tools that ensures an open and transparent network. The Open System Interconnection (OSI) model is one of the best tools available today to describe and catalog the complex series of interactions that occur in the Internet and allows vendor-independent protocols to eliminate monolithic protocol suites.

The OSI model is a layered approach to networking. Some of the layers may not even be used in a given protocol implementation, but the OSI model (see Figure 2.5) is broken up so that any networking function can be represented by one of its seven layers. As mentioned before, the OSI model allows the definition of "lower-level" functions (Layer 1 to Layer 3), which are strictly related to the physical network architecture being deployed, and "upper-level" functions (Layer 4 to Layer 7), more related to the applications being developed and used. In the following we provide a very high-level overview of the major functions provided by each layer. For more details, please refer to ref. [191].

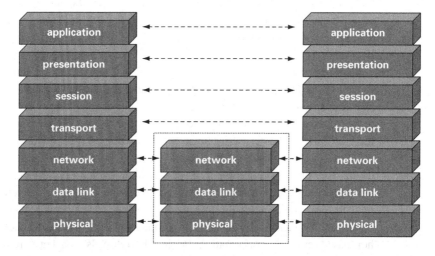

Fig. 2.5.   OSI layered architecture. The *physical*, *data link* and *network* layers are also known as *media layers*, while the remaing top four layers are referred to as *host layers*. Depiction of two hosts communicating through a single link.

- Layers 1 and 2 – the physical and datalink layers. The *physical layer* is responsible for the most substantial of all functions, e.g. the physical interconnection between two network elements. The physical layer defines the electrical, mechanical, procedural and functional specifications for activating, maintaining and deactivating the physical link between end systems. At this level the information is fragmented into small Internet units, named *bits*. The *datalink layer* deals with arbitration (e.g. how to negotiate access to a single data channel when multiple hosts are attempting to use it at the same time), physical addressing (e.g. all devices connected to the Internet must have a physical address, Medium Access Control – MAC, and must be uniquely identified globally), error detection (e.g. determines whether problems were introduced during transmission) and framing (e.g. organization of bits of information into more structured entities, named *frames*, that must be correctly interpreted by machines/technology sitting at both ends of the communication link).
- Layer 3 – the network layer. The *network layer* deals with logical addressing and path determination, e.g. routing. While the methods used for logical addressing vary with the protocol suite used, the basic principles remain the same. Network-layer addresses are used primarily for locating a host geographically. This task is generally performed by splitting the 32-bit address into two parts: the *network id* field and the *host id* field. The network id, also known as network address, identifies a single network segment within a larger TCP/IP internetwork (a network of networks). All the systems that attach and share access to the same network have a common network id within their full IP address. The host id, also known as the host address, identifies a TCP/IP node (a workstation, server, router, or other TCP/IP device) within each network. The host id for each device identifies a single system uniquely within its own network. To make IP addressing easier, IP addresses are expressed in dotted decimal notation. The 32-bit IP address is segmented into 8-bit octets. The octets are converted to decimal (base-10 numbering system) and separated by periods. For instance, the IP address 131.107.16.200 uniquely identifies a specific host with the last two octets of the IP address, e.g. 16.200, within a larger network identified by the first two octets of the IP address, e.g. 131.107.

  Information is then organized in IP packets that represent the bundles of data that form the foundation of the TCP/IP protocol suite. Every packet carries a 32-bit source and destination address, option bits, a header checksum and a payload of data. A typical IP packet is a few hundred bytes long. These packets flow by the billions across the world over Ethernet, serial lines, Fiber-Distributed Data Interface (FDDI) rings, ATM, packet radio connections, etc. There is no notion of *virtual circuit* at the IP level; every packet stands alone. IP is an unreliable *datagram* service. No guarantees are made that packets will be delivered, delivered only once, or delivered in any particular order. A packet traveling a long distance will travel through many IP links, e.g. *hops*. Each hop terminates in a host or a router, which forwards the packet to the next hop based on routing information. During these travels a packet may be *fragmented* into smaller pieces – each of which has its own IP header, but only a portion of the payload – if it is too long for a hop. A router may drop packets if it is too congested. Packets may arrive out of order, or even duplicated, at the far end. There is usually no notice of these actions: higher-layer protocols (i.e. TCP) are supposed to deal with these problems and provide a reliable circuit to the higher application.

Another very much used protocol living in this layer is the *Internet Control Message Protocol* (ICMP). ICMP is the low-level mechanism used to influence the behavior of TCP and UDP connections. It can be used to inform hosts of a better route to a destination, to report trouble with a route, or to terminate a connection because of network problems. It also supports the single most important low-level monitoring tool for system and network administrators: the *ping* program (see Section 3.3.2).

- Layer 4 – the transport layer. The *transport layer* performs a number of functions, the most important of which are error checking, error recovery and flow control. The transport layer is responsible for reliable internetwork data transport services that are transparent to upper-layer programs. There are two major protocols co-living at this layer: the *Transfer Control Protocol* (TCP) and the *User Datagram Protocol* (UDP). The TCP protocol provides reliable *virtual circuits* or *connections* to user processes. Lost or damaged packets are retransmitted; incoming packets are shuffled around, if necessary, to match the original order of transmission. The ordering is maintained by *sequence numbers* in every packet and each packet successfully received is acknowledged (the only exception being the very first packet of the conversation). Every TCP message is marked as being from a particular host and port number and to a destination host and port. The 4-tuple ⟨localhost,localport,remotehost,remoteport⟩ uniquely identifies a particular circuit.

  The UDP protocol extends to application programs the same level of service used by IP. Delivery is on a best-effort basis or *connectionless*, e.g. no virtual circuit is established between two hosts before any user data is sent; there is no error correction, retransmission, or detection of lost, duplicated, or re-ordered packets. Even error detection is optional with UDP. As a result, UDP imposes significantly less overhead compared with TCP. In particular, there is no connection setup. This makes UDP well suited to query/response applications, where the number of messages exchanged is small compared with the connection setup/teardown costs incurred by TCP. Note that UDP may still be used for reliable service provision as long as the missing functionality is assumed by higher-layer protocols or the application itself.

- Layer 5 – the session layer. The *session layer* is responsible for the establishment, management and termination of sessions between applications. Detailed tasks performed at this layer further include the synchronization of communication, the control of dialogs and the "graceful close" of sessions, session checkpointing and recovery, which are typically not used in the Internet protocols suite. A session is a series of related connection-oriented transmissions between two networked hosts. The session layer determines the type of communication that will take place, whether it is full-duplex (bidirectional), half-duplex (non-simultaneous bidirectional) or simplex (unidirectional), and the protocols that need to be used at the lower layers. For instance, without the support of the session layer a user would not be able to transfer data after logging into the system. The session layer ensures that the relevant credentials carry over to the data transfer itself. In addition, in the presence of a network failure an application making use of the session layer could resume a lengthy download from the previous checkpoint without needing to start over. Some services and protocols that live at this layer are Network File Services (NFS), Remote Procedure Call (RPC), Structured Query Language (SQL), and the X Window System and X Terminal.

- Layer 6 – the presentation layer. The *presentation layer* is one of the easiest layers to understand because we can easily see its effects. The presentation layer modifies the format of the data in order to present the data in a proper way to the end user. For instance, we might send an email via Simple Mail Transport Protocol (SMTP) to somebody including an attached image. SMTP cannot support anything beyond plain text (7-bit ASCII characters). To support the use of this image, our application needs a presentation-layer protocol to convert the image to plain text at the local host and convert it back into an image at the remote host. The presentation layer is also responsible for compression and encryption, and pretty much everything else related to the format of the data.

- Layer 7 – the application layer. The *application layer* is responsible for interacting with the actual user application. Note that it is not (generally) the user application *itself*, but rather the network applications used by the user application. For instance, in Web browsing, the user application might be represented by Microsoft Internet Explorer, but the network application is Hyper Text Transfer Protocol (HTTP). Some common application-layer protocols are HTTP, File Transfer Protocol (FTP), Telnet, SMTP, Post Office Protocol (POP3), SQL and Internet Message Access Protocol (IMAP).

2.4.1   Intra-domain and inter-domain routing protocols

Routing protocols are mechanisms for the dynamic discovery of the proper paths through the Internet. They are fundamental to the operation of TCP/IP. The intended result of a routing system is quite impressive: at every decision point within the entire Internet, the local router has adequate information to switch any IP packet to the "correct" output port, e.g. outgoing link. In a routing sense, "correct" does not only mean "closer to the destination," but also "consistent with the best possible path from the sender to the recipient." Such an outcome requires the routing protocol to maintain both local and global state information, as the router must be able to identify a set of output ports that will carry a packet closer to its destination, but also select a port from this set which represents the best possible path to that destination overall. Again, this is not all that a routing protocol must achieve. We have to add to this picture the observation that routers and links are not perfectly reliable. Whenever a network component, such as a router or a link fails, the routing protocol must attempt to repair end-to-end connectivity by establishing a new set of paths that avoid the failed component. When a component is restored, or new routers and links are added to the network, again the routing protocol must re-evaluate the topology of the network and possibly set up a new collection of switching paths through the network. And, of course, this information must be flooded to all routers in the network as soon as possible after the event.

In a small network this can be a forbidding problem. In a large network, such as the Internet, with millions of end devices and hundreds of thousands of links and routers, it is an even more forbidding problem. The technique used by the Internet to achieve the goal of end-to-end connectivity is that of dividing the problem into more manageable tasks. In the routing domain this division of the problem nicely maps to the structure of the Internet itself: each separate network runs its own local Internet routing protocol or *intra-domain routing protocol* (or *Interior Gateway Protocol*, or *IGP*) and the collection of networks is joined into one large routing domain through the use of an *inter-domain routing protocol* (or *Exterior Gateway Protocol*, or *EGP*).

There are a number of interior routing protocols, including *RIPv2*, *EIGRP*, *OSPF* and *IS–IS*. They all perform a similar function, that of maintaining an accurate view of the current topology of the local network. For all addresses that are reachable within the network, the routing protocol computes the best path to the address from all points in the network. In the event of failure of a network component, or a change in the operational state of a component, the protocol employs mechanisms to detect changes quickly and reconstruct an up-to-date view of the network state.

The most prevalent intra-domain routing protocols in today's Internet are (i) Open Shortest Path First (OSPF) [147] and (ii) Intermediate System–Intermediate System (IS–IS) [154]. Their ability to detect topological changes quickly and adjust traffic flow accordingly makes them the most popular IGPs currently operational in the Internet [97].

Both OSPF and IS–IS are link-state routing protocols, utilizing a replicated distributed database model. They work on the basis that routers exchange information elements, called *link states*, which carry information about links and nodes in the routing domain.

More specifically, each OSPF router internal to an autonomous system generates its own view of the network consisting of its neighbors and possible *costs* involved in their inter-connection. This view of the topology obtained by a single router inside the domain is reliably distributed throughout the network, across all routers, via a flooding mechanism until all nodes in the network have obtained a copy. When the distribution is complete, every router in the network has a copy of the local view of the network topology from every other router in the network. This is usually referred to as the *link-state database*. Each router then autonomously constructs the complete topology for the entire domain. The result of this process is an identical copy of the domain's topology on each router in the network.

Once a router knows the current topology of the network, along with the *costs* involved in using specific links across the network, it applies the Shortest Path First (SPF) algorithm and calculates a tree of shortest paths to each destination, placing itself as the root. At that point, we say that the routing protocol has *converged*. The *cost* associated to each link in the network is also known as the OSPF/IS–IS weight and is configured by the network operator. Decreasing the weight of a link in the network is likely to attract more traffic toward it. Consequently, link weights express administrative preference toward specific paths throughout the network.

Depending on the link weights configured across the network, the SPF algorithm may compute multiple equal-cost paths toward specific destination networks. For equal-cost IP paths, traffic is split among the different available IP paths; traffic is also described as being *load-balanced*. Such an approach leads to comparable utilization levels across equally weighted links in the same IP path and therefore a better spread of the carried traffic across the network. As presented earlier in this chapter, backbone networks are likely to feature multiple links between two adjacent PoPs. Those links are usually configured with equal link weights. As a consequence, when all links are operational, traffic is spread across all links, and, when a link fails, the affected traffic can revert to the remaining links without impacting the performance observed.

As with IGPs, there are a number of exterior routing protocols. The most common in the Internet today is the Border Gateway Protocol (BGP). When two networks exchange information using BGP they do not tell each other the precise path to a particular destination. Instead, they simply inform the neighboring network that if they receive a packet addressed to a particular destination, then they will be able to deliver it. This does not necessarily mean that the address is part of the local network. The destination network may have been learned from another neighboring network via BGP, and the network in question is willing to transit traffic between the two neighboring networks. When an ISP network connects to multiple ISPs, it is often the case that a destination address is reachable via two or more neighboring networks. Left to its own devices, BGP will select paths that traverse the fewest possible number of providers (also termed as Autonomous Systems (ASs)) to reach each destination. But BGP does not have to run in such a fully automated mode. BGP has an additional function not found in IGP – that of policy specification and enforcement. One upstream provider may be cheaper than another, or one neighbor may be a peer, while the other may be a customer. BGP allows a network to express preferences regarding which neighbor to use when choosing a path for external addresses. For more details please refer to Chapter 3. Path selection is not only possible for outgoing traffic. A network may attempt to bias incoming traffic to use particular networks over others, and do so for particular ingress points.[2] It may sound somewhat clumsy, but it is on these foundations that traffic engineering and load balancing is constructed in today's Internet.

2.5 Specifying requirements in managed IP networks: design and traffic engineering, monitoring and security operational criteria

IP networks have typically been designed to satisfy the requirements of data services (such as Internet access, content hosting and IP Virtual Private Networks (IP VPNs)). These services are principally ones that tolerate delay and packet loss: often they involve

---

[2] Such an outcome is reached when the network in question advertises particular network prefixes with a path that includes its own AS multiple times. This practice is called *AS-path prepending* and is used to inflate the AS-path to particular destinations artificially so that the network in question is not selected in the BGP route selection process.

no fast user interaction and use TCP to adjust bandwidth consumption and to retransmit in the event of packet loss.

The design of these networks has been based on the "best-effort" service model and has required no significant extension to TCP or IP, as has been demonstrated by the immense growth in the offering of Internet data services. All of the traffic is carried across the IP network as rapidly as possible but without assurances for timeliness or even reliability. Congestion due to inadequate capacity, route changes due to routing updates, link or node failures, or malicious threats targeting hardware or software network components all result in poorer response times, which users accept to some extent.

Even within the intended range of data services, there are now applications that are not always served well by the best-effort service model. Although the applications themselves may appear to tolerate delay and packet loss, *unless adequate capacity is assured*, the packet rate reduction or packet retransmission resulting from congestion can be unacceptable to users. In addition, there are now new emerging services posing requirements that are very different from those of data services, such as Voice-over-IP (VoIP), IP Television (IPTV) and multimedia. For example, VoIP is naturally contrasted to the traditional Packet Switched Data Network, and as such it needs to operate with comparable performance, despite being relayed over an otherwise "unreliable" IP infrastructure.

This objective gives rise to multiple customer satisfaction requirements that must be taken into account and summarized into SLAs contracted between ISPs and their customers. Typical terms included in SLAs regard the following aspects.

(1) *Quality of Service (QoS)*. Service providers must deliver services to the end user with excellent quality targets, e.g. delay and packet loss. Beyond the baseline delay due to the speed of light and other irreducible factors, delays in the Internet are caused by queues, which are an intrinsic part of congestion control and sharing capacity. Congestion occurrs in the Internet whenever the combined traffic that needs to be forwarded onto a particular outgoing link exceeds the capacity of that link, a condition that may be either transient or sustained. When congestion occurs in the Internet, a packet may be delayed, sitting in a router's queue while waiting its turn to be sent on, and will arrive later than a packet not subject to queuing, resulting in latency. Jitter results from variations in the queue length. If a queue fills up, packets will be dropped.

(2) *Reliability and robustness*. Service providers must ensure to their customers the delivery of the expected level of service independently of specific network conditions. For example, an ISP may claim to its customers to deliver services with 99.999% reliability, meaning that a customer should expect the service to fail no more than 5 minutes per year. Reliability is typically achieved by the combination of component reliability, component redundancy and a robust network design against failures. Moreover, it is important that, upon a failure, any service will not be subject to catastrophic collapse; instead, it must degrade gracefully to a level that is still considered acceptable by the end user and recover fast to its standard level.

(3) *Consistency and predictability* represent the hallmark of a well designed and implemented network. Well designed networks allow user expectations about the availability and quality of service to be routinely and constantly met.

(4) ISP customers have recently requested increased visibility into details of their *service and applications*. Customers are requesting more detailed metrics such as routing-paths, end-to-end delay, jitter and bandwidth breakdown by application. In turn, ISPs have introduced more sophistication and complexity into their traditional management systems, historically based on Simple Network Management Protocol (SNMP) data, by adopting packet- and flow-based monitoring infrastructures.

(5) *Security* is becoming a fundamental consideration in ensuring the stable operation of a service provider network and the robust delivery of connectivity and reachability to its customers. Historically, ISPs' customers have been protecting their network perimeter using firewalls, intrusion detection/prevention systems. As Internet attacks are becoming more complex, distributed and sophisticated, customers have realized that their countermeasures are not robust enough to protect their own infrastructure. These days, customers are turning toward their ISPs to shoulder a significant portion of the security burden, and SLAs contracted with their ISPs are now reflecting those demands as well.

All the above business requirements are translated into specific technical tasks that are usually handled by engineers operating at the *Network Operation Center* (NOC) and the *Security Operation Center* (SOC). These requirements can be classified into three major families:

(i) *monitoring and management*;
(ii) *network design and traffic engineering*; and
(iii) *security*.

Each one of the three themes of the book focuses on one of the aforementioned areas and presents ways in which theoretical tools, complemented by actual network measurements, can address some of the needs in the respective area.

Part I

Network monitoring and management

3 The need for monitoring in ISP network design and management

As networks continue to grow rapidly in size and complexity, it has become increasingly clear that their evolution is closely tied to a detailed understanding of network traffic. Large IP networks are designed with the goal of providing high availability and low delay/loss while keeping operational complexity and cost low. Meeting these goals is a highly challenging task and can only be achieved through a detailed knowledge of the network and its dynamics.

No matter how surprising this may seem, IP network management today is primarily reactive in nature and relies on trial and error when problems arise. Network operators have limited visibility into the traffic that flows on top of their network, the operational state of the network elements and the behavior of the protocols responsible for the routing of traffic and the reliable transmission of packets from end to end. Furthermore, design and planning decisions only partially rely on actual usage patterns. There are a few reasons behind such a phenomenon.

First, the designers of IP networks have traditionally attached less importance to network monitoring and resource accounting than to issues such as distributed management, robustness to failures and support for diverse services and protocols [57]. Thus, IP network elements (routers and end hosts) have not been designed to retain detailed information about the traffic flowing through them, and IP protocols typically do not provide detailed information about the state of the underlying network. Second, IP protocols have been designed to respond automatically to congestion (e.g. TCP) and failures (e.g. routing protocols such as IS–IS/OSPF). This makes it hard for a network administrator to track down the cause of a network failure or congestion before the network itself takes corrective action. Finally, the Internet is organized as a loose interconnection of networks (autonomous systems) that are administered independently. Hence the operator of a single network has no control (or even visibility) over events occurring in other networks it exchanges traffic with.

In contrast to traditional telecommunication networks, an IP network features no such thing as a circuit where end points are capable of counting the amount of activity from end to end. In addition, the fact that each packet in a flow traverses each network element independently and multiplexes with other traffic makes tracking of resources across the network an almost impossible task. As a result, traditional techniques for the design and planning of telecommunication networks do not apply to the design and management of the connectionless IP data networks.

An Internet service provider's monitoring requirements are diverse, and range from network design and capacity planning to traffic engineering and customer feedback. Currently some of these requirements are fulfilled using standard tools and techniques, such as ping, traceroute and Simple Network Management Protocol (SNMP), supplemented by a few commercial and proprietary tools. However, the problem of building a comprehensive and integrated monitoring infrastructure to address all ISPs' needs is far from solved. Several questions remain about ways in which network measurements can assist in everyday tasks, suitable timescales for their collection, the granularity of collected information, the need for router-level support, etc.

This chapter articulates the monitoring needs and challenges of an ISP. Although most of the discussion was inspired by experiences with a specific large ISP in North America, i.e. Sprint, we believe that a significant portion can be generalized to other IP backbones of similar size and design. We start with the description of the state of the art in the monitoring of operational networks and proceed to identify specific recent efforts that intend to shed more light into previously unchartered areas. The sheer amount of data flowing on top of large-scale IP networks renders exhaustive monitoring infeasible. Nonetheless, advanced measurement support on well defined locations inside the network can indeed significantly enhance current practices, as will be seen later in this book. We describe the elements of such an architecture and present solutions that can make use of such data sources towards the end of the chapter. The rest of the book will make use of the different sources of monitoring data discussed in this chapter and will demonstrate the benefits one can reap when such data are available.

3.1   Current measurement capabilities

Network traffic has been analyzed since the original development of the ARPANET [130]. Measurements collected in operational networks provide useful information about their performance and the characteristics of the traffic they carry. Measurement systems can be classified into two categories: (i) the *passive* and (ii) the *active* measurement systems. *Passive measurement systems* observe the traffic flowing through a selected point inside the network. Analysis of the collected measurements leads to workload characterization and gives insight into its characteristics. This method is non-intrusive in that it is designed in a way that the collected measurements are not significantly affected by the presence of the measurement system. *Active measurement systems*, on the other hand, inject probe traffic into the network and extrapolate the network performance based on the performance experienced by the traffic injected. Such active measurement techniques are usually considered intrusive, in the sense that probe traffic is likely to influence the network operation and hence the collected results. For instance, if an active measurement application aims at measuring the queuing delay observed on a particular link, then the existence of the packet probes themselves may alter the queuing behavior observed. Therefore, active measurement tools have to be designed carefully in order to provide correct information on the targeted metric. Efforts toward the standardization of performance metrics and active measurement techniques for their evaluation are carried out within the framework of the Internet Protocol Performance Metrics (IPPM) working group of the IETF (Internet Engineering Task Force). Already proposed standards include metrics for connectivity [132], one-way delay [21], one-way packet loss [22], round-trip delay [23] and bulk transfer capacity [135].

Understanding the network state and the performance experienced by data traffic on top of an IP network may require both passive and active measurement techniques in order to assess: (i) the state of the network devices themselves, (ii) the state of the protocols that enable the flow of traffic across the network, and (iii) the amount of traffic flowing on top of links and its behavior over several timescales. We now look into the state of the art in each one of the above categories. Note that what follows is not an exhaustive list of tools for the measurement of network state, but tools that will be later used in this book for the purposes of network design, planning and management. A more extensive description of the state of the art for Internet measurement can be found in ref. [63].

3.2 Monitoring the network devices

Network element specific information can be retrieved in a number of ways. Such retrieval can be done through the *Command Line Interface* (CLI) provided by the vendor, automated logging of events through syslog, or even standardized protocols such as SNMP (see Section 3.2.1), and Remote Monitoring MIB (RMON) (Section 3.2.2). Depending on the level of support provided by the vendor, the output produced by the aforementioned tools can provide information on the state of the physical layer (optical level failures), the state of the network layer (whether an IP link is up or down), the state of the routing protocols in use (whether a route exists to a particular destination) and also aggregate counters on the amount of activity seen on an interface. Further visibility into the configuration of the device itself can allow a network operator to ensure that appropriate Access Control Lists (ACLs) are in place or whether the routing protocol adjacencies are set up as planned.

Real-time tracking of the health of the network is achieved through the use of commercial or custom-made tools that periodically poll specific statistics from the network elements and collect them at a central location, typically called a *Network Management Station (NMS)*. Example products in this space are *Hewlett-Packard OpenView*, *Concord Network Health*, *InfoVista VistaView* and Multi-Router Traffic Grapher (MRTG) to name a few. These tools are not only deployed for the management of ISP networks, but are also in common use for the management of enterprise networks. Since, all these tools intend to capture the network-wide health state, they typically rely on SNMP, a protocol that is widely supported by today's network equipment manufacturers.

3.2.1 Simple Network Management Protocol

The SNMP is an evolution of the Simple Gateway Monitoring Protocol [43]. The first version of SNMP was defined in 1988, was updated in 1989, and became an Internet standard in 1990 [176]. Most network elements in today's Internet support the second version of SNMP, which was defined by the IETF in 1996 [44]. A third version was proposed in 1999 [45], primarily adding security and remote configuration capabilities.

SNMP is a very simple protocol, based on a client-server model and implemented over UDP. It has been designed to allow implementation in low-cost systems with little memory and computing resources. The simplicity of SNMP made it possible to require its support by all Internet systems, which in turn greatly facilitates the management of the Internet.

SNMP enables network managers to monitor and control the health of network entities remotely, e.g. routers or transmission equipment. The state of each entity is represented by a number of variables and tables. The structure of this data is defined by a "Management Information Base" (MIB). Monitoring is performed by reading these variables through SNMP commands (i.e. through *snmpget*). Control is performed by "setting" some essential parameters to a new value (i.e. through *snmpset*). There are MIBs defined for a very large number of "managed objects," notably the generic MIB which describes an IP router [126], also known as MIB-II, and specific MIBs, which describe the variables and tables for each routing protocol [30, 206].

MIB-II is particularly interesting in its ability to capture coarsely the state of a network element and the amount of traffic it sends/receives on top of each one of its interfaces, along with losses, errors, etc. The rest of the book will make extensive use of the MIB-II table that captures information on the different interfaces of a router, i.e. *MIB-II.interfaces*. Such information includes the number of bytes sent/received on an interface (MIB-II.interfaces.ifTable.ifEntry.if{In,Out}Octets), packets dropped (MIB-II.interfaces.ifTable.ifEntry.if{In,Out}Discards), as well as errors (MIB-II.interfaces.ifTable.ifEntry.if{In,Out}Errors).

Note that all such information is captured in the form of counters that are only reset by the network operator or during the initialization of the network device itself. As such, these counters will keep increasing until they reach their maximum allowable value and then wrap around. Frequent polling of the counters kept for total incoming/outgoing traffic is typically used to compute the average utilization of a link across a polling interval. If an automated SNMP agent polls ifInOctets every 5 minutes, then subtracting two sequential values allows one to compute the amount of traffic received during that particular 5-minute interval. Further division with the polling interval duration can provide the network operator with the 5-minute average value, in bytes per second, for the link utilization (incoming or outgoing). This kind of 5-minute average utilization value constitutes the fundamental input to most IP network design and planning tools.

SNMP read and write operations are triggered by the network operator. However, SNMP also offers a small number of unsolicited messages, better known as *SNMP traps*, that may be generated by the network elements themselves when particular operator-defined criteria are met, such as when communication links fail, or become active [176]. The destination of these messages is configured by the network operator. The Network Management Station receives those traps and records the health status information, while offering a graphical user interface to the network operator.

Probably the most important disadvantage of SNMP lies in the timescale of information, e.g. information can be obtained only on the timescale of minutes or longer and in the form of aggregate counters. Note that more frequent polling is feasible but is bound to introduce additional overhead on the routers and the network itself, since all information needs to be transported to the Network Management System (NMS). Consequently, the frequency of collection in large-scale IP networks is typically set to once every 5 or 10 minutes. Moreover, SNMP does not provide any information about the source and destination of the traffic flowing through the network, its dynamics at finer timescales, nor its nature. For instance, the aggregate counters reported by SNMP cannot provide information on the point-to-point traffic demands, the delay experienced across the network, the mix of protocols and applications, etc. As will be shown later in this chapter, such information is essential for a variety of tasks. Despite the aforementioned shortcomings, the lightweight nature of SNMP casts it as the de facto protocol in large-scale network management, allowing for the continuous monitoring of thousands of network devices at minimal overhead.

3.2.2 Remote-monitoring MIB

RMON is another standardized SNMP MIB, whose goal is to facilitate remote monitoring of LANs [199]. In the RMON context, a single agent is capable of monitoring a complete shared LAN. The RMON agent is endowed with local intelligence and memory and can compute higher-level statistics and buffer these statistics in case of outage. Alarm conditions are defined, as well as actions that should be taken in response, in the spirit of the SNMP traps. Lastly, filtering conditions and the corresponding actions are defined, so that the content of the filtered packets can be captured and buffered.

RMON offers great flexibility in combining the above primitives into sophisticated agent monitoring functions. However, this flexibility makes a full RMON agent implementation costly. Thus, RMON has been only implemented (at least partially) for LAN router interfaces, which are relatively low-speed. Implementations for high-speed backbone interfaces have proved to be infeasible or prohibitively expensive. Instead, router vendors have opted to develop more limited monitoring capabilities for high-speed interfaces [59]; these are described in Section 3.4.1.

RMON is a bright example on the feasibility of extensive monitoring on routing elements. Even though the best vantage points for the collection of network measurements are the routing elements themselves, cost and performance considerations would probably deem any exhaustive solution infeasible. Consequently, any additional measurement support at the routers or switches would need to constrain the associated memory and processing overhead. In the rest of the book we make a number of recommendations with respect to such functionality, but turn to packet monitors (see Section 3.5) when the amount of processing or collected information is bound to strain a router's resources.

3.3 Monitoring the state of the routing protocols

Apart from their operational state, routers can also log a variety of events that concern the behavior of the routing protocols themselves, such as the loss of IGP link adjacencies and changes in the paths selected to particular destinations, say through BGP. Despite the importance of such information for the troubleshooting of the network and the state of the routing protocols themselves, collection of such logs on a continuous basis has not been a common practice among ISPs. Only recently have operators turned their interest to the continuous collection of such information using primarily in-house tools, or open source software such as *zebra* and *PyRT*.

3.3.1   Routing information

A *routing protocol listener* is a system establishing routing sessions with the operational routers inside a network with the intent to record all the messages exchanged in the routing domain. To the best of our knowledge, there are only two publicly available implementations of such systems: (i) zebra[1] and (ii) PyRT.[2] In fact, zebra is more than just a routing protocol listener. A system with the zebra software acts as a dedicated router. It can exchange routing information with other routers and update its kernel routing table accordingly. Currently supported routing protocols include RIPv1, RIPv2, OSPFv2 and BGP-4.

The Python Routing Toolkit (PyRT) is a tool written in Python [68] that supports the passive collection and off-line analysis of routing protocol data forming minimal router peering sessions and dumping the routing PDUs (Packet Data Units) received over these sessions. PyRT currently supports BGPv4 and IS–IS. Compared with full implementations of routing protocols, such as zebra, an advantage of PyRT is that no routing information is ever advertised on these sessions and only a minimal amount of information is injected into the network.

Only recently were such systems deployed within operational networks (and they still are). However, BGP routing information has been collected since 1989 in the form of periodical BGP routing table dumps. The collected information is analyzed in ref. [99]. As we will see later in the book, routing protocol listeners are priceless, not only in troubleshooting routing protocol dynamics, but also in the computation of the traffic matrix of an IP network, an essential input to any IP network design and planning tool.

3.3.2   Path-level information

To test the availability and performance of an end-to-end path, which may cross multiple administrative domains, network operators use active measurement approaches and tools such as ping and traceroute.

Ping is a widely used facility that relies on the ability of an Internet node to send an "ICMP echo reply" when it receives an "ICMP echo request" packet from another host. (ICMP stands for Internet Control Message Protocol.) Its only use is in determining reachability from one node to another and to get a rough estimate of the round-trip time between the two nodes. It can be further used by an operator to determine whether a router in the network (or one of its interfaces) is "alive."

---

[1] The official distribution site for zebra is at http://www.zebra.org.
[2] The official distribution site for PyRT is at http://ipmon.sprint.com/pyrt/.

Traceroute is also well known and widely used. It is somewhat more powerful than ping in that it reveals the route that a packet takes from one Internet host to another. Traceroute also provides a rough estimate of the round-trip time to every network node along the path. An operator may use traceroute to trace the path between any two given points in a network. However, access is required at one of the two end points.

Ping and traceroute are certainly useful in getting a ready and basic sense of routing and reachability across a backbone. However, they provide a very limited set of abilities and are clearly insufficient to meet all the monitoring requirements of an ISP. Moreover, ping and traceroute are essentially tools to measure reachability, not performance. While they may be used to obtain rough estimates of latency and loss rate along certain paths in a backbone, such estimates may not be statistically significant. First, when active probes are used to measure the quality (e.g. delay/loss) of a network path, the interval between successive probes has to be set carefully in order to allow for statistically meaningful interpretation of the results [195]. Second, routers usually throttle the rate at which they respond to ICMP echo requests and this can interfere with a measurement method that is based on a series of pings or traceroutes. Moreover, traceroute does not provide any information about the reverse path back to the source. Given that Internet routing is largely asymmetric, e.g. the path taken by a packet from the source to the destination may be different than the path taken by a packet from the destination to the source, there is no guarantee that successive traceroutes out from the same host traverse the same round-trip path back to the host. This makes it hard to interpret the measurements obtained. Finally, network operators often respond unfavorably to a large number of pings/traceroutes directed to hosts in their network since it could be the indication of a denial-of-service attack.

Despite the issues identified, ping and traceroute are irreplacable in their ability to provide up to date information on the availability of a backbone network from end to end. Actually, most large-scale ISPs do collect similar measurements periodically but on a smaller scale. Typically, those measurements are based on a series of UDP probes between selected locations inside the network, say one router inside each PoP. Even though the functionality of UDP probes is very similar to that of ping, the fact that these probes are relayed over UDP implies that they are less likely to face rate limiting than their ICMP counterparts. The output of these measurements is processed to produce the average delay and loss experienced between any two PoPs in the network, which could be used as a health indicator and an input to SLAs.

3.4   Monitoring the traffic flow

Proactive management of an IP backbone network involves studying the traffic flow across the operational network in order to determine appropriate routing protocol configurations. Any redesign task needs to incorporate knowledge about the amount of traffic customers inject into the network and its most likely exit points. Further, the impact of peering link traffic needs to be taken into account when provisioning the core of the network. All these tasks, which will be discussed in Section 3.6, require visibility into traffic statistics at finer spatial and temporal granularity than the one currently supported by SNMP.

3.4.1   Flow-level information

Interest in additional visibility into the amount of traffic flowing on top of a network at a granularity finer than aggregate link count has led to advances in flow monitors integrated with commercial routers. State-of-the-art routers provide support for gathering information about traffic flowing through them at the level of flows. Flows can be flexibly defined by the network operator and typically correspond to all data packets that share a common key defined as common source, destination IP addresses, source, destination ports and protocol number (termed as "5-tuple"). More recent generations of flow collectors (such as Cisco NetFlow v8) can further aggregate information, taking into account the state of the BGP routing protocol, defining flows as all traffic destined to particular BGP prefixes, essentially requiring a longest prefix match of the IP address with the BGP routing table. The type of information exported by the router depends on the version of the software and the definition of what constitutes a flow. As an example we provide the flow format used in NetFlow v5 in Table 3.1: each NetFlow record lists the start and end time of the flow and the number of packets or bytes seen; it could also list the source and destination ASs involved.

At predefined intervals (typically set to 5 or 10 minutes) routers may export all flow records or all records of expired and completed flows (depending on the configuration option selected) to a well known location, where data are collected. There are several publicly available pieces of software chartered with the transport and storage of Net-Flow data. Some examples include *cflowd*, *flowd*, *Plixer* and *nnfd*. Further manipulation of the collected information can allow a network operator to identify what applications are making use of the network resources, its largest consumers, the amount of traffic received from peering links and its destinations. Later in this book we will see that flow-level information can play a very important role in network design and management.

Given the usefulness of flow-level information in network management and operation, the IETF has made efforts to standardize flow-level measurements. The Real-Time Traffic Flow Meter (RTFM) working group [41, 91] provides a formal definition of flows and describes techniques for collecting the measurement data. More recently, the IP Flow Information Export (IPFIX) working group [164] defined a format and protocol for delivering flow-level data from the measurement device to other systems that archive and analyze the information.

Flow-level data yield information at a higher aggregation level than individual packets and lower than SNMP counters. As a result, its storage and processing requirements are greater than SNMP, but significantly lower than collecting per-packet information (described in Section 3.5). However, a number of issues remain open in collecting flow-level information on routers.

Table 3.1. NetFlow v5 flow record format

Bytes	Contents	Description
0–3	srcaddr	source IP address
4–7	dstaddr	destination IP address
8–11	nexthop	IP address of next hop router
12–13	input	SNMP index of input interface
14–15	output	SNMP index of output interface
16–19	dPkts	packets in the flow
20–23	dOctets	total number of Layer-3 bytes in the packets of the flow
24–27	first	SysUptime at start of flow
28–31	last	SysUptime at the time the last packet of the flow was received
32–33	srcport	TCP/UDP source port number or equivalent
34–35	dstport	TCP/UDP destination port number or equivalent
36	pad1	unused (zero) bytes
37	tcp_flags	cumulative OR of TCP flags
38	prot	IP protocol type (e.g. TCP = 6; UDP = 17)
39	tos	IP type of service (ToS)
40–41	src_as	autonomous system number of the source, either origin or peer
42–43	dst_as	autonomous system number of the destination, either origin or peer
44	src_mask	source address prefix mask bits
45	dst_mask	destination address prefix mask bits
46–47	pad2	unused (zero) bytes

First, the implications of turning these features on the routers are poorly understood. Anecdotal evidence suggests that flow-level monitoring may severely impact the essential functions of a router (packet forwarding, updating routing information, etc.) by increasing CPU load and memory consumption. As a consequence, operators of large ISPs typically make very limited use of existing flow-based monitoring capabilities. These facilities are usually turned on for short periods of time on a few routers under special circumstances.

Second, in order to reduce complexity, many high-end routers form flow statistics from only a sampled substream of packets in order to limit the consumption of memory and processing cycles involved in flow cache lookups. As a side benefit, the rate at which flow statistics are produced is reduced in most cases, lowering the requirement for bandwidth to transmit flow statistics to a collector and for processing and storage costs at the collector. However, it is well understood that sampling entails an inherent loss of information. For some purposes, loss is easy to correct for. Assuming that one in N packets are selected, the total number of packets in the stream can be estimated by multiplying the number of sampled packets by N. The total number of bytes can be similarly inferred under specific assumptions. However, more detailed characteristics of the original traffic are not easily estimated, such as total number of flows, duration of a flow, or number of packets per flow. A more detailed treatment of sampled statistics inversion and relevant references are provided in Appendix A.

3.4.2   Packet-level information

Collecting information on traffic flows allows one to monitor traffic demands across a network, while further processing can lead to the computation of meaningful entities, such as a traffic matrix. However, despite the increased resolution into the source and destination of traffic demands and their nature (as exposed by the port numbers), no information is preserved on the timing of the individual packets in a flow or their individual content. The only way one can gain visibility into the explicit timing of each data packet and its content is through the collection of information on a per-packet basis. Measuring traffic at the granularity of individual packets yields a list of all packets (or a subset thereof, if sampling is employed) seen on an interface, their timing and can include their entire payload. Challenges in such a process involve the lossless capture of large volumes of data at very high speeds, their storage and subsequent manipulation. In particular, collecting every packet on every interface on a modern high-speed router is a daunting task, as will be seen in Chapter 4.

Two approaches have been used so far for packet monitoring. The first is to use "port-mirroring," where every packet on an incoming interface can be written out to a "monitoring" interface, in addition to being routed to the output interface. However, this approach potentially requires one additional monitoring interface per router interface – a prohibitively expensive option since it would constrain an ISP to using no more than half of the available interfaces on a router to connect to customers and peer networks. If, instead, a single monitoring interface were added for every group of, say, N interfaces, then the monitoring interface would have to support a packet-forwarding speed that is equivalent to the aggregate capacity of all N interfaces. The problem becomes tractable only if a subset of packets on each interface is captured. This introduces the requirement for sampling, a common practice also employed by router manufacturers in the collection of flow-level information, as seen in the preceding section.

The second approach is exemplified by ref. [81], where a special-purpose monitoring device is used to tap the optical signal on a link and capture a fixed part of every packet. Each packet is stamped with a highly accurate Global Positioning Satellite (GPS)-synchronized timestamp at the time of capture. While there are several benefits in capturing this information, the greatest challenges are the infrastructural cost and the dynamic nature of operational networks. These monitoring systems have to be installed inside PoPs, where space is extremely limited and prohibitively expensive. Furthermore, extreme care has to be taken when installing and operating these systems so as not to disrupt network operation accidentally. Finally, operational networks are in constant evolution, with links and routers being reconfigured, commissioned, or decommissioned. This makes the maintenance and management of the monitoring systems an operational nightmare.

As a result, network providers typically deploy such systems at a small select number of locations in order to support specific tasks. As will be seen later in this section, recent efforts have focused on similar functionality being supported by operational routers, but on an as-needed basis and for limited periods of time. Incidentally, the driving application behind most such deployments is security, e.g. the identification of malicious or attack traffic, and the need to be able to trace particular traffic streams within the context of Lawful Intercept (LI). Special-purpose packet capture systems tend to be at the heart of most Internet security products in today's market.

In the next chapter we describe packet-capture systems in more detail, since their data form a fundamental part of a lot of the analyses presented later in this book.

3.5   Packet-capture equipment

A packet-capture system typically comprises two different components: (i) a capture mechanism that collects the data off a physical link and (ii) a high-end system that can store the data fast enough to avoid loss. As mentioned above, packet capture can simply rely on port-mirroring, whereby traffic on a specific set of interfaces is mirrored to a specific port on a router. The monitoring system then observes the traffic on the mirroring port, decodes the captured packets and saves them in the appropriate format (*libpcap* tends to be the format of choice). A second option, in terms of capture, is to use a special-purpose monitoring card, such as the Endace DAG card,[3] used in several packet-capture prototypes, such as *OCXMON* [25], *Gigascope* [102] and *IPMON* [81]. Commercial products in this space typically use their own proprietary technology.

All packet-capture systems have strict requirements in terms of processing power, access time to hard disk, memory and acceptable packet-drop rates. In addition, they need to feature a well calibrated clock in order to allow for the detailed analysis of the timing information captured across packets (e.g. to study traffic burstiness) and across systems (e.g. to compute one-way delay). We discuss these requirements in what follows. Our discussion revolves around the choices made in the Sprint IPMON project. Despite minor differences between packet-capture systems, we believe that the following requirements are shared by all systems in this space. Having said that the detailed design choices listed below are indeed specific to the IPMON implementation and may be slightly different to what is on the market at the time of this publication.

3.5.1   Packet-capture system architecture

Packet-capture systems are high-end computers with a large disk array and a packet-capture card, such as the DAG card. For optical links, the DAG card decodes the SONET payloads and extracts the IP packets. When the beginning of a packet is identified, the DAG card generates a timestamp for the packet, extracts the first 48 bytes of the Packet-Over-SONET (POS) frame, which contains 4 bytes of POS header and 44 bytes of IP data, and transfers the packet record to the main memory in the PC using DMA (Direct Memory Access). The format of the packet record is shown in Figure 3.1. If the packet

---

[3] http://www.endace.co.nz/our-products/dag-network-monitoring-cards/

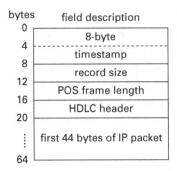

Fig. 3.1.   DAG Packet-record format. POS = Packet-Over-SONET; HDLC = High-Level Data Link
Control.

contains fewer than 44 bytes, the data are padded with zeros. Once 1 MB of data have
been copied to the main memory, the DAG card generates an interrupt, which triggers
an application to copy the data from the main memory to the hard disk. It would be
possible to transfer the data from the DAG card directly to the hard disk, bypassing
the main memory. The main memory, however, is necessary to buffer bursts of traffic,
as described later in this section. Note that the above description refers to version 3 of
the DAG architecture, which may differ to that of newer cards. Later generations of the
DAG card can also support full packet capture, where the entire packet payload can be
decoded and saved on disk, thus facilitating application-layer analysis.

The duration of the packet trace (list of packet records on disk) for the aforementioned architecture depends on the amount of available storage capacity. As link speeds
increase, the same amount of storage is bound to accommodate shorter duration traces.
To address such a potential limitation, newer generations of packet-capture prototypes
have opted for dual operation. Instead of collecting data until they run out of disk space,
new generations of packet monitors employ a "circular buffer" approach, by which they
emulate continuous operation by overwriting the oldest data. At the same time, in order
to avoid total loss of older information, they provide customizable summaries of older
data with adjustable time granularity. In other words, the network operator can obtain
a summary of all traffic seen by a packet monitor for the recent, or not so recent, past,
but can only recover detailed packet information for a limited number of hours in the
immediate past, the number of which depends on the link utilization and the available
storage capacity.

The five basic design requirements of a typical packet-capture system can be
summarized as follows. It must

- support data rates ranging from 155 Mbps (OC-3) to 2.5 Gbps (OC-48), and even
  10 Gbps (OC-192) and higher;
- provide synchronized timestamps;
- occupy a minimal amount of physical space;
- prevent unauthorized access to trace data;
- be capable of remote administration.

Table 3.2. Data-rate requirements from packet capture equipment

	OC-3	OC-12	OC-48	OC-192
Link rate (Mbps)	155	622	2480	9953
Peak capture rate (Mbps)	248	992	3968	15872
One-hour trace size (GB)	11	42	176	700

In what follows, we describe how the IPMON infrastructure addressed each one of these requirements. Any packet-monitor infrastructure will need to address the afore-mentioned challenges, and the recommendations below provide point solutions to the choices a network operator would need to make.

Data-rate requirements

The data rate of a network link determines the speed at which a DAG card must be able to process incoming packets. We report the data rates for OC-3, OC-12 and OC-48 links in the first row of Table 3.2. After the DAG card has received a packet and extracted the first 44 bytes, the timestamp and additional header information is added to the packet record and copied to main memory. If there is a sequence of consecutive packets whose size is less than 64 bytes, then the amount of data that is stored to main memory is actually greater than the line rate of the monitored link. The amount of internal bandwidth required to copy a sequence of records corresponding to minimum-size TCP packets (40-byte packets) from the DAG card to main memory is shown on the second row of Table 3.2. While for OC-3 and OC-12 link monitoring a standard 32-bit, 33-MHz PCI bus would have sufficient capacity (1056 Mbps or 132 MB/sec), monitoring an OC-48 link would probably require a faster PCI bus. A system with a 64-bit, 66-MHz PCI bus with a capacity of 4224 Mbps (528 MBps) has proven to be able to keep up at such speeds. Note that it is possible to have non-TCP packets which are smaller than 40 bytes resulting in even higher bandwidth requirements, but the system is not designed to handle extended bursts of these packets as they do not occur very frequently. It is assumed that the small buffers located on the DAG card can handle short bursts of packets less than 40 bytes in size.

Once the data have been stored in main memory, the system must be able to copy the data from memory to disk. The bandwidth required for this operation, however, is significantly lower than the amount of bandwidth needed to copy the data from the DAG card to main memory, as the main memory buffers bursts of small packets before storing them to disk. Only 64 bytes of information are recorded for each packet that is observed on the link. As reported in prior studies, the average packet size observed on backbone links ranges between 300 and 400 bytes during the busy periods of the day [193]. Assuming an average packet size of 400 bytes, the disk Input/Output (I/O) bandwidth requirements are therefore only 16% of the actual link rate. For OC-3 this is 24.8 Mbps; for OC-12, 99.5 Mbps; and for OC-48, 396.8 Mbps. To support these data rates, a three-disk RAID[4] array which has an I/O capacity of 240 Mbps (30 MBps) is

---

[4] The RAID array uses a software RAID controller available with Linux.

sufficient for the OC-3 and OC-12 systems. To support OC-48 link speeds, a five-disk RAID array with higher performance disks is needed, capable of supporting 400 Mbps (50 MBps) transfers. The final choice of how many disks and their type primarily depends on the market availability and the expected transfer rate in MBps. To minimize interference with the data being transferred from the DAG card to memory, the disk controllers should use a separate PCI bus.

Timestamp requirements

In order to be able to correlate traces collected in diverse geographical locations, the packet timestamps generated by each packet monitor need to be synchronized to a global clock signal. For that reason each DAG card features a dedicated clock on board. This clock runs at a rate of 16 MHz, which provides a granularity of 59.6 ns between clock ticks. Packets are not timestamped immediately when they arrive at the DAG card. They first pass through a chip, which implements the SONET framing and which operates on 53 bytes ATM cells. Once this buffer is full, an interrupt is generated and the packet is timestamped. In other words, timestamping happens on the unit of 53 bytes, thus introducing a maximum timestamp error of $2\,\mu s$ (the time needed for the transmission of 53 bytes on an OC-3 link; higher-speed links will introduce smaller timestamp errors [67]).

Due to room temperature and the quality of the oscillator on board the DAG card, the oscillator may run faster or slower than 16 MHz. For that reason, it is necessary to discipline the clocks using an external stratum 1 GPS receiver located at the PoP. The GPS receiver outputs a one-pulse-per-second (PPS) signal, which needs to be distributed to all of the DAG cards at the same geographical location. The most cost-effective way of distributing the GPS signal across all monitors at the same location is through a daisy chain. The longer the chain the greater the synchronization error that may be suffered by the last monitor on the chain. However, note that, even for an 8-meter long chain, that synchronization error does not exceed 28 ns, the propagation delay of the signal across the 8-meter long cable.

In addition to synchronizing the DAG clocks, the monitoring systems must also synchronize their own internal clocks so that the DAG clock is correctly initialized. This can be accomplished using the Network Time Protocol (NTP). A broadcast NTP server installed on the LAN, and connected to the monitoring systems, is capable of synchronizing the system clocks to within 200 ms. This is sufficient to synchronize the beginning of the traces, and the 1-PPS signal can be used to synchronize the DAG clock. Given that there will always be an initial period in the packet traces when the DAG cards will be adjusting their initial clock skew, it is preferrable to ignore the first few seconds of each trace.

To summarize, in order to achieve high synchronization accuracy between packet monitors, the monitoring infrastructure needs to feature access to an accurate global clock and an NTP server. More details on the potential errors introduced by this architecture are provided in Chapter 4.

Physical requirements

In addition to supporting the bandwidth requirements of high-speed links, the packet monitors must also have a large amount of hard disk storage capacity to record the traces. As the systems may be installed in a commercial network facility, where physical space is a scarce resource, this disk space must be contained in small form factor. Using a rack-optimized system, OC-3 and OC-12 systems are able to handle 108 GB of storage in only 4 U of rack space.[5] This allows the system to record data for 9.8 hours on a fully utilized OC-3 link or 2.6 hours on a fully utilized OC-12 link. The OC-48 systems described in Section 3.5.1 have a storage capacity of 360 GB, but in a slightly larger 7 U form factor. This is sufficient to collect a 2-hour trace on a fully utilized link. Fortunately, the average link utilization on most links is less than 50%, allowing for longer trace collection.

The physical size constraint is one of the major limitations of packet monitors. While technological advances are bound to lead to higher capacity storage systems in smaller form factor, there will always be a tradeoff between the amount of data that can be collected and the associated occupied physical space. Collecting packet-level traces requires *significant* amounts of hardware, making such an approach *non-scalable* for operational monitoring of entire networks.

Security requirements

Packet monitors collect proprietary data about the traffic flowing on top of the monitored Internet links. Preventing unauthorized access to this trace data is an important design requirement of all such systems. This includes preventing access to trace data stored on the systems, as well as preventing access to the systems in order to collect new data. To accomplish this, the systems should be configured to accept network traffic only from specific applications (and potentially only from specific hosts), such as *ssh* and NTP. Note that *ssh* is an authenticated and encrypted communication program, similar to *telnet*, that provides access to a command line interface to the system. This command line interface can be chosen as the only way to access trace data that has been collected by the system and to schedule new trace collections. The *ssh* daemon running on the monitor should accept connections from well defined servers and use a secure authentication mechanism (such as an RSA key-based mechanism to authenticate users).

The second type of network traffic accepted by packet-capture systems is NTP traffic for the synchronization of their internal clock. To make such a configuration more secure, one can configure packet-capture systems to only accept NTP messages that are transmitted as broadcast messages on a local network used exclusively by the packet monitors. All broadcast messages that do not originate on this network should be filtered.

Remote administration requirements

Finally, packet monitors need to be robust against failures, since they are installed, in some cases, where there is no human presence. Continuous polling from a specific

---

[5] 1 U is a standard measure of rack space and is equal to 1.75 inches or 4.45 cm.

server can accommodate the detection of failures in a timely fashion. Query responses should indicate the status of the DAG cards and of the NTP synchronization. If the response indicates the failure of either of these components, the server can attempt to restart the component through an *ssh* tunnel. If the server is not able to correct the problem, it can notify the system administrator that manual intervention is required. In some cases, even the *ssh* connection will fail, and the systems will not be accessible over the network. To handle this type of failure, the systems should be configured with a remote administration card that provides the capability to reboot the machine remotely. The remote administration card could further provide remote access to the system console during boot time. In cases of extreme failure, the system administrator could choose to boot from a write-protected floppy installed in the systems and completely reinstall the operating system remotely.

The one event that cannot be handled remotely is hardware failure. Hardware redundancy is the only way to handle such failures, and should be implemented in the monitoring infrastructure if packet-monitor uptime is vital to the operation of the network itself.

3.5.2   Packet capture on demand

Given the importance of packet capture in the debugging of network operation and its ability to allow greater visibility into the traffic carried on top of an IP network, there have been recent efforts to standardize its support by routers. Within such proposals, routers can be instructed to intercept packets that meet specific criteria for a specific period of time and forward them to a dedicated monitoring station. During such periods of time, the router replicates the packets that match the specified criterion, while continuing to forward the packets to their original destination.

Although one of the reasons behind such functionality is the desire to understand the nature of "unusual" traffic, Cisco has provided such functionality within the grounds of *lawful intercept* (LI). The Cisco Service Independent Intercept (SII) architecture was developed in response to the need for service provider and Internet service providers to comply with LI legislation and regulations. The basic premise behind such a need is that using SII ISPs can respond to court orders to report particular communication taking place through their network, on a par with traditional phone company practices. The main design challenge behind such an architecture is the need for such intercept actions to be invisible to network operators and potential offenders. Also, intercept should not impact the quality of service provided by the network. When capture of complete content is required, a network operator needs to ensure that there is enough bandwidth and router processing power.

A generic protocol for the dynamic request of packet capture from network elements has been further proposed to the IETF. The DTCP (Dynamic Tasking Control Protocol) is a message-based interface by which an authorized client may connect to a server (usually a network element or security policy enforcement point) and issue dynamic requests for data. As with SII, such requests lead to the replication of packets that meet specific criteria, as identified in the request. Additional actions include redirection, by which the packets are not replicated but redirected to the collection station without being allowed to reach their intended destination. In certain cases, such an action may involve blocking of traffic. The protocol itself contains a security architecture to address client or server spoofing as well as replay prevention. Since controlled network elements may need to respond to automated requests at a potentially high rate (during the onset of a security incident), the protocol imposes a fast response time on the network elements, especially given that some of the monitored events may actually last for a limited amount of time.

One advantage of "on-demand" capture is that the network operator does not need extra equipment close to the monitored router. A second advantage is that storage requirements are significantly smaller than with a full packet-capture system. The apparent disadvantage of such a solution is that it may overload the monitored network elements and may have significant bandwidth implications, as all packets need to be transmitted to the collection station. Nonetheless, it appears to be a definite step in the right direction.

3.6 The need for measurements in network operations

From the previous section it becomes evident that network operators have a range of tools at their disposal. All these tools pose tradeoffs between the quality of the collected information, their requirements, overheads and cost. In this section we showcase the types of tasks that could benefit from data sources like the ones mentioned above. One conclusion the reader may draw at the end of this section is that no single data source is sufficient to address all the tasks at hand. In addition, providing answers to the questions network operators pose may require access to more than one data source. In most cases, the decision on which measurements to collect is going to depend on the associated cost and overhead, rather than the usefulness of the data itself. What will become apparent by the end of the book is that a monitoring infrastructure in support of network operators' tasks needs to be heterogeneous and feature a combination of slower-timescale SNMP statistics, aggregate flow statistics, real-time packet capture and coordinated router-assisted measurements. The rest of the book will focus on the tasks listed below and present methods for the processing of different network measurements to achieve the objective at hand.

There are three primary areas where network measurements may be used in network operations: (1) within a research framework, to understand network behavior and design new metrics that can concisely capture effects of interest (Chapters 4–5); (2) in the design and planning of an IP network (Chapters 6–11); and (3) in the management of a network to support secure, high-performance services (Chapters 12–16). In what follows we briefly outline the type of measurements needed in each area and why.

3.6.1 Research

The research area of network measurements is relatively new. The first packet-capture prototypes were deployed in large Tier-1 IP networks in the early 2000s. Since that point in time, the research community has been primarily interested in collecting vast amounts of information that would allow it to understand the behavior of IP networks and ways in which one could formalize tasks typically carried out by network operators, such as routing optimization, buffer dimensioning, etc. In parallel, a lot of attention was focused on the formalization of network measurement techniques and the design of network monitors. One of the goals of such an effort was the definition of new metrics, previously elusive to capture. Examples in that latter category are those of the IP traffic matrix at different granularities (from router to router, PoP to PoP, etc.), and the queue occupancy of routers at timescales that may be meaningful in network operation, departing from 5-minute average SNMP counts.

Buffer dimensioning

Today's IP backbones are designed to meet very stringent loss, delay and availability requirements. The typical provisioning rule in effect in most large-scale IP networks is that of *overprovisioning*. Links are not supposed to exceed 50% utilization. The motivation behind such a choice is not only to limit queue occupancy and cross-router delay, but also to provide robustness to failures; for instance, if a link fails, then even if all its traffic is rerouted to a single neighboring link, that latter link will not reach prohibitive utilization levels that will cause packet drops. In reality, link utilization across operational Tier-1 networks is far lower than 50%.

Despite such low utilization levels, however, there are still no guarantees that the delays experienced by packets end-to-end are acceptable. Apart from the occasional end-to-end delay measurements performed by a network operator, there is little visibility into the delays experienced by customer traffic. SNMP is capable of reporting average or maximum queue occupancy values across 5-minute intervals on a per link basis, but these metrics can hardly relate to the performance experienced by the user.

Packet-level measurements can be used to measure the end-to-end delay experienced by packets in the Internet and across routers. In both cases, one would need to track individual packets across monitors and compute the time elapsed between their appearance at the monitoring points. In Chapter 4, we use packet monitors to measure the delay experienced by packets while crossing a single high-performance router in an operational backbone network. The analysis of the collected results sheds light onto the dynamics of router queues and appropriate metrics for delay performance. The understanding of the burstiness of traffic and a router's ability to handle traffic bursts is essential for dimensioning buffer capacities. The best current practice on sizing buffers without explicit knowledge of their behavior across time can be found in refs. [24,72,207].

Data requirements: Packet measurements.

Traffic matrix

The traffic matrix of a network captures the amount of traffic flowing between any ingress and egress point in the network. Traffic may be defined in bytes or packets and its value is bound to evolve with time. As such, a traffic matrix is a dynamic entity, which typically expresses the *average* amount of traffic seen between two locations during a pre-specified period of time, say 5 minutes. The ingress and egress points can be further flexibly defined to correspond to PoPs, routers, or even links, and customers.

Despite the fact that a traffic matrix is a vital input to most network design and planning tasks, its measurement in an IP setting is not a trivial task. One needs not only to keep track of the source and destination IP address of traffic as it enters the network, but also to correlate this information with real-time routing information that can identify its egress point, i.e. the final location inside the network before the packet is delivered to its final destination or to a peering network or customer. In Chapter 5 we show how one can measure or estimate the traffic matrix of an IP backbone network and how we can minimize the associated overhead.

Data requirements: Flow information and routing information.

Designing new metrics/developing new capabilities

Both Chapters 4 and 5 study the behavior of two network quantities that are rather elusive to capture with today's technology: the instantaneous queue length and the traffic matrix. The underlying problem is the need for fine-grain information, such as packet timing information and real-time routing updates. Both chapters recommend ways in which the measurement of these quantities could become commonplace. It is debatable whether such support will ever become an integral part of router technology. Nonetheless, the research community has been instrumental in identifying metrics of interest and recommending efficient methods for their computation. Only in the presence of fine-grain measurements can one achieve such a result.

Data requirements: Packet measurements and routing information.

3.6.2   Network design

The design of an ISP network entails a number of tasks aimed at providing a network with predictable performance, stable behavior and inherent robustness to potential failures. Fundamental inputs to all such tasks are detailed measurements about the traffic flowing on top of the network and the historical behavior of its network elements, such as failures etc. Network design could be further partitioned into the following tasks: (i) topology design, which deals with the positioning of the network elements and the interconnections between them, (ii) traffic engineering, which deals with the management of traffic flow and its predictable treatment in case of failure, and (iii) the evolution of the network, in terms of addition or upgrades of networking equipment, and the links used in their interconnection.

Topology design

ISP backbones typically consist of a set of edge/core routers interconnected by high-speed links. The connectivity between a given set of routers (and hence the router-to-router topology) should be based on the traffic exchanged between every pair of routers. For example, there is little justification for adding a direct link between two routers that exchange a negligible amount of traffic. On the other hand, adding a direct link between two routers that exchange large volumes of traffic may actually improve end-to-end latency. The availability of accurate traffic matrices is essential in such a task. Additional information on the physical layout of fiber paths allows an operator to evaluate alternative designs in case direct interconnection is not possible due to the excessive cost of adding a new fiber.[6]

Further information on the reliability of the network elements and communication links is needed to design a robust topology. Using information on the short-and long-term failure statistics of different network elements can allow a network operator to introduce the appropriate amount of redundancy. In Chapters 7 an 8, such information is used to allow a network operator to map IP links on physical-layer paths in a way that IP layer links between locations exhibit sufficient diversity at the physical layer (they feature at least two different physical-layer paths).

Data requirements: Flow information, routing information, failure information, network element configuration and maps.

Traffic engineering

The goal of traffic engineering is to transport traffic across a network in order to optimize resource utilization and achieve good performance. There is an intrinsic coupling between the traffic that a network has to carry and the routing policies that determine how this traffic is carried across the network. For example, consider an ISP that uses a link-state protocol such as IS–IS/OSPF for intra-domain routing. Knowledge of the traffic matrix is crucial for setting IS–IS link weights and for evaluating the suitability or performance of any routing protocol. The reason is that the link weights will attract or detract traffic from specific parts of the network, achieving appropriate utilization levels across links and acceptable end-to-end delays primarily determined by the length of the path followed by different flows. Detailed understanding of the amount of traffic impacted by particular routing configuration changes can allow a network operator more flexibility in achieving the desirable outcome through specific changes in IS–IS/OSPF link weights. In addition, an understanding of the interaction between BGP and IGP can further enable a network operator to decide on appropriate BGP policies and their impact on their own as well as neighboring networks (see Chapters 9 and 10).

Data requirements: Flow information, routing information, Maps and SNMP data.

Capacity planning and forecasting

Traffic measurements and routing information are key to effective capacity planning. Measurements can identify traffic bottlenecks, which may then be removed by upgrading the capacity of some links and/or creating new paths by adding links or routers. If an ISP can successfully predict the links that would be affected by adding a new customer or a peering link, it can plan ahead and upgrade those links and/or adjust router configurations to avoid congestion.

Successful capacity planning can be achieved only by accurately forecasting the growth of traffic in the network. Inaccurate forecasts can cause a network to oscillate between having excessive capacity and inadequate capacity, which in turn affects the

---

[6] Adding a fiber in an existing fiber conduit is far cheaper than digging a new conduit.

predictability of performance. Statistical forecasting techniques that predict growth in traffic based on past traffic data should be a central component of the forecasting process. In Chapter 11 we describe a measurement-based method for the forecasting of when and where upgrades will be needed in the core of an IP network given its historical growth trends. Use of this information, along with marketing predictions, is likely to enhance significantly current best practices in this area.

Data requirements: SNMP data, network element configuration and maps.

Application classification

The ability of a network operator to classify traffic accurately into network applications directly determines the success of many network management tasks and the definition of strong and reliable business strategies. Network planning, traffic engineering, service level agreement and quality of service monitoring, peering relationship reports, application-content billing and real-time service marketing are just a few examples. Unfortunately the classification of Internet traffic into network applications is not an easy task. Many techniques have been proposed that explore the classification of applications based on *ports*, *strings*, *numerical properties* and *pattern behavior* of the data stream. Unfortunately, the proliferative growth of new applications and the extensive usage of encryption techniques mean that these techniques are not reliable candidates when used as a stand-alone method. In Chapter 12 we explain the reasons for this, and in Chapters 13 and 14 we propose two promising novel methods that may accomplish this important task.

Data requirements: Flow information and packet measurements.

Data- and control-plane security

The primary challenge faced by today's service providers is maintaining service predictability, availability and high quality of delivery (i.e. QoS) in the presence of an outbreak of malicious traffic source from one/multiple end points spread across one/multiple network boundaries. DDoS, Internet worms, viruses, Spam over email and VoIP, phishing and Click-Fraud are just a few examples of malicious threats that service providers have to face every single day. While the above threats are focused on degrading application and network performance and availability, making their effect clearly visible at the flow and packet level, i.e. *data plane*, new types of threats aimed at gaining control of the entire data communication between two end hosts have recently also become the focus of network operators, i.e. *control plane*. In contrast to data-plane threats, the latter act on the routing infrastructure and their activity is completely transparent to the data plane. Examples of such are *prefix hijacking*, *routing hijacking* and *AS path-spoofing*. Due to the great amount of attention received by network security these days, in Chapter 15 we present a novel methodology aimed at detecting a large set of data-plane anomalies, and in Chapter 16 we look more deeply into control-plane anomalies.

Data requirements: Flow information, packet measurements and BGP information.

3.7   Summary

In this chapter, we have discussed the requirements of a monitoring infrastructure that forms part of the design, management and operation of an IP backbone network. We have discussed the need for a monitoring infrastructure comprising a coarse-grained component for continuous network-wide monitoring and a fine-grained component for on-demand monitoring.

We discussed some key implementation challenges for the proposed two-level monitoring system. This includes support of monitoring functionality on routers and the design of special monitoring systems capable of capturing and analyzing packet-level information.

Furthermore,we described the diverse applications of network measurements in the context of network provisioning. It is evident that network management and planning for large ISP networks is a very challenging task. Current market conditions dictate optimal use of the network resources and performance within the bounds defined in rather competitive SLAs. Network measurements can provide the necessary knowledge that could allow the formalization of different network provisioning and planning tasks. Moreover, the development of sound methodological approaches toward the estimation of different performance metrics can serve as a framework, according to which ISPs, and customers, can evaluate the performance offered by packet-switched networks such as the Internet.

The next two chapters will elaborate on ways in which network operators can measure and quantify instantaneous queue sizes on routers, and ways in which they can measure or infer their network's traffic matrix.

4   Understanding through-router delay

End-to-end packet delay is an important metric to measure in networks, both from the network operation and application performance points of view. An important component of this delay is the time for packets to traverse the different switching elements along the path. This is particularly important for network providers, who may have SLAs specifying allowable values of delay across the domains they control. A fundamental building block of the path delay experienced by packets in IP networks is the delay incurred when passing through a single IP router. In this chapter we go through a detailed description of the operations performed on an IP packet when transitting an IP router and measurements of their respective time to completion, as collected on an operational high-end router. Our discussion focuses on the most commonly found router architecture, which is based on a cross-bar switch.[1]

To quantify the individual components of through-router delay, we present results obtained through a unique set of measurements that captures all packets transmitted on all links of an operational access router for a duration of 13 hours. Using this data set, this chapter studies the behavior of those router links that experienced congestion and reports on the magnitude and temporal structure of the resulting packet delays. Such an analysis reveals that cases of overload in operational IP links in the core of an IP network *do* exist, but tend to be of *small magnitude* and *low frequency*. One would expect that such behavior may be singificantly different at the edge of the Internet, where overload may occur more frequently.

Using the knowledge acquired through the analysis of the collected data, the second part of the chapter looks into the derivation of a physical model for router-delay performance. Throughout, we confirm the prevailing assumption that the bottleneck of current router architectures is in the output queues, and justify the commonly used fluid output queue model for an IP router. We go further to demonstrate that refinements to the original model are possible and lead to excellent accuracy. One of the main conclusions of this analysis is that the gap between queueing theory and practice is not that big, as long as one is capable of incorporating enough reality into the model.

Portions of this chapter are reprinted or adapted from N. Hohn, D. Veitch, K. Papagiannaki, and D. Veitch (2004). "Bridging router performance and queueing theory." In *ACM Sigmetrics*, pp. 355–366; http://doi.acm.org/10.1145/1005686.1005728.
[1] An alternative architecture utilizes shared memory instead of a cross bar, thus introducing different through-router-delay behavior.

Having derived a simple model for through-router delays, the third part of the chapter combines the insight from the data analysis and the model derivation to address the question of how one could derive appropriate delay statistics that can be effectively summarized and reported while revealing enough detail about the delay process itself. The current best practice relies on the inference of delay from SNMP utilization. In other words, a network operator assumes that through-router delay is acceptable as long as the *5-minute average* link utilization (provided through SNMP) does not exceed 60% or 70% (where the exact utilization number is determined according to prior experience with performance at those utilization levels). Using a simple counterexample, the chapter argues that such a technique is fundamentally flawed, since queueing delay strongly depends on the structure of the traffic arriving at the router, which is hard to predict. On the other hand, direct on-the-router measurements are shown to convey the appropriate information, while requiring limited processing and memory resources.

Finally, we take a look into the localized periods of congestion, as measured in the collected data. We call such short-lived periods of congestion *micro-congestion episodes*, and we study their causes and the way they manifest themselves in operational large-scale IP networks.

4.1   A router under the microscope

We begin our discussion with a brief description of the virtual output queued router architecture and the measurements that will be used for its study.

4.1.1   Router architecture

The discussion focuses on one particular type of router architecture; a *store & forward* router that is based on a *cross-bar* switching fabric and makes use of *Virtual Output Queues* (VOQs). What this means is that (i) a packet needs to be fully stored in memory before being processed and forwarded to the appropriate output link (store & forward), (ii) a packet is broken down into smaller units, called *cells*, and transmitted to its output link across a cross-bar switch that connects each input link with each output link, and (iii) each input link implements a virtual output queue for each destination link. The router itself features a number of linecards that can be of different technology, with each linecard controlling the input and output directions of the same link. Such an architecture can be found in the Cisco GSR routers. For further details, please see ref. [129].

A typical datapath followed by a packet crossing this kind of router is as follows. First, when a packet arrives at the input link of a linecard, its destination address is looked up in the forwarding table. This does not occur, however, until the packet completely leaves the input link and fully arrives in the linecard's memory – the "store" part of store & forward. Virtual output queuing means that each input interface has a separate First In First Out (FIFO) queue dedicated to each output interface. The packet is stored in the appropriate queue of the input interface, where it is decomposed into fixed-length cells. When the packet reaches the head of line, it is transmitted through the switching fabric cell by cell (possibly interleaved with competing cells from VOQs at other input interfaces dedicated to the same output interface) to its output interface and reassembled before being handed to the output link scheduler – the "forward" part of store & forward. The packet might then experience queuing before being serialized without interruption onto the output link. In queuing terminology it is "served" at a rate equal to the bandwidth of the output link, and the output process is of fluid type because the packet flows out gradually instead of leaving in an instant.

In the above description the packet might be queued both at the input interface and the output link scheduler. However, in practice such routers have the capacity to switch packets at a much higher rate than their input bandwidth, which means that any significant queuing should only take place in the output link scheduler.

Overheads of different encapsulation layers

To study the behavior of through-router delays, we present findings from the instrumentation of an operational VOQ router at the access of a large-scale IP network; this router facilitates the connection of four customers to the backbone network. The router is a Cisco GSR with interfaces that use the High-level Data Link Control (HDLC) protocol as a transport layer to carry IP datagrams over a Synchronous Optical NETwork (SONET) physical layer. Packet-over-SONET (POS) is a popular choice to carry IP packets in high-speed networks because it provides a more efficient link layer than classical IP-over-ATM, and faster failure detection than other broadcast technologies. However, given the HDLC-over-SONET encapsulation, a link's effective bandwidth is no longer equal to its nominal value.

The first level of encapsulation is the SONET framing mechanism. A basic SONET OC-1 frame contains 810 bytes and is repeated with an 8 kHz frequency. This yields a nominal bandwidth of exactly 51.84 Mbps. Since each SONET frame is divided into a transport overhead of 27 bytes, a path overhead of 3 bytes and an effective payload of 780 bytes, the bandwidth accessible to the transport protocol, also called the effective bandwidth, is in fact 49.92 Mbps. OC-n bandwidth (with $n \in \{3, 12, 48, 192\}$) is achieved by merging n basic frames into a single larger frame and sending it at the same 8-kHz rate. The bandwidth accessible to the transport protocol, is therefore $(49.92 * n)$ Mbps. For instance, the effective bandwidth of an OC-3 link is exactly 149.76 Mbps.

The second level of encapsulation is the HDLC transport layer. This protocol adds 5 bytes before and 4 bytes after each IP datagram, irrespective of the SONET interface speed [179].

These layer overheads mean that, in terms of queuing behavior, an IP datagram of size b bytes carried over an OC-3 link should be considered as a $(b + 9)$-byte packet transmitted at 149.76 Mbps. The importance of these seemingly technical points will be demonstrated in Section 4.3.

Timestamping of POS packets

Measuring packet delays through operational routers requires highly accurate synchronization. To ensure such a degree of accuracy, all measurements were made using high-performance passive monitoring DAG cards [64] (described in Chapter 3). The monitoring of the OC-3 and OC-12 links was done using DAG 3.2 cards, while OC-48 links were monitored using DAG 4.11 cards. The cards use different technologies to timestamp POS packets.

DAG 3.2 cards are based on a design dedicated to ATM measurement and therefore operate with 53-byte chunks corresponding to the length of an ATM cell. The POS timestamping functionality was added at a later stage without altering the original 53-byte processing scheme. However, since POS frames are not aligned with the 53-byte divisions of the POS stream operated by the DAG card, significant timestamping errors occur. In fact, a timestamp is generated when a new SONET frame is detected within a 53-byte chunk. This mechanism can cause errors of up to $2.2\,\mu s$ on an OC-3 link [67].

DAG 4.11 cards are dedicated to POS measurement and do not suffer from the above limitations. They look past the POS encapsulation (in this case HDLC) to timestamp each IP datagram consistently after the first (32-bit) word has arrived.

As a direct consequence of the characteristics of the measurement cards, timestamps on OC-3 links have a worst-case precision of $2.2\,\mu s$. Adding errors due to potential GPS synchronization problems between different DAG cards leads to a worst-case error of $6\,\mu s$ [81]. This number should be kept in mind when we assess the router model performance.

4.1.2   Experimental setup

The data analyzed in this chapter were collected in August, 2003, at a gateway router of Sprint's Tier-1 network. The requirement behind the experimental setup was the monitoring of all links of an operational router, which translated to the search for routers that featured linecards that could be monitored using DAG cards, i.e. OC-3, OC-12 and OC-48 rates at the time of the experiment, and were located in one of the locations already monitored using packet-capture systems. Moroever, the target router needed to have no more than five or six links, such that it can be monitored using ten or twelve monitoring boxes at most. Such a search returned a single match, a router with six interfaces that, if monitored, accounted for more than 99.95% of all traffic flowing through it. A small Ethernet link carrying less than five packets per second was not monitored due to the unavailability of additional monitoring boxes and the respective DAG cards; given the small amount of traffic affected, such an omission was considered acceptable. The experimental setup is illustrated in Figure 4.1. Two of the interfaces are OC-48 linecards connecting to two backbone routers (BB1 and BB2), while the other four connect customer links: two trans-Pacific OC-3 linecards to Asia (C2 and C3), one OC-3 (C1) and one OC-12 (C4) linecard to domestic customers.

Each DAG card was synchronized with the same GPS signal and output a fixed-length 64-byte record for each packet on the monitored link. The details of the record depend on the link type (ATM, SONET, or Ethernet). In our case, all the IP packets were carried with the Packet-Over-SONET (POS) protocol, and each 64-byte record consisted of 8 bytes for the timestamp, 12 bytes for control and POS headers, 20 bytes for the IP header and the first 24 bytes of the IP payload (Figure 3.1).

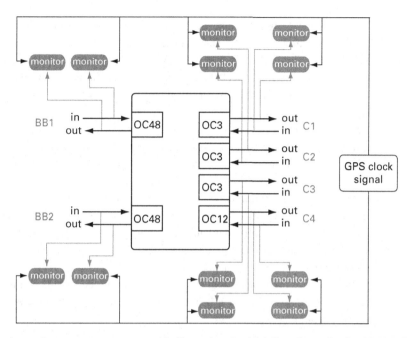

Fig. 4.1.   Experimental setup: gateway router with 12 synchronized DAG cards monitoring 99.9% of all IP traffic flowing through. ©ACM, 2004.

We captured 13 hours of mutually synchronized traces, representing more than 7.3 billion IP packets or 3 TBytes of traffic. The DAG cards were physically located close enough to the router so that the time taken by packets to go between them could be neglected.

4.1.3   Packet matching

The next step after the trace collection is the packet-matching procedure. It consists in identifying, across all the traces, the records corresponding to the same packet appearing at different interfaces at different times. In our case, the records all relate to a single router, but the packet-matching program can also accommodate multi-hop situations. We describe below the matching procedure, and illustrate it in the specific case of the customer link C2-out. Our methodology follows ref. [156].

We match identical packets coming in and out of the router by using a hash table. The hash function is based on the CRC algorithm and uses the IP source and destination addresses, the IP header identification number and, in most cases, the full 24-byte IP header data part. In fact, when a packet size is less than 44 bytes, the DAG card uses a padding technique to extend the record length to 64 bytes. Since different models of DAG cards use different padding content, the padded bytes are not included in the hash function. Our matching algorithm uses a sliding window over all the synchronized traces in parallel to match packets hashing to the same key. When two packets from two different links are matched, a record of the input and output timestamps, as well as the 44-byte POS payload, is produced. Sometimes two packets from the same link hash to the same key because they are identical; these packets are duplicate packets generated by the physical layer [162]. They can create ambiguities in the matching process and are therefore discarded; however, their frequency is monitored.

The task of matching packets is computationally intensive and demanding in terms of storage: the total size of the result files rivals that of the raw data. For each output link of the router, the packet-matching program creates one file of matched packets per contributing input link. For instance, for output link C2-out, four files are created, corresponding to the packets coming, respectively, from BB1-in, BB2-in, C1-in and C4-in (the input link C3-in has virtually no traffic and is discarded by the matching algorithm). All the packets on a link for which no match could be found were carefully analyzed. Apart from duplicate packets, unmatched packets comprise packets going to or coming from the small unmonitored link, or sourced and destined from/to the router itself. There could also be unmatched packets due to packet drops at the router. Since the router did not drop a single packet over the 13 hours (verified using the respective SNMP counters), no such packets were found.

Assume that the matching algorithm has determined that the mth packet of output link Λ_j corresponds to the nth packet of input link λ_i. This can be formalized by a *matching function* $\mathcal{M}$, obeying

$$\mathcal{M}(\Lambda_j, m) = (\lambda_i, n). \tag{4.1}$$

The matching procedure effectively defines this function for all packets over all output links. Packets that cannot be matched are not considered part of the domain of definition of $\mathcal{M}$.

Table 4.1 summarizes the results of the matching procedure. The percentage of matched packets is at least 99.6% on each link, and is as high as 99.98%, showing

Table 4.1. Trace details

Each packet trace was collected on August 14, 2003, between 03:30 and 16:30 UTC. They comprise 7.3 billion IP packets and 3 TBytes of IP traffic.

Set	Link	No. of packets	Average rate (Mbps)	Matched packets (% total traffic)	Duplicate packets (% total traffic)	Router traffic (% total traffic)
BB1	in	817 883 374	83	99.87%	0.045	0.004
	out	808 319 378	53	99.79%	0.066	0.014
BB2	in	1 143 729 157	80	99.84%	0.038	0.009
	out	882 107 803	69	99.81%	0.084	0.008
C1	out	103 211 197	3	99.60%	0.155	0.023
	in	133 293 630	15	99.61%	0.249	0.006
C2	out	735 717 147	77	99.93%	0.011	0.001
	in	1 479 788 404	70	99.84%	0.050	0.001
C3	out	382 732 458	64	99.98%	0.005	0.001
	in	16 263	0.003	N/A	N/A	N/A
C4	out	480 635 952	20	99.74%	0.109	0.008
	in	342 414 216	36	99.76%	0.129	0.008

Table 4.2. Breakdown of packet matching for output link C2-out

Set	Link	No. of matched packets	Traffic on C2-out
C4	in	215 987	0.03%
C1	in	70 376	0.01%
BB1	in	345 796 622	47.00%
BB2	in	389 153 772	52.89%
C2	out	735 236 757	99.93%

convincingly that almost all packets are matched. In fact, even if there were no duplicate packets and if absolutely all packets were monitored, 100% could not be attained because of router-generated packets, which represent roughly 0.01% of all traffic in our data set.

The packet-matching results for the customer link C2-out are detailed in Table 4.2. For this link, 99.93% of the packets can be successfully traced back to packets entering the router. In fact, C2-out receives most of its packets from the two OC-48 backbone links BB1-in and BB2-in. This is illustrated in Figure 4.2, where the utilization of C2-out across the full 13 hours is plotted. The breakdown of traffic according to packet origin shows that the contributions of the two incoming backbone links are roughly similar. This is the result of the Equal-Cost MultiPath (ECMP) policy employed in the network when packets may follow more than one (equal-cost) path to the same destination. While the utilization in Mbps (Figure 4.2(a)) gives an idea of how congested the link might be, the utilization in packets per second is important from a packet-tracking perspective. Since the matching procedure is a per packet mechanism, Figure 4.2(b) illustrates the fact that roughly all packets are matched: the sum of the input contributions is almost indistinguishable from the output packet count.

In the remainder of the chapter, we focus on link C2-out because it is the most highly utilized link and is fed by two faster links. It is therefore the best candidate for observing queuing behavior within the router. The reasons behind such queue buildup are investigated in Section 4.5.

4.2 Preliminary delay analysis

In this section we analyze the data obtained from the packet-matching procedure. We start by carefully defining the system under study, and then present the statistics of the delays experienced by packets crossing it. The point of view is that of looking at the outside of the router, seen largely as a "black box," and we concentrate on simple statistics. In the next section we begin to look inside the router, where we examine delays in greater detail.

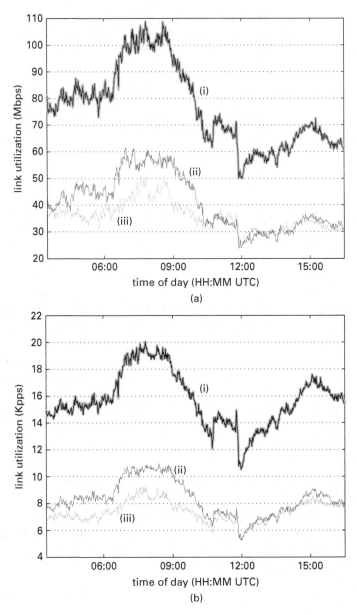

Fig. 4.2.   Utilization for link C2-out in (a) megabits per second (Mbps) and (b) kilo packets per second (kpps). The total output curve C2-out (see the shaded curve in (i)) is hidden by the sum of all input links (see the solid line in (i)), in agreement with the 99.93% packet-matching result of Table 4.2. Curves (ii) and (iii) depict input BB2-in to C2-out and input BB1-in to C2-out, respectively. The contributions from links C1-in and C4-in are too small to be visible. ©ACM, 2004.

4.2.1   System definition

Recall the notation from equation (4.1): the mth packet of output link Λ_j corresponds to the nth packet of input link λ_i. The DAG card timestamps an IP packet on the incoming interface side as $t(\lambda_i, n)$, and later on the outgoing interface at time $t(\Lambda_j, m)$. As the DAG cards are physically close to the router, one might think to define the through-router delay as $t(\Lambda_j, m) - t(\lambda_i, n)$. However, we know from Section 4.1.1 that the DAG timestamps packets in different ways. Depending on the version number, this would amount to defining the router "system" in a somewhat arbitrary way.

For self-consistency and extensibility to a multi-hop scenario, where we would like individual router delays to add, there should be symmetry between the arrival and departure instants of the packet to the router. It is natural to focus on the end of the (IP) packet for two reasons. (1) as a store & forward router, the output queue is the most important component to describe. It is therefore appropriate to consider that the packet has left the router when it *completes* its service at the output queue, that is when it has completely exited the router. (2) Again as a store & forward router, no action (e.g. the forwarding decision) is performed until the packet has fully entered, which moreover it must do at the input line rate. Thus the input buffer can be considered as part of the input link, and arrival at the router only occurs with the last bit.

The arrival and departure instants in fact define the "system," which is the part of the router which we study, and is not exactly the same as the physical router as it excises the input buffer. This buffer, being a component that is already understood, does not have to be modeled or measured. Defining the system in this way can be compared to choosing the most practical coordinate system to solve a given problem.

We now establish the precise relationships between the DAG timestamps defined earlier and the time instants $\tau(\lambda_i, n)$ of arrival and $\tau(\Lambda_j, m)$ of departure of a given packet to the system as just defined. Denote by $l_n = L_m$ the size of the packet in bytes when indexed on links λ_i and Λ_j, respectively, and let θ_i and Θ_j be the corresponding link bandwidths in bits per second. We denote by H the function giving the depth of bytes into the IP packet where the DAG timestamps it; H is a function of the link speed, but not the link direction. For a given link λ_i, H is defined as follows:

$$H(\lambda_i) = 4 \quad \text{if} \quad \lambda_i \quad \text{is an OC-48 link,}$$
$$= b \quad \text{if} \quad \lambda_i \quad \text{is an OC-3 or OC-12 link,}$$

where we take b to be a uniformly distributed integer between zero and $\min(l_n, 53)$ to account for the ATM-based discretization described earlier. We can now derive the desired system arrival and departure event times:

$$\tau(\lambda_i, n) = t(\lambda_i, n) + 8(l_n - H(\lambda_i))/\theta_i, \qquad (4.2)$$
$$\tau(\Lambda_j, m) = t(\Lambda_j, m) + 8(L_m - H(\Lambda_j))/\Theta_j.$$

These events and event times are displayed schematically in Figure 4.3.

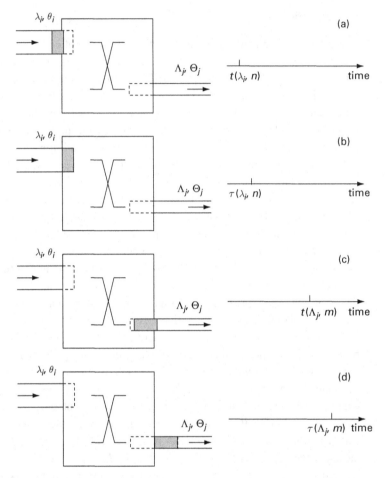

Fig. 4.3.   Four snapshots in the life of a packet crossing the router. (a) The packet is timestamped by the DAG card monitoring the input interface at time $t(\lambda_i, n)$, at which point it has already entered the router, but not yet the system. (b) It has finished entering the router (arrives at the system) at time $\tau(\lambda_i, n)$ and (c) is timestamped by the DAG at the output interface at time $t(\Lambda_j, m)$. (d) It fully exits the router (and system) at time $\tau(\Lambda_j, m)$. ©ACM, 2004.

With the above notations, the through-*system* delay experienced by packet m on link Λ_j is defined as follows:

$$d_{\lambda_i, \Lambda_j}(m) = \tau(\Lambda_j, m) - \tau(\lambda_i, n). \tag{4.3}$$

To simplify the notation, we shorten this to $d(m)$ in what follows.

4.2.2   Delay statistics

Figure 4.4 shows the minimum, mean and maximum delays experienced by packets going from input link BB1-in to output link C2-out over consecutive 1-minute intervals. As observed in ref. [156], there is a constant minimum delay across time, up to

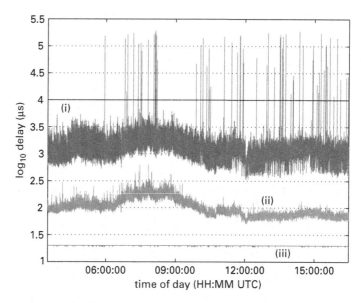

Fig. 4.4. Packet delays from BB1-in to C2-out. (i) Maximum, (ii) mean, (iii) minimum. All delays above 10 ms are due to option packets. ©ACM, 2004.

timestamping precision. The fluctuations in the mean delay roughly follow the changes in the link utilization presented in Figure 4.2. The maximum delay value has a noisy component with similar variations to the mean, as well as a spiky component. All the spikes above 10 ms have been individually studied. The analysis revealed that they are all caused by IP packets carrying options. Option packets take different, specific paths through the router since they are processed through software, while all other packets are processed with dedicated hardware. This explains why they take significantly longer to cross the router.

In any router architecture, it is likely that a component of delay will be proportional to packet size. This is certainly the case for store & forward routers, as discussed in ref. [111]. To investigate this here we compute the "excess" minimum delay experienced by packets of different sizes, that is not including their transmission time on the output link, which is a packet-size-dependent component that is already understood. Formally, for every packet size L, we compute

$$\Delta(L) = \min_m \{d(m) - 8l_m/\Theta_j | l_m = L\}. \tag{4.4}$$

Note that our definition of arrival time to the system conveniently excludes another packet-size-dependent component, namely the time interval between beginning and completing entry to the router at the input interface.

Figure 4.5 shows the values of $\Delta(L)$ found for the output link C2-out. The IP packet sizes observed varied between 28 and 1500 bytes. We assume (for each size) that the minimum value found across 13 hours corresponds to the true minimum, i.e. that at least one packet encountered no contention on its way to the output queue and no packet in the output queue when it arrived there. In other words, we assume that the

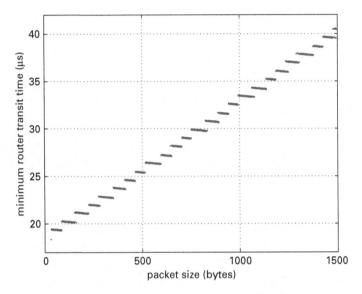

Fig. 4.5.   Measured minimum router transit times for output link C2-out. ©ACM, 2004.

system was empty from the point of view of this input–output pair. This means that the excess minimum delay corresponds to the time taken to make a forwarding decision (not packet-size-dependent), to divide the packet into cells, transmit it across the switch fabric and reassemble it (each being a packet-size-dependent operation), and finally to deliver it to the appropriate output queue. The step-like curve means that there exist ranges of packet sizes with the same minimum transit time. This is consistent with the fact that each packet is divided into fixed-length cells, transmitted through the back-plane cell by cell and reassembled. A given number of cells can therefore correspond to a contiguous range of packet sizes with the same minimum transit time.

4.3   Modeling

Having established our system definition, and defined and observed the main character-istics of what we are measuring, we are now in a position to exploit the completeness of the data set to look inside the system. This enables us to find a physically meaningful model which can be used both to understand and predict the end-to-end system delay very accurately.

4.3.1   The fluid queue

We first recall some basic properties of FIFO queues that will be central in what follows. Consider a FIFO queue with a single server of deterministic service rate μ, and let t_i be the arrival time to the system of packet i of size L_i bytes. We consider that the entire packet arrives instantaneously (which models a fast transfer across the switch), but that it leaves progressively as it is served (modeling the output serialization). Thus it is a fluid queue at the output but not at the input. Nonetheless, we will for convenience refer to it as the "fluid queue."

Let W_i be the time packet i waits before being served. The service time of packet i is simply L_i/μ, so the *system time*, that is the total amount of time spent in the system, is just given by

$$S_i = W_i + \frac{L_i}{\mu}. \tag{4.5}$$

The waiting time of the next packet $(i + 1)$ to enter the system can be expressed by the following recursion:

$$W_{i+1} = \left[W_i + \frac{L_i}{\mu} - (t_{i+1} - t_i) \right]^+, \tag{4.6}$$

where $[x]^+ = \max(x, 0)$. The service time of packet $i + 1$ is given by

$$S_{i+1} = [S_i - (t_{i+1} - t_i)]^+ + \frac{L_{i+1}}{\mu}. \tag{4.7}$$

Let $U(t)$ denote the amount of *unfinished work* at time t, that is the time it would take, with no further inputs, for the system to drain completely. The unfinished work at the instant following the arrival of packet i is nothing other than the end-to-end delay that that packet will experience across the queuing system. It is therefore the natural mathematical quantity to consider when studying delay. Note that it is defined at all real times t.

4.3.2 Simple router model

The delay analysis of Section 4.2 revealed two main features of the system delay that should be taken into account in a model: the minimum delay experienced by a packet, which is size-, interface- and architecture-dependent, and the deterministic delay corresponding to the time spent in the output buffer, which is a function of the rate of the output interface only.

The delay across the output buffer could by itself be modeled by the fluid queue as described above; however, it is not obvious how to incorporate the minimum-delay property in a sensible way.

Assume, for instance, that the router has N input links contributing to a given output link and that a packet of size b arriving on link j experiences at least the minimum possible delay $\Delta_j(b)$ before being transferred to the output buffer. A representation of this situation is given in Figure 4.6(a). Our first problem is that, given different technologies on different interfaces, the functions Δ_1, ..., Δ_N are not necessarily identical. The second is that we do not know how to measure, to take into account, the potentially complex interactions between packets which do *not* experience the minimum excess delay but instead experience some larger value due to contention with cross traffic.

We address this by simplifying the picture further, in three ways. First, we assume that the minimum delays are identical across all input interfaces: a packet of size b arriving

(a)

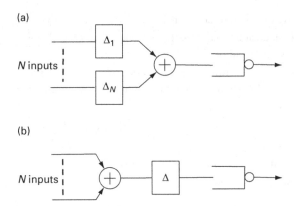

(b)

Fig. 4.6.   Router mechanisms. (a) Simple conceptual picture including VOQs. (b) Actual model with a
single common minimum delay. ©ACM, 2004.

on link j now experiences an excess minimum delay $\Delta(b) = \min_j\{\Delta_j(b)\}$. Second,
we assume that the multiplexing of the different input streams takes place before the
packets experience their minimum delay. By this we mean that we preserve the order
of their arrival times and consider them to enter a single FIFO input buffer. Finally, we
ignore all complex interactions between the input streams. Our highly simplified picture is
shown in Figure 4.6(b). We will justify these simplifications a posteriori in Section 4.3.3,
where the comparison with measurement shows that the model is remarkably accurate.

Let us now analyze the model. Suppose that a packet of size b enters the system at
time t^+ and that the amount of unfinished work in the system at time t^- was $U(t^-) >
\Delta(b)$. The following two scenarios produce the same total delay:

(1) the packet experiences a delay $\Delta(b)$, then reaches the output queue where it has to
    wait $U(t) - \Delta(b) > 0$ before being served;
(2) the packet reaches the output queue straight away and has to wait $U(t)$ before being
    served.

In other words, as long as there is more than $\Delta(b)$ amount of work in the queue when
a packet of size b enters the system, the fact that the packet should wait $\Delta(b)$ before
reaching the output queue can be neglected. Once the system is busy, it behaves exactly
like a simple fluid queue. This implies that no matter how complicated the front end of
the router is, one can simply neglect it when the output queue is sufficiently busy. The
simplicity and robustness is the main reason behind choosing the model.

A system equation for our two-stage model can be derived as follows. Assume that
the system is empty at time t_0^- and that a packet k_0 of size l_0 enters the system at time
t_0^+. It waits $\Delta(b_0)$ before reaching the empty output queue, where it immediately starts
being served. Its service time is b_0/μ, and therefore its total system time is given by

$$S_0 = \Delta(b_0) + \frac{b_0}{\mu}. \tag{4.8}$$

Suppose a second packet enters the system at time t_1 and reaches the output queue before the first packet has finished being served, i.e. $t_1 + \Delta(l_1) < t_0 + S_0$. It will start being served when packet k_0 leaves the system, i.e. at $t_0 + S_0$. Its system time will therefore be given by

$$S_1 = S_0 - (t_1 - t_0) + \frac{L_1}{\mu}. \tag{4.9}$$

The same recursion holds for successive packets k and $k + 1$ as long as the amount of unfinished work in the queue remains above $\Delta(l_{k+1})$ when packet $k + 1$ enters the system:

$$t_{k+1} + \Delta(l_{k+1}) < t_k + S_k. \tag{4.10}$$

Therefore, as long as equation (4.10) is verified, the system times of successive packets are obtained by the same recursion as for the case of a busy fluid queue:

$$S_{k+1} = S_k - (t_{k+1} - t_k) + \frac{L_{k+1}}{\mu}. \tag{4.11}$$

Suppose now that packet $k + 1$ of size l_{k+1} enters the system at time t_{k+1}^+ and that the amount of unfinished work in the system at time t_{k+1}^- is such that $0 < U(t_{k+1}^-) < \Delta(l_{k+1})$. In this case, the output buffer will be empty by the time packet $k + 1$ reaches it after having waited $\Delta(l_{k+1})$ in the first stage of the model. The service time of packet $k + 1$ is therefore given by

$$S_{k+1} = \Delta(l_{k+1}) + \frac{L_{k+1}}{\mu}. \tag{4.12}$$

A crucial point to note here is that, in this situation, the output queue can be *empty* but the system remains *busy* with a packet waiting in the front end. This is also true of the actual router.

Once the queue has drained, the system is idle until the arrival of the next packet. The time between the arrival of a packet to the empty system and the time when the system becomes empty again defines a *system busy period*. In this brief analysis, we have assumed an infinite buffer size. It is a reasonable assumption since it is quite common for a line card to be able to accommodate up to 500 ms worth of traffic.

4.3.3 Evaluation

We now evaluate our model and compare its results with our empirical delay measurements. The model delays are obtained by multiplexing the traffic streams BB1-in to C2-out and BB2-in to C2-out and feeding the resulting packet train to the model in an exact trace driven "simulation." Figure 4.7 shows two sample paths of the unfinished work $U(t)$ corresponding to two fragments of real traffic destined to C2-out. A vertical jump marks the arrival time of a new packet. The resultant new local maximum is the time taken by the newly arrived packet to cross the system, that is its delay. The black dots represent the actual measured delays for the corresponding input packets. In practice, the queue state can only be measured when a packet exits the system. Thus the

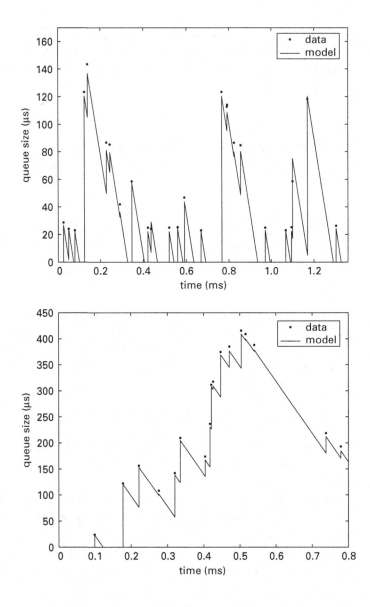

Fig. 4.7.   Comparisons of measured and predicted delays on link C2-out. Black line: unfinished work $U(t)$ in the system according to the model. Black dots: measured delay value for each packet. ©ACM, 2004.

black dots can be thought of samples of the continuous $U(t)$ function from the model, and agreement between the two seems very good.

In order to see the limitations of our model, we focus on a set of busy periods on link C2-out involving 510 packets all together. Figure 4.8(a) shows the system times experienced by incoming packets, both from the model and from measurements. The largest busy period has a duration of roughly 16 ms and an amplitude of more than 5 ms. Once

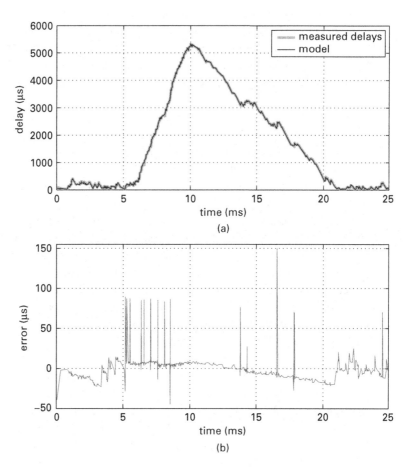

Fig. 4.8.   (a) Measured delays and model predictions. (b) Absolute error between data and model.
ⒸACM, 2004.

again, the model reproduces the measured delays very well. Figure 4.8(b) shows the error of our model, that is the difference between measured and modeled delays at each packet arrival time, plotted on the same time axis as in Figure 4.8(a).

There are three main points one can make about the model accuracy. First, the absolute error is within $30\,\mu s$ of the measured delays for almost all packets. Second, the error is much larger for a few packets, as shown by the spiky behavior of the error plot. These spikes are due to a local reordering of packets inside the router that is not captured by our model. Recall from Figure 4.6 that we made the simplifying assumption of swapping the multiplexing and delay operator in our router model. This means that packets exit our system in the exact same order as they entered it. However, in practice, local reordering can happen when a large packet arrives at the system on one interface just before a small packet on another interface. Given that the minimum transit time of a packet depends linearly on its size (see Figure 4.5), the small packet can overtake the large one and reach the output buffer first. Once the two packets have reached the output buffer, the amount of work in the system is the same, irrespective of their arrival order. Thus these local errors do not accumulate. Intuitively, local reordering requires that two packets arrive almost at the same time on two different interfaces. This is much more likely to happen when the links are busy. This is in agreement with Figure 4.8, which shows that spikes always happen when the queuing delays are increasing, a sign of high local link utilization.

The last point worth noticing is the systematic linear drift of the error across a busy period duration. This is due to the fact that our queuing model drains slightly faster than the real queue. We could not confirm any physical reason why the effective bandwidth of the link C2-out is smaller than what was predicted in Section 4.1.1. The important point to make here is that this phenomenon is only noticeable for very large busy periods, and is lost in measurement noise for most busy periods.

The model presented above has some limitations. First, it does not take into account the fact that a small number of option packets will take a "slow" software path through the router instead of being entirely processed at the hardware level. As a result, option packets experience a delay much larger than the other packets before reaching the output buffer, but, as far as the model is concerned, transit times through the router only depend on packet sizes. Second, the output queue stores not only the packets crossing the router, but also the "unmatched" packets generated by the router itself, as well as control POS packets. These packets are not accounted for in the model.

Despite its simplicity, our model is considerably more accurate than other single-hop delay models. Figure 4.9(a) compares the errors made on the packet delays from the OC-3 link C2-out presented in Figure 4.8 with three different models: our two-stage model; a fluid queue with OC-3 nominal bandwidth; and a fluid queue with OC-3 effective bandwidth. As expected, with a simple fluid model, i.e. when one does not take into account the minimum transit time, all the delays are systematically underestimated. If moreover one chooses the nominal link bandwidth (155.52 Mbps) for the queue instead of a carefully justified effective bandwidth (149.76 Mbps), the errors inside a busy period build up very quickly because the queue drains too fast. There is in fact only a 4% difference between the nominal and effective bandwidths, but this is enough to create errors up 800 μs inside a large busy period.

Figure 4.9(b) shows the cumulative distribution function of the delay error for a 5-minute window of C2-out traffic. Of the delays inferred by our model, 90% are within 20 μs of the measured ones. Given the timestamping precision issues described in Section 4.1.1, these results are very satisfactory.

We now evaluate the performance of our model over the entire 13 hours of traffic on C2-out as follows. We divide the period into 156 intervals of 5 minutes. For each interval, we plot the average relative delay error against the average link utilization. The results are presented in Figure 4.9. The absolute relative error is less than 1.5% for the whole trace, which confirms the excellent match between the model and the measurements. For large utilization levels, the relative error grows due to the fact that large busy periods are more frequent. The packet delays therefore tend to be underestimated more often due to the unexplained bandwidth mismatch occurring inside large busy periods. Overall, our model performs very well for a large range of link utilizations.

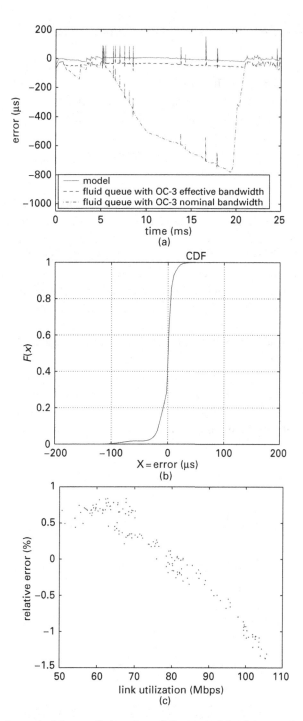

Fig. 4.9. (a) Comparison of error in delay predictions from different models of the sample path from Figure 4.8. (b) Cumulative Distribution Function (CDF) of model error over a 5-minute window on link C2-out. (c) Relative mean error between delay measurements and model on link C2-out versus link utilization. ©ACM, 2004.

4.4   Understanding and reporting delay performance

In Section 4.3 we saw that the presented router model can accurately predict delays when the input traffic is fully characterized. However, in practice, the traffic is unknown, which is why network operators rely on available simple statistics, such as curves giving upper bounds on the delay as a function of the link utilization, when they want to infer packet delays through their network. The problem is that these curves are not unique and depend heavily on the traffic characteristics. To illustrate this important point, we briefly give two examples.

First, we perform the following experiment: within each 5-minute window of aggregated traffic destined to C2-out, we replace the original packet arrival times by a Poisson process with the same rate and keep the packet sizes unchanged. The newly created Poisson traffic therefore only differs from the original in its packet arrival times. We feed both the real traffic and the Poisson traffic through our model and compare the mean delay experienced by packets within each 5-minute window. In Figure 4.10 we plot for each 5-minute window the average packet delay as a function of the link utilization. For a given utilization, packets from the Poisson traffic systematically experience smaller delays than the original packets. Although not surprising, this shows very clearly that a curve of mean delay versus link utilization depends strongly on the statistics of the input traffic, and is not universal.

Now suppose that there is a group of back-to-back packets on link C2-out. This means that the local link utilization is 100%. However, this does not imply that these packets have experienced large delays inside the router. They could in fact very well be coming back-to-back from the input link C1-in with same bandwidth as C2-out. In this case,

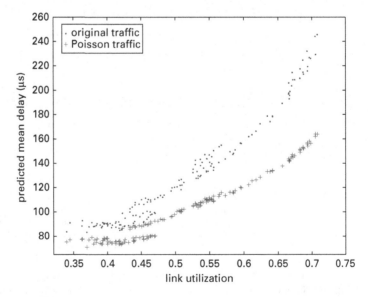

Fig. 4.10.   Predicted mean packet delays for original traffic and Poisson traffic with same average packet rate and same packet sizes.

they would actually flow through the router with minimum delay in the absence of cross traffic.

These two arguments show how misleading utilization can be as a way of inferring packet delays. To improve the inference of packet-delay performance, one needs to understand the source of large delays, that is the queue build-up events in the output buffer. Instead, we propose to study performance-related questions by going back to the source of large delays: queue buildups in the output buffer. Using our understanding of the router mechanisms obtained from our measurements and modeling work of the previous sections, we now describe the statistics and causes of busy periods, and propose a simple mechanism that could be used to report useful delay information on a router.

4.4.1 Busy periods

Definition

Recall from Section 4.3 that we defined *busy periods* as the time between the arrival of a packet in the empty system and the time when the system goes back to its empty state. The equivalent definition in terms of measurements is as follows: a busy period starts when a packet of size b bytes crosses the system with a delay $\Delta(b) + b/\mu$ and ends with the last packet before the start of another busy period. This definition, which makes full use of our measurements, is a lot more robust than an alternative definition based solely on packet inter-arrival times at the output link. For instance, if one were to detect busy periods by using timestamps and packet sizes to group together back-to-back packets, the following two problems would occur. First, timestamping errors could lead to wrong busy period separations. Second, and more importantly, according to our system definition from Section 4.3.2, packets belonging to the same busy period are not necessarily back-to-back on the output link (see equation (4.12)).

Statistics

To describe busy periods, we begin by collecting per-busy-period statistics, such as duration, number of packets and bytes, or amplitude (maximum delay experienced by a packet inside the busy period). The Cumulative Distribution Functions of busy-period amplitudes and durations are plotted in Figures 4.11(a) and (b) for a 5-minute traffic window. For this traffic window, 90% of busy periods have an amplitude smaller than $200\,\mu s$, and 80% last less than $500\,\mu s$. Figure 4.11(c) shows a scatter plot of busy-period amplitudes against busy-period durations for amplitudes larger than 2 ms on link C2-out (busy periods containing option packets are not shown). There does not seem to be any clear pattern linking amplitude and duration of a busy period in this data set, although the longer the busy period, the larger its amplitude.

A scatter plot of busy-period amplitudes against the median delay experienced by packets inside the busy period is presented in Figure 4.11(d). One can see a linear, albeit noisy, relationship between maximum and median delay experienced by packets inside a busy period. This means intuitively that busy periods have a "regular" shape, i.e. busy periods during which most of the packets experience small delays and only a few packets experience much larger delays are very unlikely.

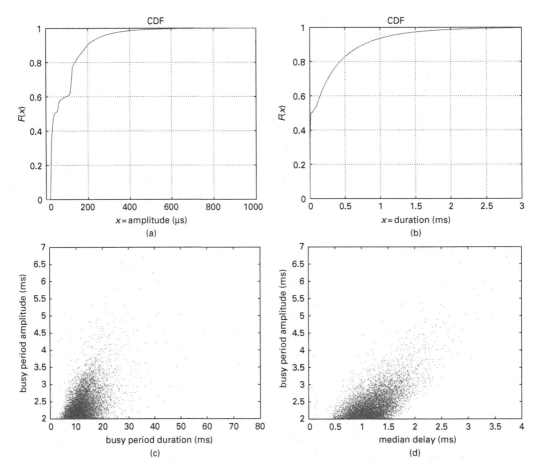

Fig. 4.11.   (a) CDF of busy-period amplitudes. (b) CDF of busy-period durations. (c) Busy-period amplitudes as a function of busy-period durations. (d) Busy-period amplitudes as a function of median packet delay. ©ACM, 2004.

Origins

Our full router measurements allow us to go further in the characterization of busy periods. In particular, we can use our knowledge about the input packet streams on each interface to understand the mechanisms that create the busy periods observed on our router output links. It is clear that busy periods are created by a local packet arrival rate larger than the output link service rate. This can be achieved by a single input stream, the multiplexing of different input streams, or a combination of both phenomena. We now illustrate these different mechanisms.

Let us first understand how the collection of busy periods shown in Figure 4.8 was created. To do so, we store the individual packet streams BB1-in to C2-out and BB2-in to C2-out, feed them individually to our model and obtain *virtual* busy periods. The delays obtained are plotted in Figure 4.12(a), together with the true delays measured on link C2-out. In the absence of cross traffic, the maximum delay experienced by packets from each individual input stream is around 1 ms. However, the largest delay for the

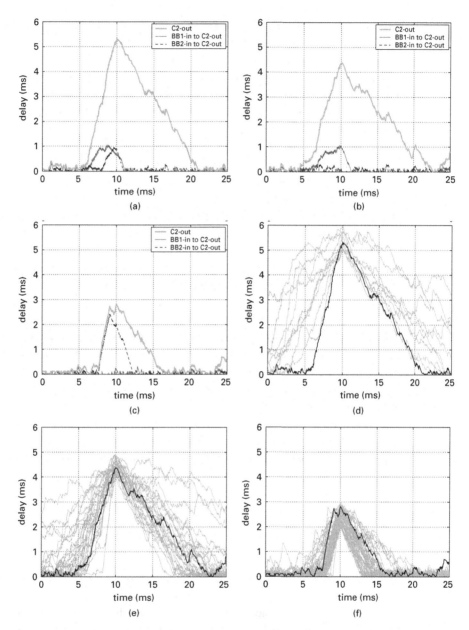

Fig. 4.12. (a)–(c) Illustrations of the multiplexing effect leading to a busy period on the output link C2-out. (d)–(f) Collection of largest busy periods in each 5-minute interval on the output link C2-out. See text for details. ©ACM, 2004.

multiplexed inputs is around 5 ms. The large busy period is therefore due to the fact that the delays of the two individual packet streams peak at the same time. This non-linear phenomenon is the basic ingredient explaining all the large busy periods observed in our traces. We go into more detail on the causes behind busy periods in Section 4.5. A more surprising example is illustrated in Figure 4.12(b), which shows one input stream creating at most a 1 ms packet delay by itself and the other a succession of 200 μs delays. The resulting congestion episode for the multiplexed inputs is again much larger than the individual contributions. A different situation is shown in Figure 4.12(c), where one link contributes almost all the traffic of the output link for a short time period. In this case, the measured delays are almost the same as the virtual ones caused by the busy input link.

It is interesting to note that the three large busy periods plotted in Figures 4.12(a)–(c) all have a roughly triangular shape. Figures 4.12(d)–(f) show that this is not due to a particular choice of busy periods. Of course, as expected, the larger the amplitude of the busy period, the smaller the number of such busy periods, indicating that congestion in C2-out is not persistent. For each 5-minute interval, we detect the largest packet delay, we denote by t_0 the corresponding packet arrival time, and we plot the delays experienced by packets between 10 ms before and 15 ms after t_0. The resulting sets of busy periods are grouped according to the largest packet delay observed: Figure 4.12(d) corresponds to busy periods with a largest amplitude between 5 ms and 6 ms, Figure 4.12(e) between 4 ms and 5 ms; and Figure 4.12(f) between 2 ms and 3 ms. For each of Figures 4.12(d), (e) and (f), the black line highlights the busy period detailed in Figures 4.12(a), (b), and (c), respectively. The striking point is that most busy periods have a roughly triangular shape. The largest busy periods have slightly less regular shapes, but a triangular assumption can still hold. We are using this interesting empirical finding to describe a novel performance reporting scheme that can capture the structure of busy periods on a router interface.

These results are reminiscent of the theory of large deviations, which states that rare events happen in the most likely way. Some hints on the shape of large busy periods in (Gaussian) queues can be found in ref. [19], where it is shown that, in the limit of large amplitude, busy periods tend to be antisymmetric about their midway point. This is in perfect agreement with what we see here.

4.4.2   Modeling busy periods

Although a triangular approximation may seem very crude at first, we now study how useful such a model could be. To do so, we first illustrate in Figure 4.13 a basic principle: any busy period of duration D seconds is bounded above by the busy period obtained in the case where the D seconds worth of work arrive in the system at maximum input link speed. The amount of work then decreases with slope -1 if no more packets enter the system. In the case of the OC-3 link C2-out, fed by the two OC-48 links BB1 and BB2 (each link being 16 times faster than C2-out), it takes at least $D/32$ s for the load to enter the system. From our measurements, busy periods are quite different from their theoretical bound. The set of busy periods shown in Figures 4.8 and 4.12(a) is again plotted in Figure 4.13 for comparison. One can see that the busy-period amplitude A is much lower than the theoretical maximum, in agreement with the scatter plot of Figure 4.11(c).

In the rest of the chapter, we model the shape of a busy period of duration D and amplitude A by a triangle with base D, height A and same apex as the busy period.

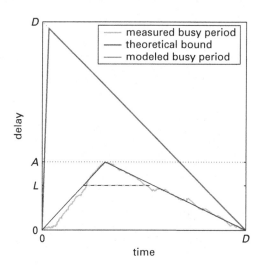

Fig. 4.13. Modeling of busy-period shape with a triangle. ©ACM, 2004.

This is illustrated in Figure 4.13 by the triangle superposed over the measured busy period. This very rough approximation can give surprisingly valuable insights on packet delays. We define our performance metric as follows. Let L be the delay experienced by a packet crossing the router. A network operator might be interested in knowing how long a congestion level larger than L will last on average, because this gives a direct indication of the performance of the router.

Let $d_{L,A,D}$ be the length of time the workload of the system remains above L during a busy period of duration D and amplitude A, as obtained from our delay analysis. Let $d_{L,A,D}^{(T)}$ be the approximated duration obtained from the shape model. Both $d_{L,A,D}$ and $d_{L,A,D}^{(T)}$ are plotted with a dashed line in Figure 4.13. From basic geometry one can show that

$$d_{L,A,D}^{(T)} = \begin{cases} D(1 - L/A) & \text{if } A \geq L \\ 0 & \text{otherwise.} \end{cases} \tag{4.13}$$

In other words, $d_{L,A,D}^{(T)}$ is a function of L, A and D only. For the metric considered, a two-parameter family (A, D) is therefore enough to describe busy periods, and the knowledge of the position of the busy-period maximum does not improve our estimate of $d_{L,A,D}$.

Denote by $\Pi_{A,D}$ the random process governing (A, D) pairs over time. The mean length of time during which packet delays are larger than L is given by

$$T_L = \int d_{L,A,D} \, d\Pi_{A,D}. \tag{4.14}$$

Note that T_L can be approximated by our busy period model as follows:

$$T_L^{(T)} = \int d_{L,A,D}^{(T)} \, d\Pi_{A,D} = \int D\left(1 - \frac{L}{A}\right) d\Pi_{A,D}. \tag{4.15}$$

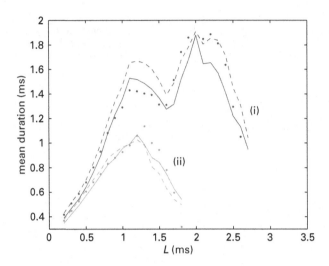

Fig. 4.14.   Average duration of a congestion episode above L ms defined by equation (4.15) for two different utilization levels, (i) 0.7 and (ii) 0.3, on link C2-out. Solid lines, data; dashed lines, equation (4.15); dots, equation (4.17). ©ACM, 2004.

We use equation (4.15) to approximate T_L on the link C2-out. The results are plotted in Figure 4.14 for two 5-minute windows of traffic with different average utilizations. For both utilization levels, the measured durations (solid line) and the results from the triangular approximation (dashed line) are fairly similar. This shows that our very simple triangular shape approximation captures enough information about busy periods to answer questions about duration of congestion episodes of a certain level. The small discrepancy between data and model can be considered insignificant in the context of Internet applications because a service provider will realistically only be interested in the order of magnitude (1 ms, 10 ms, 100 ms) of a congestion episode greater than L. Our simple approach therefore fulfils that role very well.

Let us now qualitatively describe the behaviors observed in Figure 4.14. For a small congestion level L, the mean duration of the congestion episode is also small. This is due to the fact that a large number of busy periods have an amplitude larger than L, as seen, for instance from the amplitude CDF in Figure 4.11(a). Since, however, most busy periods do not exceed L by a large amount, the mean duration of the congestion episode is small. It is also worth noting that the results are very similar for the two different link utilizations. This means that busy periods with small amplitude are roughly similar at this timescale and do not depend on the link utilization.

As the congestion level L increases, the mean duration first increases as there are still a large number of busy periods with amplitude greater than L on the link. Most busy periods are larger than L, and therefore the mean duration increases. With an even larger congestion level L, however, fewer and fewer busy periods are considered. The ones that do cross the threshold L do so for a smaller and smaller amount of time, up to the point where there are no busy periods larger than L in the trace.

4.4.3 Reporting busy-period statistics

The study presented above shows that one can obtain useful information about delays by using both the amplitude and duration of busy periods. Now we look into ways in which such statistics could be concisely reported.

We start by forming busy periods from buffer occupancy measurements and collecting (A, D) pairs during 5-minute intervals (a feasible number for SNMP reporting). This is feasible in practice since the buffer size is already accessed, for example, by active queue management schemes. Measuring A and D is easily performed on-line. In principle, we need to report the pair (A, D) for each busy period in order to recreate the process $\Pi_{A,D}$ and evaluate equation (4.15). Since this represents a very large amount of data in practice, we instead assume that busy periods are independent and therefore that the full process $\Pi_{A,D}$ can be described by the joint marginal distribution $F_{A,D}$ of A and D. Thus, for each busy period we simply need to update a sparse 2-D histogram. The bin sizes should be as fine as possible, consistent with available computing power and memory. We do not consider these details here; they are not critical since, at the end of the 5-minute interval, a much coarser discretization is performed in order to limit the volume of data finally exported. We control this directly by choosing N bins for each of the amplitude and the duration dimensions.

As we do not know a priori what delay values are common, the discretization scheme must adapt to the traffic to be useful. A simple and natural way to do this is to select bin boundaries for D and A separately based on quantiles, i.e. on bin populations. For example, a simple equal population scheme for D would define bins such that each contained $(100/N)\%$ of the measured values. Denote by M the $N \times N$ matrix representing the quantized version of $F_{A,D}$. The element $p(i, j)$ of M is defined as the probability of observing a busy period with duration between the $(i - 1)$th and ith duration quantile, and amplitude between the $(j - 1)$th and jth amplitude quantile. Given that, for every busy period, $A < D$, the matrix is triangular, as shown in Figure 4.15. Every five minutes, $2N$ bin boundary values for amplitude and duration, and $N^2/2$ joint probability values, are exported.

The 2-D histogram stored in M contains the 1-D marginals for amplitude and duration, characterizing, respectively, packet delays and link utilization. In addition, however, from the 2-D histogram we can see at a glance the relative frequencies of different busy-period *shapes*. Using this richer information, together with a shape model, M can be used to answer performance-related questions. Applying this to the measurement of T_L, and assuming independent busy periods, equation (4.15) becomes

$$T_L^{(T)} = \int d_{L,A,D}^{(T)} \, dF_{A,D} = \int_{A>L} D\left(1 - \frac{L}{A}\right) dF_{A,D}. \quad (4.16)$$

To evaluate this, we need to determine a single representative amplitude A_i and average duration D_j for each quantized probability density value $p(i, j)$, $(i, j) \in \{1, ..., N\}^2$, from M. One can, for instance, choose the center of gravity of each of the tiles plotted in Figure 4.15. For a given level L, the average duration T_L can then be estimated by

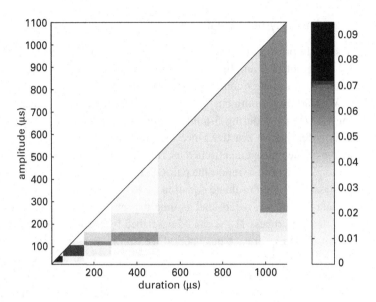

Fig. 4.15.   Histogram of the quantized joint probability distribution of busy-period amplitudes and durations with $N = 10$ equally spaced quantiles along each dimension for a five-minute window on link C2-out.

$$\widetilde{T_L^{(T)}} = \frac{1}{n_L} \sum_{j=1}^{N} \sum_{\substack{i=1 \\ A_i > L}}^{j} d_{L,A_i,D_j}^{(T)} p(i, j), \qquad (4.17)$$

where n_L is the number of pairs (A_i, D_j) such that $A_i > L$. Estimates obtained from equation (4.17) are plotted in Figure 4.14. They are fairly close to the measured durations, despite the strong assumption of independence.

The above gives just one example, for the metric T_L, of how the reported busy-period information can be used; however, other performance-related questions could be answered in a similar way. In any case, our reporting scheme provides far more insight into packet delays than currently available statistics based on average link utilization, despite its simplicity.

4.5   Micro-congestion episodes and their causes

From the previous results, it is evident that, while core routers feature output buffers capable of holding on the order of 500 ms to 1 s of data, delays rarely exceed millisecond levels. When examined on finer timescales, however, during localized periods of congestion, or *micro-congestion episodes*, delays can still reach levels which are of concern to core network providers bound by SLAs. Typically, backbone networks are structured in a hierarchy, where link bandwidths decrease as one moves from the long-haul links connecting different PoPs (OC-192 as of 2008), through those interconnecting core routers within a PoP (OC-48 to OC-192), down to access links connecting customers to access routers (OC-3, OC-12 or gigabit Ethernet). The access links, being closer to the edge of the network, are more interesting to study from the delay perspective for two reasons. First, the list of potential causes of delays in a network widens as we move toward the edge. Second, an access link is typically managed by the customer. SLAs therefore do not apply, and the link may be run at higher load levels to lower costs, again increasing the potential for congestion. The aim of this final part of Chapter 4 is to examine in detail the causes of micro-congestion episodes in an access router leading away from the core, with a particular emphasis on delays. Although a full separation is not possible, there are, nonetheless, different generic "causes" or mechanisms of congestion in general, and delay in particular, which can be identified. Briefly, these are related to:

- reduction in link bandwidth from core to access
- multiplexing of multiple input streams
- degree and nature of burstiness of traffic input stream(s).

In particular, we seek to answer the question "What is the dominant mechanism responsible for delays?" in such a context. More generally, a knowledge of the relative importance of different causes of higher than usual delay, and their interactions, gives insight into how delays may evolve in the future, potentially extending our findings to environments beyond the single router studied here.

4.5.1 Congestion mechanisms

Fundamentally, all congestion is due to one thing: *too much traffic*. The different mechanisms above relate to different ways in which traffic can be built up or concentrated, resulting in a temporary shortage of resources in the router. To explain the mechanisms precisely, we must have a model of router operation, as it is the router which will multiplex traffic arriving from different high-speed links and deliver it (in this case) to the lower-speed output link. For this purpose, we use the model of Section 4.3.

In the framework of the model, micro-congestion can now be precisely understood as the statistics of delays suffered during busy periods, which are time intervals where the system is continuously busy, but idle to either side. Here, by "system" we mean a given output interface and the portion of the router, leading from the input interfaces, related to it. Note, however, that packets are deemed to arrive at the system only after they have fully arrived at one of the input interfaces involved. For an input link, we will use "busy period" in a different, but related, sense, to refer to a train of back-to-back packets (corresponding to a busy period of the output link of the router upstream). We can now discuss the three mechanisms.

Bandwidth reduction

Clearly, in terms of average rate, the input link of rate μ_i could potentially overwhelm the output link of rate $\mu_o < \mu_i$. This does not happen for our data over long timescales, but locally it can, and does, occur. The fundamental effect is that a packet of size p bytes, which has a width of p/μ_i seconds on the input wire, is stretched to p/μ_o seconds at the output. Thus, packets which are too close together at the input may be "pushed" together and forced to queue at the output. In this way, busy periods at the input can only worsen: individually they are all necessarily stretched, and they may also then meet and merge with others. Furthermore, new busy periods (larger than just a single packet) can be created that did not exist before. This stretching effect also corresponds, clearly, to an increase in link utilization; however, it is the small-scale effects, and the effect on delay, that we emphasize here. Depending on other factors, stretching can result in very little additional delay or significant buildups.

Link multiplexing

For a given output link, input traffic will typically arrive from different traffic streams over different input links. This is particularly the case given the ECMP routing currently deployed by network providers for load balancing purposes. Whether these streams are correlated or not, the superposition of multiple streams increases the packet "density" at all scales and thereby encourages both the creation of busy periods at the output and the inflation of existing ones. To first order this is simply an additive increase in utilization level. Again, however, the effect on delays could either be very small or significant, depending on other factors.

Burstiness

It is well known that traffic is highly variable or bursty on many timescales [160]. The duration and amplitude (the highest degree of backlog reached) of busy periods will depend upon the details of the packet spacings at the input, which is another way of saying that it depends on the input burstiness. For example, packets which are already highly clustered can more easily form busy periods via the bandwidth-induced "stretching" above. To put it in a different way, beyond the first-order effect of utilization level, effects at second order and above can have a huge impact on the delay process.

4.5.2   Methodology and results

In order to study micro-congestion episodes, we continue our analysis on the full router data set that has been the focal point of this chapter. As mentioned before, our measurements cover all the links of an operational IP router at the edge of the network, serving four customers and connecting to two different backbone routers at OC-48 rate. With one exception, every link on the router has an average link utilization below 50% and thus experiences low congestion, and in particular low delays (99.26% were below 0.5 ms). The exception, which we study in detail here, is an access link at OC-3 rate fed from the two OC-48 backbone links, where average utilizations measured over 5-minute intervals reached as high as 80%. Busy periods on this link lasted up to 15 ms, and resulted in maximum through-router delays as high as 5 ms. In what follows we

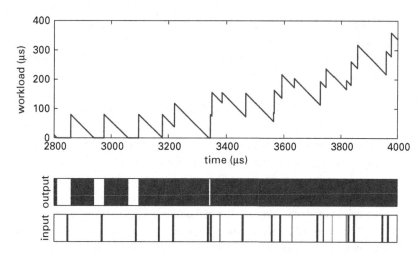

Fig. 4.16. The bandwidth reduction effect. Bottom: input/output bar plots: busy periods on the OC-48 input and OC-3 output links. Top: resulting queueing workload process. The output busy periods are longer and far fewer than at the input.

look into 13 hours of data, comprising more than 129 000 busy periods. The completeness of this data set, and the analysis methodology, allows us to measure in fine detail the evolution of busy periods both on the input links and in the router itself. We can therefore empirically answer essentially any question regarding the formation and composition of busy periods, or on the utilization and delay aspects of congestion, that we wish. In queueing terminology we have full access to the entire sample path of both the input and output processes and the queueing system itself.

An example of the sample path of the queue workload at the output covering over three busy periods is given in Figure 4.16. The first packet arrives at the output link and finds an empty queue. The amount of time to clear the output queue is equal to the amount of time it takes to transmit a packet of that size at the output rate (OC-3). The second packet is also transmitted without facing any queueing delay. However, the spacing between the third, fourth and fifth packet on the input link is not sufficiently large for the packets to be transmitted at the lower speed OC-3 link without delaying the packets behind them. We can clearly see the workload building up, forming a micro-congestion episode.

In order to quantify the contribution of the three mechanisms in the formation of micro-congestion episodes, we are going to use the concept of "semi-experiments" (see ref. [95] for details of this concept). Using the actual recorded input traffic we are going to explore virtual scenarios by feeding data through the physical router model of Section 4.3 and observing the output queue. The experiments take the following form. First, a "total" traffic stream S_T is selected. It is fed through the model with output rate μ and the locations and characteristics of all the resulting busy periods are recorded. Note that even in the case when S_T is the full set of measured traffic, we still must use the model to determine when busy periods begin and end, as we can only measure the system when packets arrive or depart, whereas the model operates in continuous time.

For a given busy period we denote its *starting time* by t_s, its *duration* by D, its *amplitude*, that is the maximum of the workload function (the largest of the delays $\{d_j\}$ suffered by any packet), by A, and we let t_A be the time when A occurred. When we need to emphasise the dependence on the stream or the link bandwidth, we write $A(S_T, \mu)$ and so on.

We next select a substream S_S of traffic according to some criteria. We wish to know the extent to which the substream contributes to the busy periods of the total stream. We evaluate this by feeding the substream into the model, since the detailed timing of packet arrivals is crucial to their impact on busy-period shape and amplitude. The focus remains on the busy periods of the total stream, even though the substream has its own busy-period structure. Specifically, for each busy period of S_T we will look at the contribution from S_S appearing in the interval $[t_s, t_A]$ during which it was building up to its maximum A. Exactly how to measure the contribution will vary depending upon the context. It is in fact not possible in general to separate fully the congestion mechanisms, as the busy-period behavior is a result of a detailed interaction between all three. The extent to which separation is feasible will become apparent as the results unfold.

4.5.3 Reduction in bandwidth

Let us illustrate the first mechanism in Figure 4.16. The two "bar plots" visualize the locations of busy periods on the OC-48 input link (lower horizontal bar) and the resulting busy periods following transmission to the smaller OC-3 output link (upper horizontal bar). For the latter, we also graph the corresponding system workload induced by the packets arriving at the busy period, obtained using the model. We clearly see that the input busy periods – which consist typically of just one or two packets – have been stretched and merged into a much smaller number of much longer busy periods at the output.

In order to quantify the "stretching" effect, we perform a virtual experiment where the total traffic stream S_T is just one of the main input OC-48 streams. By restricting to just a single input stream, we can study the effect of link bandwidth reduction without interference from multiplexing across links. In this case our "substream" is the same as the total stream ($S_S = S_T$), but evaluated at a different link rate. We quantify "stretching and merging" using the normalized *amplification factor* (AF):

$$\text{AF} = \frac{A(S_T, \mu_o)}{\max_k A_k(S_T, \mu_i)} \frac{\mu_o}{\mu_i},$$

where $\text{AF} \geq 1$. The amplitude for the substream is evaluated across all k busy periods (or partial busy periods) that fall in $[t_s, t_A]$. In simple cases where packets are well separated, so that all busy periods at both the input and output consist of just a single packet, stretching is purely linear and $\text{AF} = 1$. If queueing occurrs so that non-trivial busy periods form at the output, then $\text{AF} > 1$. The size of the AF is an indication of the extent of the delay increase due to stretching. If the utilization at the output exceeds unity then theoretically it will grow without bound.

We present the cumulative distribution function for the AF in Figure 4.17 for each of the main input streams separately. Less than 5% of the busy periods are in the 'linear'

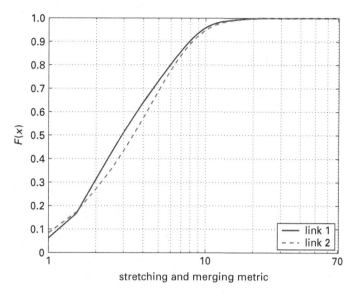

Fig. 4.17. Empirical distribution functions of the amplification factor for the OC-48 input streams.

regime with minimal delay detected via AF = 1. The majority are significantly ampli-
fied by the non-linear merging of input busy periods into larger-output busy periods.
If instead we had found that in most cases the AF was close to unity, it would have
been an indication that most of the input traffic on that link was shaped at OC-3 rate
upstream.

To get a feeling for the size of the values reported in Figure 4.17, note that a realistic upper bound is given by AF = 240 000, corresponding roughly to a 500 ms buffer
being filled (in a single busy period) by 40 byte packets well separated at the input, that
would induce a maximum workload of 129 s when served at OC-48 rate. A meaningful
value worthy of concern is AF = 1030, corresponding to delays of 20 ms built up from
375 byte packets, the average packet size in our data.

4.5.4 Link multiplexing

To examine the impact of multiplexing across different input links, we let the total
stream S_T be the full set of measured traffic. The ramp-up period, $[t_s, t_A]$, for two busy
periods of S_T are shown as curves labeled (i) in Figures 4.18 and 4.19. We select our
substreams to be the traffic from the two OC-48 backbone links, S1 and S2. By looking
at them separately, we again succeed in isolating multiplexing from the other two mechanisms in some sense. However, the actual impact of multiplexing is still intimately
dependent on the "stretch-transformed" burstiness structure on the separate links. What
will occur cannot be predicted without the aid of detailed traffic modeling. Instead, we
will consider how to measure what does occur and see what we find for our data.

Figures 4.18 and 4.19 show the delay behavior (consisting of multiple busy periods)
due to the separate substreams over the ramp-up period. The non-linearity is striking:
the workload function is much larger than the simple sum of the workload functions

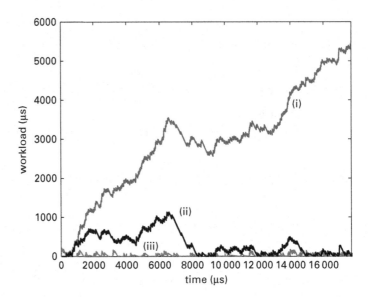

Fig. 4.18.   Effect of multiplexing on the formation of busy periods (from t_S to t_A). (i) Busy period; (ii) link 1; (iii) link 2.

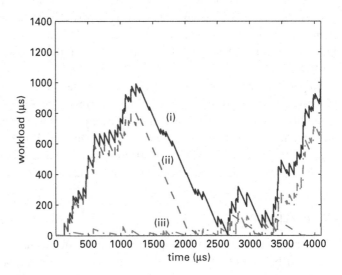

Fig. 4.19.   A "bimodal" busy period, where the contribution to A is ambiguous. (i) Busy period; (ii) link 1; (iii) link 2.

of the two input substreams, although they comprise virtually all of the total traffic. Each individual link contributes less than 1 ms of workload at worst. Similar results were briefly presented in Section 4.4.2. Nonetheless, the multiplexed traffic leads to a significantly longer ramp-up period that reaches more than 5 ms of maximum workload at t_A on the right of the plot.

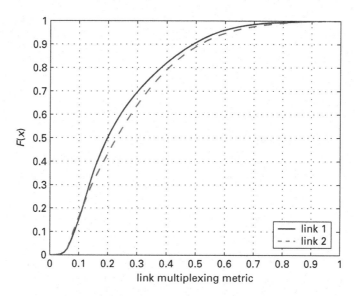

Fig. 4.20.   Empirical distribution functions of the link multiplexing ratio for the OC-48 input streams.

To quantify this effect we define the "link multiplexing" ratio as follows:

$$LM = \frac{\max_k A_k(S_i, \mu_o)}{A(S_T, \mu_o)},$$

which obeys $LM \in [0, 1]$. Values close to zero indicate that the link has a negligible individual contribution. Therefore if all substreams have small values, the non-linear effect of multiplexing is very strong. In contrast, if $LM \approx 1$ for some link, then it is largely generating the observed delays itself, and multiplexing may not be playing a major role. Large values of LM are in fact subject to ambiguity, as illustrated in Figure 4.19, where the ratio is large for both links. The busy period has a bimodal structure. The first mode is dominated by link 1; however, its influence has died off at time t_A, and so is not significantly responsible for the size of A.

The results on LM are presented in Figure 4.20. In more than 95% of the busy periods, traffic from each individual link contributes to less than 60% of the actual busy-period amplitude. Therefore, it appears that multiplexing is an important factor overall for the delays experienced over the access link.

4.5.5   Flow burstiness

There are many definitions of burstiness. It is not possible to address this issue fully without entering into details requiring traffic models, which is beyond the scope of this book. A look at burstiness of Internet traffic can be found in ref. [63]. We therefore focus on burstiness related to 5-tuple flows to investigate the impact that individual flows, or groups of flows, have on overall delays. We begin by letting the total traffic S_T be that from a single link, to avoid the complications induced by multiplexing. In order to obtain an insight into the impact of flow burstiness, we first select as a substream the "worst"

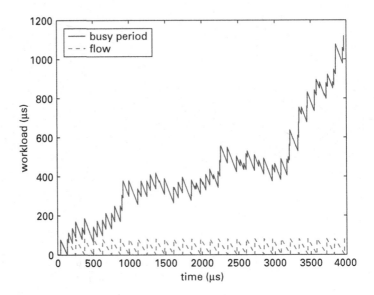

Fig. 4.21.   Flow with multiple packets and no significant impact on the queue buildup.

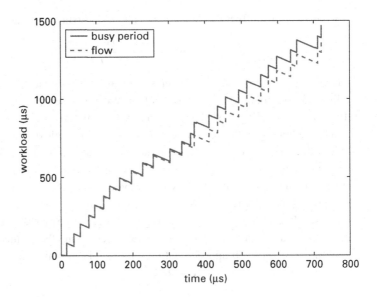

Fig. 4.22.   Flow with multiple packets that dominates the busy period.

individual flow in S_T in the simple sense of having the largest number of packets in the ramp-up period $[t_s, t_A]$. Two extreme examples of what can occur in the ramp-up period are given in Figures 4.21 and 4.22. In each case the busy-period amplitude is large; however, the flow contribution varies from minimal in Figure 4.21 to clearly dominant in Figure 4.22.

To refine the definition of worst flow and to quantify its impact we proceed as follows. For each busy period in the total stream we classify traffic into 5-tuple flows. We then use each individual flow S_j as a substream and measure the respective $A(S_j, \mu_o)$. The worst or "top" flow is the one with the largest individual contribution. We define "flow burstiness" (FB) as follows:

$$FB = \max_j \frac{\max_k A_k(S_j, \mu_o)}{A(S_T, \mu_o)},$$

where, as before, the inner maximum is over all busy periods (or partial busy periods) of the relevant substream falling in $[t_s, t_A]$. The top flow may or may not be the one with the greatest number of packets.

Our top flow metric takes values $FB \in (0, 1]$. If the FB is close to zero, then we know that all flows have individually small contributions. Alternatively if the FB is large, then, similarly to the LM, there is some ambiguity. We certainly know that the top flow contributes significant delay, but in the case of bimodality this flow may not actually be responsible for the peak defining the amplitude. In addition, knowledge about the top flow can say nothing about the other flows.

We present the cumulative distribution function for the FB in Figure 4.23 for each of the OC-48 links. For more than 90% of the busy periods the contribution of the top flow was less than 50%. In addition, for 20% of the busy periods the contribution of the top flow was minimal (for example, as in Figure 4.21), that is it was the smallest possible, corresponding to the system time of a single packet with size equal to the largest one appearing in the flow. If the top flow has little impact, it is natural to ask if perhaps a small number of top flows together could dominate. One approach would be to form a stream of the n largest flows in the sense of the FB. However, as the choice of n is arbitrary, and it is computationally intensive to look over many values of n, we first change our definition to select a more appropriate substream. We define a flow to be bursty if its substream generates a packet delay which exceeds the minimal delay (as defined above) during the ramp-up period. Note that only very tame flows are not bursty

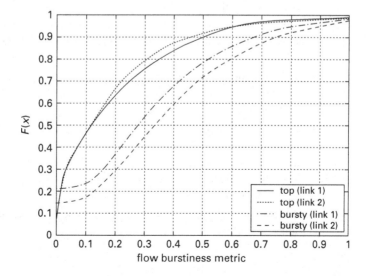

Fig. 4.23.   Empirical distribution functions of flow burstiness for the OC-48 input streams.

by this definition! We denote by S_b the substream of S_T that corresponds to all bursty flows, and we compute the new flow burstiness metric as follows:

$$FB' = \frac{\max_k A_k(S_b, \mu_o)}{A(S_T, \mu_o)}$$

As before, $FB' \in [0, 1]$, and its value can be interpreted in an analogous way to before. The difference is that, as the substream is much larger in general, a small value is now extremely strong evidence that individual flows do not dominate. Note that it is possible that no flow is bursty, in which case $FB' = 0$, and therefore that the top flow is not necessarily bursty. This has the advantage of avoiding the classification of a flow as dominant, thereby giving the impression that it is bursty in some sense, simply because a trivial flow dominates a trivial busy period. Our results are presented in Figure 4.23. As expected, the contribution of S_b to the busy period is more significant: in 20% of cases it exceeds 60%. On the other hand, 20% of the busy periods had FB' equal or close to zero, indicating that they had no bursty flows. Indeed, we found that only 7.7% of all flows in our dataset were classified as "bursty" according to our definition. Only in a very small number of cases does the top or the subset of bursty flows account for the majority of the workload (for example, as in Figure 4.22). Consequently, it seems that in today's network flow dynamics have *little impact* on the delays experienced by packets in core networks.

4.6   Lessons learned

In this chapter we looked into the characterization of congestion episodes, their frequency, duration and amplitude. We clearly demonstrated that the current network operator practice of using 5-minute average utilization values to "predict" user-experienced delay is fundamentally flawed. Using actual measurements collected from operational routers we studied the delays experienced by packets as they cross commercial routers, and we derived a model that allowed us to propose a highly accurate method to report on delay performance through a router. Using the derived model, we further studied the main causes of micro-congestion in today's networks. Interestingly, we saw that flow burstiness has little impact on the formation of micro-congestion episodes; probably due to limited capacity at the last mile. In contrast, flow multiplexing, and in particular the reduction of link capacity from the core to the edge, appear to be significant causes of queue buildups in today's highly overprovisioned backbone networks. We should however note that even though queue buildups were observed in our data, they were never followed by packet loss and they never exceeded 6 ms in amplitude. Given that the monitored link featured 80% utilization on a 5-minute average, it constituted probably one of the highest utilized links across the entire network, reducing the likelihood that other links around the network exhibited similar queueing behavior.

5 Traffic matrices: measurement, inference and modeling

The traffic matrix (TM) of a telecommunications network measures the total amount of traffic entering the network from any ingress point and destined to any egress point. The knowledge captured in the TM constitutes an essential input for optimal network design, traffic engineering and capacity planning. Despite its importance, however, the TM for an IP network is a quantity that has remained elusive to capture via direct measurement. The reasons for this are multiple. First, the computation of the TM requires the collection of flow statistics across the entire edge of the network, which may not be supported by all the network elements. Second, these statistics need to be shipped to a central location for appropriate processing. The shipping costs, coupled with the frequency with which such data would be shipped, translate to communications overhead, while the processing cost at the central location translates to computational overhead. Lastly, given the granularity at which flow statistics are collected with today's technology on a router, the construction of the TM requires explicit information on the state of the routing protocols, as well as the configuration of the network elements [77]. The storage overhead at the central location thus includes routing state and configuration information. It has been widely believed that these overheads would be so significant as to render computation of backbone TMs, through measurement alone, not viable using today's flow monitors.

This assumption has been one of the main motivations behind recent research targeted toward estimation techniques that can infer the TM from readily available SNMP link counts [42, 122, 137, 182, 215, 216]. The SNMP link counts constitute only partial information, and thus basic inference methods are limited in how low they can drive the error rates. Hence, these previous efforts have explored different avenues for extracting additional information from the network. Research efforts, such as those contained in refs. [42], [137], [217] and [218], usually postulate some underlying model for the TM elements and then use an optimization procedure to produce estimates that are consistent with the link counts. In ref. [182], the authors propose changing the IGP link weights in order to obtain more information to reduce the uncertainty in the estimates.

Portions of this chapter are reprinted or adapted from (1) A. Soule, A. Lakhina, N. Taft *et al.* (2005). "Traffic matrices: balancing measurements, inference and modeling." In *ACM Sigmetrics Performance Review* **33** : 1, 362–373 (see http://doi.acm.org/10.1145/1071690.1064259) and (2) K. Papagiannaki, N. Taft and A. Lakhina (2004). "A distributed approach to measure IP traffic matrices." In *Proceedings of 4th ACM SIGCOMM Conference on Internet Measurement*, Taormina, Italy, October, 2004, pp. 161–174 (see http://doi.acm.org/10.1145/1028788.1028808).

While this technique is powerful in collapsing errors, it requires carriers to alter their routing in order to obtain a TM. It is not clear that carriers are willing to do this. In ref. [122] the authors recognize that some of the optimization approaches may not scale to networks with large numbers of nodes (such as TMs at the link-to-link granularity level), and hence they propose a method for partitioning the optimization problem into multiple subproblems. This improves the scalability, but at the expense of some accuracy. Most of these studies have come from carriers whose interest lies in backbone TMs at larger timescales for the purposes of improving network traffic engineering.

There has been some debate as to whether or not inference techniques are really needed. Some researchers believe that the communication, storage and processing overheads of direct measurement are prohibitive, thus rendering it impractical. Other researchers believe that the TM problem is an implementation issue, which can be solved by advances in flow monitoring technology. While many of the inference techniques perform quite well, monitoring capabilities on the network elements have made noticeable progress, and technologies for the collection of flow statistics have been made available on a wide variety of router platforms.

In this chapter we describe what a traffic matrix is and investigate its different forms in a large-scale IP network. We then describe ways in which it can be measured and estimated using readily available SNMP statistics. Note that the entire discussion is focused on pure IP-based networks (as is the remainder of the book); the problem of TM measurement is significantly simpler in a MPLS network [90]. The chapter covers the three different generations of TM inference techniques and provides their comparison on the same data set.

5.1   What is an IP traffic matrix?

The elements of a TM are Origin–Destination (OD) flows, where the definition of the origin and destination object (i.e. node) can be selected as per the needs of the application using the TM. The granularity of a TM is determined by the choice of definition for the source or destination "object." The most typical objects are links, nodes and PoPs. In this chapter we will primarily consider TMs at the granularity of "link-to-link," "router-to-router" and "PoP-to-PoP". Other definitions are possible, accommodating representations where the level of traffic aggregation at the source is not necessarily the same as that for the destination, for example a TM can capture all the traffic from one PoP and destined to any AS. For a taxonomy of IP TMs and their potential applications in traffic engineering and network planning, please refer to ref. [136].

In addition to selecting the flow granularity of a TM, a network operator also needs to specify its time granularity. As mentioned in Chapter 3, NetFlow statistics are typically collected every 5 minutes. If TMs are built using NetFlow, then the smallest time granularity is that of 5 minutes. Coarser time granularities can average traffic across smaller time intervals and define the TM to capture the total amount of traffic exchanged across, for instance, one hour or one day. In this chapter we will focus on hourly traffic matrices, since most traffic engineering applications are targeted toward longer timescales.

One hour is actually small for such applications, but we believe this timescale is useful because it allows us to observe diurnal patterns [159] and busy periods that may last for a small number of hours. Note that, according to the above problem definition, the presented TMs cannot be used within an anomaly detection setting, which would require finer timescale measurements.

A TM is typically organized in two different ways: (1) an $N \times N$ matrix (Figure 5.2(a)), where N is the number of nodes in the network, or (2) a row (or column) vector $1 \times N^2$ vector ($N^2 \times 1$), as shown in Figure 5.2(b). For instance, for the small network depicted in Figure 5.1, the *router-to-router* TM can be represented as in Figure 5.2.

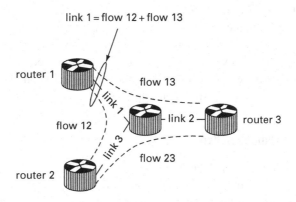

Fig. 5.1.   Example network flow 12 captures the amount of traffic from router 1 to router 2. The traffic on link 1 is the sum of all traffic sourced at router 1.

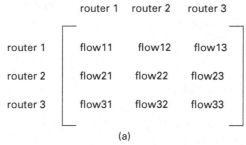

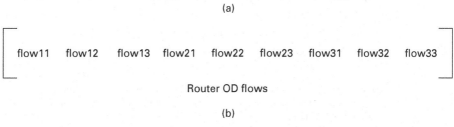

Fig. 5.2.   (a) Matrix and (b) vector representation of a traffic matrix.

5.1.1   How to measure an IP traffic matrix

There are three steps needed to obtain a TM from measurements. The first is to gather information about the traffic *source* by collecting measurements using NetFlow, or a similar monitor, at the ingress of the network. Packets are observed and statistics are stored at the granularity of flows, which can be flexibly defined as shown later. The second step is to identify the *destination* for each flow. For this particular operation, and as will be seen later in this chapter, access to routing information is required. The third step is to assemble all the information at the right granularity level (link, router or PoP) in a way that is consistent with the network topology.

In this section we describe a generic algorithm for doing steps 2 and 3. The Net-Flow data, gathered for step 1, serves as input to this algorithm. We describe the state of the art today for implementing such an algorithm. This is based on NetFlow v8 and essentially requires a (semi-)centralized approach.[1] Cisco's most recent release of Net-Flow v9, makes initial steps toward enabling the TM to be computed using a distributed approach. However, we will see that this version does not go far enough to enable a truly distributed approach. We quantify the storage and communication overheads for both centralized and distributed approaches.

5.1.2   The backbone network

Before proceeding, let us present a sample topology for which a network operator would like to measure the IP TM. This topology is consistent with the topology of the European Sprint backbone network and will help clarify concepts in what follows. A similar collection was not possible for the North American Sprint backbone due to the fact that not all edge routers supported NetFlow.

To derive the IP TM of the Sprint European IP backbone network we needed to activate NetFlow at the full edge of the network, which comprises 13 PoPs, one for each major European city. The number of routers in each PoP ranges from five to ten. The routers are organized in a hierarchy as depicted in Figure 5.3. Customers connect to the network by being directly attached to gateway (gw) routers. Backbone (bb) routers aggregate the traffic of multiple-gateway routers and forward it to the core of the network. The backbone routers are used for connecting peers to the backbone and also to inter-connect the PoPs.

In order to obtain a TM by direct measurement, one needs to collect all packets entering the backbone. To achieve that, NetFlow was enabled on all incoming peering links and all the links going from gateway routers to backbone routers. This latter set of links captures nearly all customer traffic. It only misses traffic that enters and leaves the network at the same gateway router.[2] The advantage of enabling NetFlow on the gateway

---

[1] Throughout the remainder of this chapter, we will use the term *centralized* to describe fully centralized or semi-centralized approaches. The latter would employ multiple collection stations, say one in each PoP, where a subset of the network routers will report their flow statistics.

[2] This implies that the only elements that may be impacted from this configuration choice are those that feature the same source and destination node.

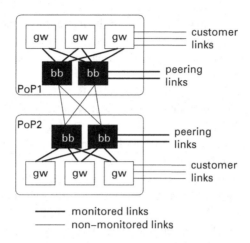

Fig. 5.3.   Setup for NetFlow data collection; gw = gateway router, bb = backbone router. ©ACM, 2004.

to backbone links, instead of the gateway routers themselves, is that we need to monitor a much smaller number of links while missing only local traffic. Such a choice
significantly reduces the complexity of our implementation.

   We used NetFlow v8, which is usually referred to as "sampled aggregated NetFlow"
[59]. Each record for NetFlow v8 is 40 bytes long, and features a flow identifier, its
source and destination network prefix, its associated load in bytes and packets, as well
as its starting and ending time. Rather than examine every packet, NetFlow v8 employs
periodic sampling in which one sample is collected every Nth packet (set to 250 in
what follows). Traffic statistics are not computed over each packet but are based on this
subset of the packets.

   Each router ships the collected NetFlow statistics to a configurable node (the collection station) every 5 minutes. We instrumented our European backbone with a single
collection station for all 27 routers in this backbone. We used 5 minutes as our reporting
interval since it coincides with the default SNMP reporting time interval. The collection station stores all flow statistics from *all* backbone routers in the IP network. We
collected three weeks of data during the summer of 2003.

5.1.3   Identifying the egress node

A TM is typically computed for a single domain or Autonomous System (AS). As
described above, NetFlow v8 statistics collected at each router are collected at the granularity level of source and destination network prefixes. These source and destination
prefixes will often reside outside the AS whose traffic matrix is computed. Thus the
source and destination of each packet need to be mapped onto the entry node and exit
node within the given AS. Identifying the entry node is simple, as it is defined as the
link or node where a packet enters a given domain (i.e. the location where NetFlow
sees the packet). The exit point for a particular source/destination network prefix flow
will depend both on its entry point as well as its actual destination. To identify this exit

point one needs to obtain a view of the forwarding table of the router that recorded the flow. Consolidation of intra- and inter-domain routing (from the vantage point of the router), as well as topological information, can resolve each prefix flow into its egress node inside the network.

The task at hand is to map accurately the destination network prefix in each flow record to (i) a backbone–gateway egress link, (ii) an egress backbone router and (iii) its egress PoP. To do this we need the BGP routing table from each router (or each route reflector [33] if deployed in the network), the ISIS/OSPF link weights and the router-level topology [77].

If a network prefix is advertised through BGP, then the BGP table specifies the last router within the AS that needs to forward traffic for this destination prefix, usually referred to as "BGP next hop." However, often the BGP next hop corresponds to the IP address of the first router outside the AS. In this case, configuration files can be used to map this address to a router interface within the network of interest.

This identified router interface will very likely correspond to the interface of a gateway router in our setting. At this point, we can identify the egress PoP, but we need additional information to be able to map this router interface (for the given destination prefix) to a particular egress backbone router and backbone–gateway link. Here we use the router topology, along with the associated link weights, to compute the shortest paths across the network per the intra-domain routing protocol in use. Using these paths we can find the egress backbone router(s) and backbone–gateway link(s) that will forward traffic toward the previously identified gateway router. The number of such routers and links may be more than one if the PoP topology is densely meshed and equal-cost multipath is enabled by the provider. In that case, we apportion the total flow equally toward each one of the routers or links selected using the process mentioned above.

5.1.4   Computing the traffic flow

A generic algorithm for the computation of the TM of an IP network can be summarized as in Figure 5.4. At the heart of this algorithm is a routine called $find\_egress\_node(f)$ that returns the egress node at the desired level of granularity (link, node or PoP) according to the method described above. There are four nested loops in this algorithm, one for each time interval n, one for each router r, one for each link l and one for each flow f. The $find\_egress\_node(f)$ routine operates at the level of a flow because that is the form of the NetFlow input. After the egress node is identified, the flows are aggregated so that the algorithm yields a TM at each of the granularity levels. In this pseudocode, $L(r)$ denotes the number of links at router r, and $F(l)$ denotes the number of flows on link l. This sketch of an algorithm makes it easy (i) to see how the overheads are computed, (ii) to identify all the additional routing/configuration data needed and (iii) to clarify what the change from a centralized to a distributed approach implies.

Centralized approach

Because NetFlow does not implement a procedure such as $find\_egress\_node(f)$, all of the flow data need to be shipped by each router to a specific collection station that

Algorithm: COMPUTETM($data, T, R, L, F$)

> **for** $n \leftarrow 1$ **to** T
> $ISIS = isis(n); \%the\ same\ topology\ network - wide$
> $configuration = \cup_{r=1}^{R} configuration file(r,n);$
>   **for** $r \leftarrow 1$ **to** R
> $routingtable = BGProutingtable(RR(r), n);$
> $\%BGP\ routing\ table\ of\ the\ route\ reflector$
> $in\ r's\ PoP.$
>   **for** $l \leftarrow 1$ **to** $L(r)$
>   **for** $f \leftarrow 1$ **to** $F(l)$
> $EN(f) = find\_egress\_node(f, routingtable,$
> $configuration, ISIS);$
> $TM(l, EN(f)) = TM(l, EN(f)) + data(f, t);$
>   **return** (TM)

Fig. 5.4.   Pseudocode for the computation of the TM.

can carry out the above algorithm. Thus, today's state of the art essentially mandates a centralized solution. The collection station needs to have explicit information about each PoP's BGP routing table and the IS–IS/OSPF weights in effect at each time interval n. In addition, it needs to have an accurate view of the network topology, in terms of the configuration of each router inside the network. For an implementation of such a centralized solution, one needs to download router configuration files and BGP routing tables on an as-continuous basis as possible. Given that router configuration files do not change that often, downloading such information once a week will be sufficient. BGP routing information is bound to change more often. As a result, one needs to rely on a continuous BGP feed, utilizing routing protocol listeners (Section 3.3.1), or one must use snapshots of BGP routing tables, collected, say, once a day from each PoP. Previous work has shown that BGP routing table changes occurring during a single day rarely affect large amounts of traffic [171], thus providing an acceptable compromise. Note that in networks employing route reflectors, all routers are bound to have a similar view of the BGP routing domain as the route reflector they are assigned to, and downloading the BGP routing table from each PoP route reflector will be sufficient. Certainly there are inaccuracies incurred by not having perfectly up to date routing information. However, obtaining more frequent updates of this information greatly increases the communication overhead. (The only way to avoid these inaccuracies completely is to use a distributed approach as described below.)

Toward distributed approaches

A process similar to $find\_egress\_node(f)$ is already performed by the router itself before switching the packet to its destination. Therefore a truly distributed approach would be one in which each router saved the information on the egress point of each network prefix while performing the lookup in its routing tables. Since one router constitutes one source that sends potentially to all other routers in the network, saving traffic statistics on the amount of traffic destined to each egress point is equivalent to the router computing one row of the TM. With this approach the only data that need to be shipped to a collection station is the TM row itself.

In order to do this, the router would need to change the flow record to include fields such as egress link, egress router and/or egress PoP. If flow records were kept at the level of links or routers rather than prefixes, this would dramatically reduce the on-router storage. The communications overhead is also greatly scaled back since sending one row, of even a link-to-link TM, is far smaller than shipping individual prefix-level flow records. Furthermore, the computational overhead at the collection station has now been reduced to that of assembling the rows without any egress node identification activity.

Recent advances in the area of flow monitoring have led to new definitions for flow records that incorporate explicit routing information defining flows such that the destination field captures the BGP next hop address. Such a change can be found in NetFlow v9 [58], which thus constitutes a significant movement toward the distributed solution described above. This improvement is not yet sufficient for the following reason. When a particular 5-tuple flow is mapped onto a "BGP-next-hop" flow, there is always the risk that the destination network prefix may not be advertised through BGP. For ease of implementation, NetFlow v9 addressed this issue by using "0.0.0.0" as the BGP next hop. Such a design choice implies that all the traffic that may be going to internal customers is missed if these customers are not advertised through BGP. For ISPs with a large number of customers, this may translate to many elements of the TM being altogether missed or inaccurately estimated when a particular "unknown" flow would actually map to an existing flow in the cache.

The feasibility of direct measurement approaches is dependent upon the ability to implement a routine such as *find_egress_node(f)* at a router. We distinguish two factors regarding the implementation of this routine. First, we point out that the information needed is already available in today's routers in the Routing Information Base (RIB). The RIB contains (i) the mapping between each destination prefix and its BGP next hop (as dictated through BGP), (ii) the mapping between the BGP next hop and its egress node (as identified through the intra-domain protocol in use) and (iii) the mapping between the egress node and the appropriate local interface that should be used to reach that node. The second factor regarding the implementation is the need to gain efficient access to this information. This could require changes in the software architecture, and is the main challenge to implementation.

5.1.5   TM overheads: a concrete example

In order to provide the reader with a sense of how much data would need to be stored, shipped and processed for the computation of the TM of a large-scale IP network, we use the three weeks of data collected from the European Sprint IP backbone network using today's technology (prefix-to-prefix NetFlow).

Table 5.1. Overheads (in number of records) for computing the TM in *one* time interval

Scheme approach	Storage (router)	Communications (network)	Computational (collection station)
SrcPrefix2DstPrefix	$L/R \cdot \bar{F}(l)$	$L \cdot \bar{F}(l)$	5.5 million lookups, aggregation to the required granularity (additional storage: 13 BGP routing tables, one IS–IS routing table, topology)
Centralized	$3 \cdot 67\,000$	$81 \cdot 67\,000 =$ 5.5 million	
l2l row *Distributed*	$L/R \cdot L$ ($3 \cdot 81 = 243$)	L^2 ($81 \cdot 81 = 6561$)	matrix assembly
r2r row *Distributed*	R (27)	R^2 ($27 \cdot 27 = 729$)	matrix assembly
p2p row *Distributed*	P (13)	$R \cdot P$ ($27 \cdot 13 = 351$)	matrix assembly

Note that p2p = PoP-to-PoP matrix; r2r = router-to-router matrix; l2l = link-to-link matrix.

5.1.6   Storage, communications and computational overheads

The overheads involved in computing a TM at each of the granularity levels are listed in Table 5.1. The storage overhead per router refers to the amount of flow statistics stored at the router. These elements are updated and shipped to the collection station on an ongoing basis. The communication overhead refers to the total amount of information that needs to be transmitted through the network domain toward the collection station. This includes the inputs from all domain routers. The computational overhead we show is that of the activity at the collection station.

First we consider the centralized approach that is available today. Because the basic granularity of the collected and processed flow statistics is that of a network prefix, we have labeled the centralized solution as prefix-to-prefix in the first row of Table 5.1. Table 5.1 lists the associated overheads for the centralized solution for *one* time interval, that is the derivation of *one* TM, regardless of its actual granularity. Let L denote the total number of links, R the number of routers, and P the number of PoPs in the network. The average number of links per router is thus L/R. Based on the actual configuration parameters of the European network under study and the current state of the art, the collection station had to perform 5.5 million egress node identifications (last column). This required 13 BGP tables, since each PoP has a different vantage point, and one IS–IS routing table residing at the collection station.

The next three rows of this table correspond to the overheads for a distributed solution for cases where the end goal is a TM at a specific granularity level. The notation "p2p" refers to a PoP-to-PoP matrix, "r2r" indicates the router-to-router and "l2l" is link-to-link. With fully distributed solutions, the collection station merely needs to assemble the rows it receives as input and build the matrix. Letting F denote the total number of flows, then clearly in any typical network we will have $P < R < L \ll F$. As an example, in a typical large Tier-1 backbone for a geographic region the size of the USA, P is on the order of tens, R is on the order of hundreds, L is on the order of a few thousands, and F is on the order of millions or even billions.

The advantages of moving from a centralized approach to a distributed one are clear: (1) the router storage overhead is reduced from $O(F)$ to $O(R)$ or $O(L)$, where R and L can be many orders of magnitude smaller than F and (2) the communication overhead is reduced by two or three orders of magnitude (depending upon the target TM). In addition, recall that a problem with the centralized approach is that the routing table information at the collection station will regularly become out of date. Distributed solutions do not suffer from this problem since the information on the routers is essentially always up to date (as fast as the protocol can perform updates).

The amount of storage and communication overhead partly depends upon what the ISP wants to collect. For example, if an ISP is sure it only ever wants to collect the r2r matrix, then each router should compute one row of this matrix and the communication overhead is limited to R^2. If, however, a carrier prefers to leave open the flexibility of looking at a TM at any of the granularity levels, then they should seek the l2l matrix, which can be aggregated into a r2r and p2p by checking which link belongs to which router, which router belongs to which PoP, and so on. In this case the communication overhead is L^2.

There is one other very critical piece to the overhead issue, and that is the frequency with which one wants to collect a TM. Suppose carriers decide they want the TM to be updated K times each day. Then, all of the numbers in Table 5.1 would be multiplied by K for each day. In the case we focus on, these overheads would be incurred once every hour.

Clearly one required step for direct measurement of a TM to become practical is for flow monitoring at routers to increase their functionality so as to compute rows of TMs, thereby enabling a distributed approach. However, we believe that this is still not enough. A communications overhead of L^2 or R^2 records incurred every hour may still be regarded as excessive. Exploiting temporal information could enable lower communication overheads, as shown in ref. [158].

In what follows, we describe alternative approaches to estimate the TM in the presence of different kinds of input data. Given that SNMP data are collected by most operational networks on a continuous basis, an interesting question to ask is "how accurate would a TM be in the absence of flow data?" We address this question and describe alternative approaches to reduce the error associated with the inference of a TM from the readily available SNMP data.

5.2   TM estimation from SNMP link counts

The TM estimation problem posed as an inference problem can be briefly stated as follows. The relationship between the TM, the routing and the link counts can be described by a system of linear equations $y = Ax$, where y is the vector of link counts, x is the TM organized as a vector and A denotes a routing matrix in which element A_{ij} is equal to one if OD pair j traverses link i and zero otherwise. The elements of the routing matrix can have fractional values if traffic splitting is supported. In networking environments today, y and A are readily available; the link counts y can be obtained through standard SNMP measurements and the routing matrix A can be obtained by computing shortest paths using IGP link weights together with the network topology information. The problem at hand is to estimate the TM x. This is not straightforward because there are many more OD pairs (unknown quantities) than there are link measurements (known quantities). This is manifested by the matrix A having less than full rank. Hence the fundamental problem is that of a highly under-constrained, or ill-posed, system.

By looking back on the development of TM estimation techniques, we see that most techniques have been motivated by trying to tackle the fundamental issue of the problem being ill-posed. The idea of introducing additional side information is a common approach to solving ill-posed inverse problems. A first generation of techniques was proposed in refs. [42] and [194]. Their approach to handling the highly under-constrained problem was to introduce additional constraints related to the second-order moment of the OD pairs. These methods used simple models for OD pairs (e.g. Poisson, Gaussian) that contained neither spatial nor temporal correlations in the *OD flow model*. A comparative study of these methods [137] revealed that these methods were highly dependent upon an initial prior estimate of the TM. This illustrated the need for improved OD flow models.

Shortly thereafter, a second generation of techniques [151, 182, 215, 216] emerged that made use of side information coming from additional sources of SNMP data. These extra measurement data were used to calibrate a model of OD flows. The calibrated model was then incorporated inside some type of estimation procedure. The OD flow model used in refs. [215] and [216] is that of a gravity model that captures the fraction of traffic destined to an egress node as a portion of the total traffic arriving at an ingress node. They assume that ingress and egress nodes are independent and calibrate the gravity model using SNMP data from access and peering links. These data are different from the SNMP data used inside the estimation, which come from inter-router links. The method in ref. [182] tackles the issue of an ill-posed system via another approach. The key idea is to change explicitly the link weights, thereby changing the routing and moving OD flows onto different paths. By doing this enough times, the rank of the routing matrix is increased. The inter-router SNMP data are then collected from each of the different routing configurations used. The side information here can be thought as this additional SNMP data coming from altered routing configurations. This route change method uses all of the collected data to calibrate a model for OD flows that is based on a Fourier expansion plus noise model. In fact, this method attacks the ill-posed nature of the problem two ways simultaneously: that of increasing the rank and of using an improved OD flow model. The model is a purely temporal one for OD flows, while the tomogravity approach uses a purely spatial model for the OD flows.

These second-generation methods yielded improvements on first-generation techniques and also provided useful insights. However, these methods have not been able to push error rates sufficiently low for carriers. With recent advances in flow monitoring techniques, some of the issues involved in their use are dissipating. Moreover, it is becoming easier to understand which sorts of changes to flow monitors are required in order to make the flow data required for TMs less cumbersome to obtain. Even assuming these changes will take place in the future, we showed in Chapter 5.1 that the communications overheads will remain large. One potential solution to reduce the communications overhead would be to turn flow monitors only when needed and only on the affected nodes.

In what follows we investigate this idea further, generalize it to other methods, and explore a much broader range of questions that arise when one has the possibility of considering using partial flow measurements for TM estimation.

Starting with the premise that we can do partial flow measurement, many important questions arise. Since such measurement data will contain information about spatial and temporal correlations, we can ask which OD flow models are optimum at exploiting the temporal, spatial or spatio-temporal correlations? How many measurements should be carried out? What are the improvements in error-rate reduction that can be achieved for a given level of measurement overhead? How well do these methods handle the dynamic nature of TMs? What are the strengths and weaknesses of different approaches with respect to other traffic engineering applications that rely upon TM data as an input? What are the tradeoffs in and limitations of inference methods coupled with partial flow measurement?

We address these questions by studying the behavior of five TM estimation methods, described in detail in Section 5.3. We include two second-generation methods: that of ref. [216], called the *tomogravity* method, and that of ref. [182], which we refer to as the *route change* method. We label methods that use partial flow measurements as third-generation methods, and we consider three such methods. One method, called the *fanout* method hereafter, is the one proposed in ref. [158]. We introduce a method that relies upon Principal Component Analysis (PCA), called the *PCA* method. The final method makes use of Kalman filtering techniques and is hence referred to as the *Kalman* method. The Kalman method was originally introduced in ref. [183]. The version included herein is an extended version of the original method, which includes adaptive recalibration of the underlying model.

If flow monitors are turned on network-wide for a period of time, they yield an exact copy of the TM for that period of time. If all the flow monitors at a single PoP (router) are turned on for a period of time, this yields an exact copy of one row of a p2p (r2r) TM. Because previous work [118, 158, 182] has found strong diurnal patterns in TMs, the third-generation methods we study assume that when flow monitors are turned on, they are left on for periods of 24 hours. This measurement of OD flows is then used to calibrate a model. With this kind of rich side information, powerful models that capture spatial and temporal correlations via measurement (not via assumption) can be designed.

A TM is a dynamic entity that includes many sources of variability, including short-lived variations, long-lived variations (e.g. resulting from multi-hour equipment failures) and permanent sources of change (addition or removal of customers/routers/links). It is thus natural to expect that any underlying model of OD flows will need to adapt to these changes by being recalibrated. Hence each of the three flow-based methods studied herein include a mechanism to check for change. When changes are detected, flow monitors are turned on again for 24 hours, and the corresponding model is recalibrated.

This chapter looks into a number of dimensions in the problem area of TM estimation.

(1) First, we classify each method based on the type of correlation (spatial, temporal or spatio-temporal) they draw on to model OD flows. We clarify the amount, type and cost of supporting information they require. This taxonomy helps us better understand and explain the performance of each method.

(2) Second, we evaluate all of the methods in a unified fashion – on the same set of data, using the same error metrics. We find that the third-generation methods (Fanout, Kalman and PCA) outperform the second-generation methods overall, and can yield accurate estimates for many of the smaller OD flows that are traditionally more difficult to handle. This is somewhat expected because the third-generation methods rely on rich priors obtained from direct measurements. However, what is more surprising is that with only 10–20% additional measurements, we can reduce the errors by more than one-half. Thus, large gains in accuracy can come at a low measurement cost, which underscores the feasibility of the next generation of combined measure-and-infer techniques for computing TMs.

(3) Third, we will examine how each method responds to dynamic changes in the TM. We observe that the tomogravity method, because of its memoryless model, accomodates large and sudden changes best. However, because the spatial correlations it uses in modeling are less accurate than those obtained from measurement, the tomogravity method suffers from biased estimates, producing less accurate estimates of the overall mean behavior of OD flows. Conversely, the third-generation methods are able to capture the long-term mean behavior of OD flows, but track large and sudden changes in an OD flow less well. Finally, we study the methods' estimation bias versus the resulting error variance. We find that, in stark contrast to the third-generation methods, the tomogravity estimates are biased, but have low error variance. This suggests that tomogravity may be better suited for applications requiring responsive change detection, such as anomaly detection, whereas the hybrid measure-and-infer methods are well suited for capacity planning, which requires faithful estimates of the overall mean estimates.

Recent work in the area [217] presents ways in which the task of TM estimation can also be made robust to erroneous/inaccurate input. The choice of technique for the estimation of the TM of an IP network will ultimately lie with the operator, and is likely to depend primarily on the associated overhead and the targetted accuracy required by the intended application.

5.3   Methods

Let us first describe each of the five methods considered. In what follows, we will high-light three key aspects of each method: the type of OD flow model used, the type of data (or side information) brought in to calibrate the model, and the method of estimation. Focusing on these three aspects of each method is helpful in understanding the differences and similarities between various methods without getting lost in the details. We classify each model as being either spatial, temporal or spatio-temporal. A *spatial model* is one that captures dependencies among OD flows, but has no memory. *Temporal models* are those where an OD flow model is dependent on its past behavior, but inde-pendent of other OD flows. Spatial models thus capture correlations across OD flows, while temporal models capture correlations in time. Clearly, *spatio-temporal* models are those that incorporate both types of correlation.

The three third-generation methods presented here use different underlying OD flow models. The common feature of these methods is that they rely on data from flow monitors to calibrate their underlying models. All of these methods assume that flow monitors are initially turned on network-wide for a period of 24 hours for initial model calibration. The flow monitors can then be turned off until further notice. All of these methods include simple schemes for change detection, and when changes are detected flow monitors are turned back on for another period of 24 hours.

In the data we use for validation, we had an estimate of our TM at each 10 minute time interval. Hence all methods estimate the TM on a timescale of 10 minutes (the underlying time unit t).

5.3.1   Notation

To facilitate subsequent discussion, we introduce the relevant notational convention used here. As mentioned in Section 5.2, x represents the TM at a specific point in time, organized as a column vector with N elements. Likewise, y is the column vector of traffic on L links at a point in time, and A is the $L \times N$ routing matrix. We will also frequently refer to the TM over time, and we use X to denote this structure; X is a $T \times N$ matrix, where each column j corresponds to the time series of OD flow j. Similarly, Y is the $T \times L$ multivariate time series of link traffic. To represent the TM (or link traffic) at a particular time t, we use x_t (or y_t). To identify the jth OD pair at time t, we use $x_t(j)$, and, when needed, $x(i, j, t)$ represents the traffic at time t for the OD pair sourced at node i and destined to node j. We will reserve the terms $\hat{x}$ and $\hat{X}$ for the estimated traffic demands. In general, we will use uppercase letters to denote matri-ces, lowercase letters to denote vector quantities and italic lowercase letters to denote scalars. Finally, all vectors are column vectors unless otherwise stated.

5.3.2   Tomogravity method

In refs. [215] and [216], the authors developed a method for estimating TMs that starts by building a prior TM using a gravity model. The basic principle of the gravity model is to assume proportionality relationships. For ease of exposition, we omit the time index in our notation. Let $x(i, *)$ denote the total traffic entering an ingress node i. If node i is a router, this corresponds to the incoming traffic on all the access and peering links. Let $x(*, j)$ denote the total traffic departing the network from node j. Again, if node j is a router, this includes all the traffic departing the AS in question on either access or peering links. The gravity model postulates that

$$x(i, j) = x(i, *)\frac{x(*, j)}{\sum_j x(*, j)}. \tag{5.1}$$

This implies that the total amount of data node i sends to node j is proportional to the amount of traffic departing the network at j relative to the total amount of traffic departing the entire network. The authors of ref. [215] call this the *simple gravity model*. This model essentially assumes complete independence between sources and destinations. However, in typical IP backbones, there are two policies that lead to deviations from pure independence. Due to hot-potato routing policies, traffic from a customer edge traveling towards a peer will be sent to the nearest exit point. The second policy is that there should be no traffic transiting the network from one peer to another.

Hence, the authors define the generalized gravity model to capture these policies. This is interpreted as a conditional independence of source and destination, conditioned on additional side information identifying link types. Let $\mathcal{A}$ denote the set of access links and let $\mathcal{P}$ denote the set of peering links. The generalized gravity model is defined as follows:

$$\hat{x}(i, j) = \begin{cases} 0, & \text{for } i \in \mathcal{P}, | \in \mathcal{P} \\ \dfrac{x(i, *)}{\sum_{i \in \mathcal{A}} x(i, *)} x(*, j) & \text{for } i \in \mathcal{A}, j \in \mathcal{P} \\ x(i, *)\dfrac{x(*, j)}{\sum_{j \in \mathcal{A}} x(*, j)} & \text{for } i \in \mathcal{P}, | \in \mathcal{A} \\ \rho \dfrac{x(i, *)}{\sum_{i \in \mathcal{A}} x(i, *)} x(*, j) & \text{for } i \in \mathcal{A}, j \in \mathcal{A}, \end{cases} \tag{5.2}$$

where ρ is a normalization constant (see ref. [215]).

Using an information theoretic perspective, the authors show that a gravity model is a prior capturing complete independence between sources and destinations and therefore is equivalent to a maximum entropy formulation [125]. The gravity model is then used inside a convex optimization problem that combines a minimum-mean-square-error approach with a maximum-entropy approach. This combination is achieved using a regularization strategy that is a common approach for dealing with ill-posed problems. The optimization problem is given by

$$\min ||y - Ax||_2^2 + \lambda^2 \ K(x||\hat{x}) \ \text{subject to } x > 0, \tag{5.3}$$

where $||\cdot||_2$ denotes the L_2 norm, $\lambda > 0$ denotes a regularization parameter and $K(x||\hat{x})$ is the Kullback–Leibler divergence, which is a well known measure of distance between probability distributions. The Kullback–Leibler divergence is used here as a way to write the mutual information, as the goal is to minimize the mutual information. The minimization of mutual information is equivalent to the maximization of entropy. The idea is thus that, among all the TMs that satisfy the link constraints, the method picks the one closest to the gravity model.

The model used in this method is a spatial one that describes a relationship between OD flows. The gravity model essentially captures a node's fanout for each node. The *fanout* for a source node is simply the fraction of its total traffic that it sends to a given destination. For each source node, the sum of its fanout must equal one. So, although the gravity model assumes independence across *nodes*, it does not assume independence among *OD flows*. All the OD flows sharing a common source are interdependent because of this requirement that the fanouts sum to one; similarly for OD flows sharing a destination node. The gravity model is not a temporal model because at any moment in time the calculation of the gravity model does not depend on history. The data used to calibrate the gravity model come from SNMP data on access and peering links. The estimation part of this method uses SNMP link counts from inter-router links Y and a routing matrix A, and solves the minimization problem in equation (5.3) to produce an estimate for X.

5.3.3   Route change method

We now summarize the method developed in refs. [151] and [182], where the latter paper uses the algorithm in the first as part of its overall methodology. In ref. [151], the authors proposed the idea of changing the IGP link weights, and then taking new SNMP measurements under this new routing image, in order to decrease the ill-posedness of the original estimation problem. This strategy increases the number of constraints on the system because additional link counts, collected under different routing scenarios, yield new equations into the linear system that will increase the rank of the system if they are linearly independent of the existing equations. In ref. [151] the authors proposed a heuristic algorithm to compute the weight changes required to obtain a full rank system. The advantage here is that, with a full rank system, there is a huge potential to reduce errors. However, for such systems to be practical the number of times carriers have to change link weights needs to be kept small. Note that the problem is not yet solved by obtaining a full rank system because the additional measurements will be collected over timescales of multiple hours and thus the TM itself will have changed.

In ref. [182] the authors assume the diurnal pattern to be cyclo-stationary, while the fluctuation process, i.e. random noise around the diurnal pattern, to be a zero mean stationary process with covariance matrix Q. According to such assumptions, the authors propose a Fourier expansion of the diurnal pattern, and the OD flow model $x(i, j, t)$ is formalized as follows:

$$x(i, j, t) = \sum_h \theta_h(i, j)b_h(t) + w(i, j, t), \tag{5.4}$$

where the first term is the Fourier expansion for the diurnal trends and the second term captures the stationary fluctuations. Here $b_h(t)$ denotes the hth basis function while

$\theta_h(i, j)$ refers to the coefficients corresponding to an OD pair sourced at node i and destined to node j.

By coupling route changes with temporal models for OD flows, the authors devise a new method for estimating the first- and second-order statistics of OD flows under both a stationary and a cyclo-stationary condition. In order to do this, the authors develop an expanded system description of $y = Ax$, where the routing matrix, now a function of time, is modified to include many block matrices that incorporate the different routing images. The linear system is also modified to incorporate the models for X; this has an impact on the pseudo-inverse solution used on the expanded full rank system to obtain a TM estimate. In this context, the estimation of the first-order statistics collapses to the estimation of the diurnal pattern, i.e. θ_h in equation (5.4), while the estimation of the second-order statistics becomes that of estimating the covariance matrix Q of the fluctuations process.

The OD flow model used here is a temporal model in which each OD flow is dependent on its past. This is not a spatial model in that it is assumed that OD flows are independent of one another. This model operates on a different timescale than other methods. Let's say, for the sake of example, that it takes three days to carry out all routing configurations. In this method, all the SNMP data from inter-router links are collected during this three-day period, as well as each of the routing matrices that apply for each snapshot. The estimation step (using a pseudo-inverse method) is applied once for the three-day period, and estimates of the TM for each 10-minute slot of those three days are produced. In this method, the calibration of the OD flow model (i.e. determining the coefficients of the OD flow model) are performed simultaneously with the estimation of the TM itself. Hence the side information used for the OD flow models is the SNMP data from multiple snapshots.

To make this approach applicable for real operators, the authors coupled their ability to estimate the variance of each OD flow with the observation (from empirical data) that the flows with large variance are the flows with large average rate. By doing so, they were able to identify the heaviest flows. This in turn enabled them to reduce the number of routing configurations required for the overall process, since the focus is not on estimating the entire TM but its heaviest elements.

5.3.4 Fanout method

This method [158] is a purely data-driven method that relies on measurements alone to obtain the TM. No routing matrix is used, and no inference is performed. A node fanout is defined as the vector capturing the fraction of incoming traffic that each node forwards to each egress node inside the network, where a node can be a PoP, a router or a link. Let $f(i, j, t) = x(i, j, t)/\sum_j x(i, j, t)$ denote the fraction of traffic entering node i that will egress the network at node j at time t. A node's baseline fanout is then given by the vector $f(i, *, t) = \{f(i, j, t) \forall j\}$ at time t. In both refs. [90] and [158] the authors independently and simultaneously found that node fanouts exhibit strong diurnal patterns and are remarkably predictable over the cycle of one day. This stability of fanouts means that the fanout of node i, at say 14:00 on one day, can be used to predict row i of the traffic matrix at 14:00 on subsequent days. This direct measurement technique exploits the finding of predictable fanouts.

Initially, the method assumes that flow monitors are turned on network-wide for a period of 24 hours. Using these data, the baseline fanouts can be computed for each node. Assuming the flow monitor measures the OD flows every 10 minutes, then, in a 24-hour, period, we have 144 measurement intervals in the baseline. Given the fanout for each 10-minute window within a day, we use the fanout from this first day at a particular time t to predict the TM on another day at the same time t. For example, the measured fanouts at 15:10 on the day of NetFlow collection are used to predict the TM at 15:10 on other days. The TM estimate is given by

$$\hat{x}(i, j, t) = \hat{f}(i, j, t)x(i, *, t), \tag{5.5}$$

where $x(i, *, t)$ denotes the total incoming traffic into node i and $\hat{f}(i, j, t) = f(i, j, d + t \% 144)$, where d denotes how many days in the past the fanout was calibrated. Note that $x(i, *, t)$ can be obtained by summing the SNMP link counts on all incoming access and peering links at the node. Hence the TM is obtained by multiplying the fanouts by SNMP link counts.

The observation on the stability of the fanouts is critical, because the idea here is that each node measures its own fanout and then ships it to the Network Management System (NMS). The NMS has all the SNMP data readily available, and can thus produce the complete TM (given the fanouts from all nodes) using the above equation. If the fanouts are stable for a period of a few days, then many days can go by before a node needs to ship a new fanout to the NMS.

Because the TM is a dynamic entity, the fanouts *will* change over time. This method thus includes a heuristic scheme for detecting changes in fanouts. Once a change is detected, then the fanouts need to be recalibrated, i.e. another 24 hours of NetFlow measurements are needed. Each node randomly selects one 10-minute interval within the next 24 hours and recomputes its fanout only for that 10-minute interval. The relative change between the newly measured fanout and the fanout vector corresponding to the same 10-minute interval in the baseline captures the diversion from the baseline. We define $\Delta(i, j) = \|\hat{f}_{i,j} - f_{i,j}\|$, if $\Delta(i, j) > \delta f_{i,j}$, for a randomly choosen time interval, then the entire row $f(i, *)$ is remeasured for the following 24 hours. Otherwise another interval is randomly selected within the next 24 hours, and the process reiterates.

We point out here that this procedure for checking diversion from the baseline is performed on a per node basis. If one node detects a fanout diversion, then a new collection of flow volume data is initiated *only on that node*. This means that a network-wide recollection of flow data is not needed. If some router fanouts change frequently, they will be updated often; those exhibiting great stability may not be updated for weeks at a time. With this approach, only the portion of the TM experiencing a dynamic change needs to undergo a recalibration.

The advantage of the fanout method over a full direct measurement method, in which each node measures its own row of the TM on an ongoing basis, is the reduction in communication overhead. The savings come by shipping fanouts only once every few days, rather than TM elements every 10 minutes. Moreover, flow monitors need not be on all the time, but are instead only turned on as needed.

This method comprises modeling OD flows via their fanout behavior. The model is a spatial model in the same way the gravity model was; namely that OD flow fanouts sharing the same source are correlated (due to the requirement that fanouts at a source node must sum to one). Similarly, correlation exists among OD flows with a common destination. However, it is clear from equation (5.5) that this model is also a temporal model since an estimate depends upon the history of a fanout. The fanout method thus uses a spatio-temporal model for OD flow fanouts. The data used to compute the fanouts come from flow measurements, as will be the case for all of our third-generation methods. To summarize, this method uses no inference, but rather direct calculation to produce TM estimates. These are considered estimates because the fanouts are an approximation when used outside the measurement period.

5.3.5   Principal components method

The *principal components method* attacks the TM estimation problem by studying the intrinsic dimensionality of the set of OD flows. Recall that the central difficulty of the TM estimation problem stems from the fact that the apparent dimensionality of OD flows, X, is much larger than the available link traffic measurements, Y. In ref. [118] the authors used Principal Component Analysis (PCA) to study the intrinsic dimensionality of the set of all OD flows. PCA is a dimension reduction technique that captures the maximum energy (or variability) in the data onto a minimum set of new axes called principal components. The authors of ref. [118] found that the entire set of N OD flows, when examined over long timescales (days to weeks), can be accurately captured by low-dimensional representations. In particular, the ensemble of OD flows are very well described by about five to ten common temporal patterns (chiefly diurnal cycles) called eigenflows.

We can take advantage of this low-dimensional representation of OD flows to develop a new approach for TM estimation. The key idea is that, instead of estimating all the N OD flows, we need only estimate the k most important eigenflows. Because $k \ll N$, the problem of estimating the eigenflows from link traffic becomes well posed.

Formally, let X denote the $T \times N$ multivariate time series of OD flows, as per our convention. Then, we can use PCA to write X as follows:

$$X = USV^T, \tag{5.6}$$

where U is the $T \times N$ matrix of eigenflow time series and V is an N by N matrix with principal components as its columns. Finally, S is an N by N diagonal matrix of singular values and $S(i, i)$ is a measure of energy captured by principal component i. The low effective dimensionality of X means that the top k principal components capture the overwhelming fraction of the energy of X; the remaining principal components account for only a negligible fraction and can be omitted. We can therefore reduce the dimensionality by selecting only the top k principal components. We can also approximate all the traffic demands at a time t as follows:

$$x_t \approx V'S'u'_t \quad t = 1, ..., T, \tag{5.7}$$

where V' is $N \times k$ with only the top k principal components, S' is the corresponding diagonal matrix, of size $k \times k$, and vector u'_t has the values of the k most significant eigenflows at time t.

Traffic matrix estimation using PCA assumes that the transformation matrices (V' and S') are already known (e.g. from prior measurements of OD flows) and stable [118]. Thus the problem becomes one of estimating the top k eigenflows, u'_t, from the set of link measurements y_t:

$$y_t = AV'S'u'_t \quad t = 1, ..., T. \tag{5.8}$$

Now, because the dimensionality of u'_t is much smaller, equation (5.8) is a well-posed estimation problem (in fact, we set $k = 10$). To solve for u'_t, we use the pseudo-inverse of $AV'S'$ and obtain an estimate $\hat{u}'_t$. Then, using equation (5.7), we compute an estimate of the traffic demands, $\hat{x}_t$. We set negative entries in $\hat{x}_t$ to zero, and then rely on the iterative proportional fitting algorithm of ref. [42] to refine our estimate, subject to the constraints imposed by the link traffic.

In order to obtain the decomposition $X = USV^T$, we need the TM X. This can be viewed as a prior TM and is obtained by using 24 hours of NetFlow data. Given S and V, we now simply collect the usual SNMP link counts and solve equation (5.8) to obtain the eigenflows.

The principal components methodology also has a recalibration step, which relies on informed remeasurements of the entire TM. Recalibration ensures that the PCA model is consistent with a changing TM, thus keeping the demand estimates accurate. The recalibration step has two components: (i) deciding when to trigger remeasurements of the entire TM, and (ii) updating the PCA model (the V' and S' matrices).

Ideally, we should trigger measurements only when we are certain that the demand estimates obtained from the PCA model are erroneous. A sufficient condition to assess the accuracy of the demand estimates is to compare them with the link traffic counts. When the demand estimates do not match the link traffic counts, it is likely that the TM has substantially changed and that the PCA model is no longer accurate. One strategy to assess the accuracy of the PCA model is to examine the maximum relative error over all links:

$$\epsilon_t = \max \left| \frac{y_t - A\hat{x}_t}{y_t} \right|,$$

where y_t is the traffic on all links at time t, $\hat{x}_t$ is our estimate of the traffic demands at that timepoint and ϵ_t is the resulting maximum relative error over all links (the vector division to compute the relative error is component-wise). To detect if a remeasurement is needed, we can check if ϵ_t exceeds a threshold, i.e. if $\epsilon_t > \delta$, where δ is an error tolerance parameter. However, short-lived variations can trigger expensive measurements in this scheme. In order to tolerate such transient changes, we monitor ϵ for a period of 24 hours. New measurements are only performed when there is a sustained change in ϵ. Sustained changes are quantified by counting the total number of entries in ϵ that exceed

δ, and then checking if the result is more than some fraction κ of the monitoring period (24 hours). In general, if δ and κ are small, recalibrations will be triggered frequently; when δ and κ are large, recalibrations will be rare. For the results in this chapter, we set $\delta = 0.9$ and $\kappa = 0.5$; we therefore trigger a recalibration if the relative error is more than 90% and is sustained for more than half a day (because this combination of parameters produced a measurement overhead of 19%, which is introduced in Section 5.4, allowing comparison with the other third-generation methods). We emphasize that the recalibration strategy proposed here is one of many potential change detection methods.

Once a recalibration is triggered, new measurements of the entire TM are performed for a 24-hour period. The new set of OD flows X are then decomposed using equation (5.6) to obtain the V' and S' matrices. In this manner, we update the PCA model of the TM, which is used to estimate all subsequent traffic demands until the next recalibration is triggered.

The PCA model of the TM exploits the dependency among the ensemble of OD flows over long timescales. Therefore, like the tomogravity model, the PCA model also relies on spatial correlation in OD flow traffic. In the PCA model, each OD flow is decomposed into a weighted sum of a handful of common eigenflows, which then become the quantity to estimate. The estimation step treats each timepoint separately to determine the value of the dominant eigenflows. The weights that capture the dependency between the OD flows are specified by the V' and S' matrices (as detailed in ref. [118]) and calibrated using direct flow measurements for a period of 24 hours. Because the original TM estimation problem is transformed into a well-posed problem (equation (5.8)), the estimation step in the PCA method is simply a pseudo-inverse solution.

5.3.6   Kalman-filtering-based method

We can view the OD flows as an underlying *state* of the network traffic. Since these flows evolve in time, the states do too. Unfortunately the states are not directly observed. Instead we use the SNMP link counts Y as an indirect observation of the item we are really seeking (the TM). Estimating the system state (traffic state in this case) from indirect observations, when these observations are linear combinations of the underlying state vector, is a common environment for the application of Kalman filtering. Kalman filtering is a powerful method because it can be used not only for estimation, but also for prediction. A full version of this method and its ramifications can be found in ref. [183].

We refer to y_t as the observation vector at discrete time t, while $Y_t = \{y_t\}$ is the set of all observations up to (and including) time t. Let $x_t = \{x(i, j, t) \forall i, \forall j\}$ be a vector denoting the entire set of OD flows at discrete time t. Thus x_t refers to the state of the system at time t. We model the evolution of the traffic state according to the following linear system:

$$x_{t+1} = Cx_t + w_t. \qquad (5.9)$$

In this model C is the state transition matrix and w_t is the traffic system noise process. For a single OD flow, the diagonal elements of C capture the temporal correlations appearing in the flow's evolution. The non-diagonal elements of C describe the dependency of one OD flow on another, thus capturing any spatial correlations among the OD flows (if and when they exist). The noise process w_t captures the fluctuations naturally occurring in OD flows. This model follows that of typical linear time-invariant dynamical systems. Another way to interpret this model is to say that the term C captures the deterministic components of the state evolution while the term w_t captures the random or variable component. An attractive feature of this model is that it captures both temporal and spatial correlations in a single equation.

The observations of the traffic state are made through a measurement system; in our case this corresponds to the SNMP link count vector. The traditional linear equation relating the link counts to the TM is now rewritten here as follows:

$$y_t = Ax_t + m_t, \tag{5.10}$$

where y_t is the link count vector, x_t is the state at time t and A is the routing matrix. We have added the term m_t to represent additive measurement noise. It is well known that the SNMP measurement process has its own errors, and thus we incorporate them to allow for minor differences between link observations and linear combinations of OD flows. Using Kalman-filtering terminology we say that this equation captures the relationship between the observation vector and the state of the system.

Let $\hat{x}_{t|t-1}$ refer to the *prediction* of x_t at time t based upon all information up to time $t-1$. We use $\hat{x}_{t|t}$ to denote the *estimation* of x_t at time t. The estimation step takes the previous prediction and adds the most recent measurement (observation) to make the next estimate. We now briefly describe how the estimation process, as required by TM estimation, is carried out. Since Kalman filtering carries out both an estimation and a prediction, this implies that the method has additional power and consequences not explored in detail here (see ref. [183]).

The task is to determine $\hat{x}_{t+1|t+1}$ (written as $\hat{x}_{t+1}$ for short) given a set of observations $y_1, ..., y_{t+1}$. The estimation error at time t is given by $\hat{x}_{t|t} - x_t$. We want an estimator that is optimal in the sense that it minimizes the variance of the error. The variance of the error is given by

$$E\left[||x_{t+1} - \hat{x}_{t+1}||^2\right] = E\left[(x_{t+1} - \hat{x}_{t+1})^T (x_{t+1} - \hat{x}_{t+1})\right]. \tag{5.11}$$

Let $P_{t|t} = E[(\hat{x}_{t|t} - x_t)(\hat{x}_{t|t} - x_t)^T]$ denote the covariance matrix of errors at time t. In the following we assume both the state-noise W and the measurement-noise M to be zero-mean white-noise processes, uncorrelated and with covariance matrices Q and R. Let the initial conditions of the state of the system be denoted by $\hat{x}_{0|0} = E[x_0]$ and $P_{0|0}$. The Kalman filter estimates a process by using a form of feedback control using two types of equations.

Prediction step
This step predicts the state and variance at time $t+1$ dependent on information at time t:

$$\hat{x}_{t+1|t} = C\hat{x}_{t|t}, \tag{5.12}$$

$$P_{t+1|t} = CP_{t|t}C^T + Q. \tag{5.13}$$

Estimation step

This step is also known as the "measurement-update" step. It updates the state and variance using a combination of the predicted state and the observation Y_{k+1},

$$\hat{x}_{t+1|t+1} = \hat{x}_{t+1|k} + G_{k+1}[y_{t+1} - A\hat{x}_{t+1|t}], \tag{5.14}$$

$$P_{t+1|t+1} = (Id - G_{t+1}A)P_{t+1|t}(Id - G_{t+1}A)^T + G_{t+1}RG_{t+1}^T, \tag{5.15}$$

where Id is the identity matrix and G_{t+1} is called the Kalman gain matrix. If we let the estimation error at time t be given by $\bar{x}_{t|t} = \hat{x}_{t|t} - x_t$, then we compute the gain by minimizing the conditional mean-squared estimation error $E[\bar{x}_{t+1|t+1}^T \bar{x}_{t+1|t+1}|y_k]$. By applying basic linear algebra, we can write $G_{t+1} = P_{t+1|t}A_{t+1}^T[A_t P_{t+1|t}A_{t+1}^T + R_{t+1}]^{-1}$. The above equations (5.12)–(5.15), together with the initial conditions specific above, define the discrete-time sequential recursive algorithm for determining the linear minimum variance estimate known as the *Kalman filter*.

We make a few observations about this system. Equation (5.12) is a one-step-ahead predictor, whereas equation (5.14) serves as the estimate we can use to populate a TM. Note that equation (5.14) includes a self-correcting step. Our estimate is equal to our prediction (in the previous timeslot) plus a gain factor multiplied by the error in our *link estimate*. Using the new incoming measurement at $t + 1$, we can compare the true link count versus our estimation for the link count. We adjust our estimate for the *TM* based upon our error in link estimation, $A\hat{x}_{t+1|t}$. The Kalman gain matrix is also a quantity that is updated and adjusted at each measurement interval. It is adjusted so as to minimize the conditional mean-squared estimation error.

One of the most salient features of the Kalman filter estimation method is this self-correcting feature. It essentially provides an estimation method that continually adjusts the state prediction, over each measurement interval, based on the error in the observations.

To apply Kalman filtering to TM estimation, one more step is needed. In order to produce our estimates in equations (5.12) and (5.14), we need to know the matrices C, Q and R as well as the initial conditions $\hat{x}_{0|0}$ and $P_{0|0}$. We propose to use 24 hours of NetFlow measurements to compute these variables. This step can be viewed as calibrating the system. Estimating the matrices C, Q and R from NetFlow data is a procedure that itself requires maximum likelihood estimation. For details on this estimation, see ref. [183].

The Kalman method also has a change detection and recalibration procedure to ensure that the underlying state space model adapts to changes in the TM. Like the PCA method, Kalman computes estimates for the link counts $\hat{y} = A\hat{x}$ using its TM estimates. Comparing these to the measured SNMP link counts yields an indirect error metric for the TM estimate. In Kalman filtering terminology, this error term is typically called the *innovation* process and is given by

$$i_{t+1} = y_{t+1} - A\hat{x}_{t+1|t}. \tag{5.16}$$

Recall that, in the Kalman method, this error term is already used inside equation (5.14). The innovations process should be a zero-mean process.

Our change detection method is as follows. In each 10-minute time slot we compute i_t and check if it is above a threshold. For the threshold we use twice the error variance, i.e. $2 \times A\mathrm{diag}(P_{t|t})$. We do this throughout the day. At the end of each day (at midnight), we check to see what percentage of the errors were above this threshold. This checking of the innovations process is done on a per-link basis. If more than 10% of these errors exceeded the threshold, then we schedule 24 hours of flow measurements on a network-wide basis. Since we include the spatial correlation, it is not possible to update only a link with some simple method. At the end of the 24-hour period, we recalibrate our C, Q and R matrices using the measured TM.

The motivation for waiting until the end of the day before deciding that new measurements are needed is to avoid recalibration when transient changes occur. Because we examine a day's worth of error metrics, we are essentially smoothing out the transient changes. The downside is that the recalibration could come many hours late, thus increasing our overall errors, since we continue using the old models to generate estimates.

As mentioned earlier, the Kalman method builds a spatio-temporal model for the OD flows. It is clear from equation (5.9) that the behavior of OD flows at time $t + 1$ is dependent upon the state of the flow at time t, and the spatial dependence is captured by the off-diagonal elements of the C matrix. Similar to the other third-generation methods, this approach uses 24 hours of flow measurement data to calibrate its underlying model (in this case, the critical C matrix, as well as the error covariance matrices Q and R). Then the estimation procedure uses a Kalman filter, which is the best linear minimum variance estimator in the case of zero-mean Gaussian noise.

5.3.7   Discussion

In Table 5.2, we summarize, for each method, the type of OD flow model used, the source of data used to calibrate the model, the type of estimation used by each method and the data used for estimation. Each of these models relies on insights that have been learned from measurements over the past few years. The Fourier model used in the route change method came from the realization of the presence of strong diurnal patterns in TM data. The PCA method is built on the observation that OD flows have low intrinsic dimensionality. The fanout method relies on the stability of fanouts. The generalized gravity model includes networking insights by incorporating routing policies such as hot-potato routing and no-transit traffic forwarding. Knowing that both spatial and temporal correlations exist among OD flows, and that the system is inherently linear ($y = Ax$), the Kalman method builds a linear dynamic system that can capture both types of correlations in a single equation.

5.4   Performance analysis

As mentioned in Section 5.1.2, the performance evaluation of the five estimation algorithms is performed on the same data set, collected from the European Tier-1 Sprint IP network in the summer of 2003, introduced earlier in this chapter.

Table 5.2. Summary of data used for modeling and estimation

Method	tomogravity	route change	Kalman	PCA	fanout
Modeling					
Data used	SNMP (access, peer)	SNMP (inter-router) under multiple snapshots	Flow data, 24 hours	Flow data, 24 hours	Flow data, 24 hours
Model	gravity	cyclo-stationary (Fourier + noise)	linear dynamic system	eigenflows and eigenvectors	node fanout
Type	spatial	temporal	spatio-temporal	spatial	spatio-temporal
Estimation					
Data used	SNMP (inter-router), one routing matrix A	SNMP (inter-router), multiple A matrices	SNMP (inter-router), A matrix	SNMP (inter-router), A matrix	SNMP (access, peer), no A matrix
Method	regularized MMSE optimizaiton	pseudo-inverse	Kalman filter	pseudo-inverse	none (direct formula)

5.4.1 Spatial and temporal errors

To assess the overall performance level of each of the methods, we consider both temporal and spatial errors. Our main error metric is the relative L2 norm. By *spatial error*, we mean that we obtain an error metric per OD flow that summarizes its errors over its lifetime. For this, we use

$$RelL2_{\text{SP}}(n) = \frac{\left(\sum_{t=1}^{T} (x_t(n) - \hat{x}_t(n))^2 \right)}{\left(\sum_{t=1}^{T} x_t(n)^2 \right)}. \tag{5.17}$$

The ensemble of all spatial errors gives us a set of errors over all the OD flows. In Figure 5.5 we plot these spatial errors for each of our five methods. The x-axis represents the ordered flows, from largest to smallest, sorted according to their mean. In this plot we include all the OD flows constituting the top 95% of the load in the entire TM.

The temporal error gives an error metric for each time slot summarizing the errors over all OD flows at that instant. For this we use the following:

$$RelL2_{\text{T}}(t) = \frac{\left(\sum_{n=1}^{N} (x_t(n) - \hat{x}_t(n))^2 \right)}{\left(\sum_{n=1}^{N} x_t(n)^2 \right)}. \tag{5.18}$$

The temporal error is similar to what an ISP might see when trying to estimate the entire TM at a particular moment in time. This error metric is given in Figure 5.6 for all five

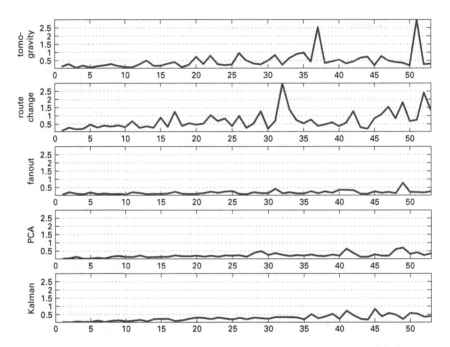

Fig. 5.5.   Spatial relative L2 norm error (x-axis is flow id; flows ordered from largest to smallest in mean) for all five estimation techniques. ©ACM, 2005.

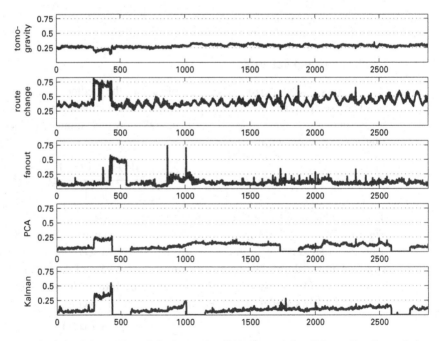

Fig. 5.6.   Temporal relative L2 error (x-axis in time units of 10 minutes) for all five estimation techniques. ©ACM, 2005.

methods. The x-axis here is in units of 10-minute slots. We include the estimation errors for 20 days, excluding the first day since it is used for calibration by the PCA, Kalman and fanout methods. (Recall that when flow monitors are turned on, we assume they are used to populate the TM and the modeling-based inference approaches are ignored in those timeslots.)

In the spatial error plots, we observe the well known phenomenon that errors increase as the size of the flow decreases. However, we also note that this effect is far less pronounced in the third-generation methods than in the second-generation methods. Since this holds across all the different third-generation techniques, we conclude that we are seeing one of the benefits of using partial flow data coupled with smart OD flow models, in that we can better estimate the smaller flows.

In the temporal error plots, note that the PCA and Kalman methods appear to have exactly zero errors at certain periods. These periods are when recalibration occurred and we thus exclude them from our error calculation. The moments of recalibration are not visible with the fanout method, because the fanout method does not recalibrate the entire TM model at the same time, but rather one or more PoPs at a time as needed.

Interestingly, we observe that the tomogravity method exhibits nearly constant temporal errors. We will examine this behavior further in later sections. The tomogravity method achieves roughly 25% temporal errors, while all of the third-generation methods achieve average temporal errors between 10 and 15%.

An alternative way of viewing the spatial and temporal errors is to examine their Cumulative Density Functions (CDFs), as given in Figures 5.7 and 5.8. We see that, in terms of both spatial and temporal errors, the third-generation methods always outperform the second-generation methods. In particular, the route change method achieves the largest errors (and is not included in the rest of the comparative evaluation); the tomogravity method experiences the second largest errors; and the three methods using partial flow measurements all exhibit errors that are roughly similar.

Because our temporal errors summarize errors across all flows for each timepoint individually, we would expect that methods that assume spatial correlation would perform best. The tomogravity and PCA methods do well in the sense that their distributions are narrower than other methods. The Kalman and fanout methods do well in that they have low errors, although they have a somewhat longer tail than the other methods. The longer tail indicates that there are some moments in time that are more difficult to estimate than others. As we will see later, this is due to a somewhat conservative approach to adaptation.

It is interesting to note that the distributions of the spatial errors tend to be much less narrow than that of the temporal errors. The TM that we obtained via measurement contained TM elements that spanned eight orders of magnitude. It is easy to understand the numerical problems introduced by trying to estimate all of these elements simultaneously. By including the OD flows that constitute the top 95% of the load, we include flows that span two orders of magnitude, from 10^5 to 10^7 bps. The large dispersion of the spatial errors indicates that some flows may simply be harder to estimate. We know from the literature (and confirmed by Figure 5.5), that, in general, smaller flows are harder to estimate.

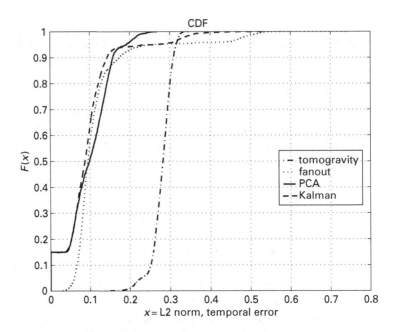

Fig. 5.7.   CDF of temporal errors. ©ACM, 2005.

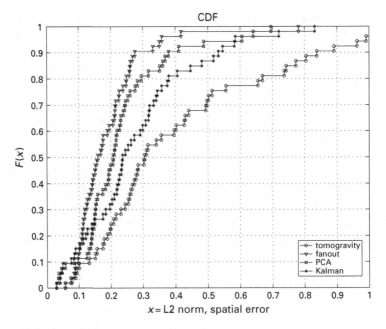

Fig. 5.8.   CDF of spatial errors. ©ACM, 2005.

5.4.2   Over-modeling

We see in both Figures 5.5 and 5.6 that the route change method performs the least well of all the methods. Before continuing we pause here to explain why this happens

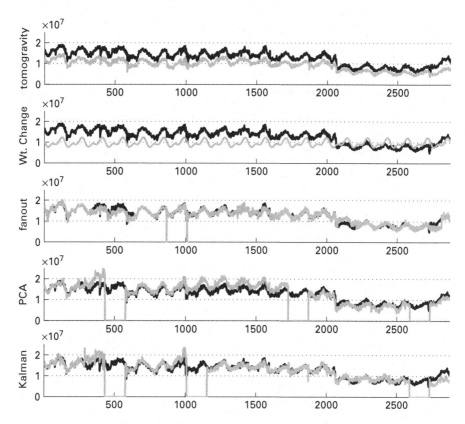

Fig. 5.9.   Eighth largest OD flow in bps. Original in black; estimates in gray. Time unit is 10 minutes.
©ACM, 2005.

and to state why we do not include this method in the remainder of our analysis. To
see how this method behaves, we use the example of our eighth largest OD flow, as
depicted in Figure 5.9. In this plot, the original traffic is depicted via the black line and
the estimate is given via the gray line. We see that the route change method yields a
perfectly cyclical pattern that does not deviate throughout the three-week period. This
is because the route change method assumes that all OD flows follow the model in
equation (5.4). The implication of this model is very strong; it assumes each OD flow
is cyclo-stationary, and any deviation from this model is not captured in the estimates.
To see this even more closely, we plot a blow-up of a short period of our fourth largest
flow in Figure 5.10. We clearly see the impact of the modeling assumption in that it is
enforced too strongly.

This is in part due to the limited set of basis functions (five) used. We have used five
here because this was the number used in ref. [182]. The performance could clearly
be improved by using a larger number of basis functions for the Fourier expansion.
Because the errors in this method are significantly larger than the other methods, we do
not include this in the remainder of our study. We wish to point out that, historically, this
method was the first to introduce a temporal model for OD flows, thus paving the way

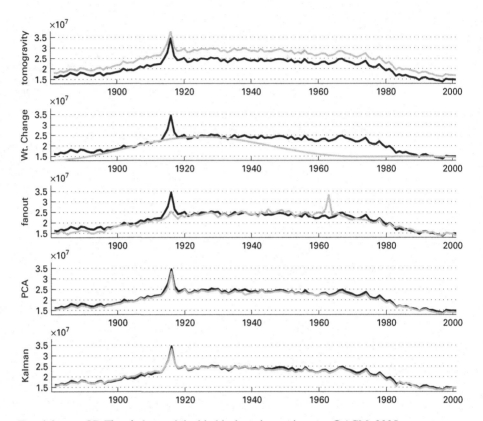

Fig. 5.10.   Fourth largest OD Flow in bps. original in black; estimates in gray. ©ACM, 2005.

for richer models based on properties learned from data. Part of the difficulty incurred by this method may be due to the fact that it does not include any spatial correlations.

5.4.3   Measurement overhead

We saw in Section 5.4.1 that the third-generation methods outperform the second-generation methods in terms of both spatial and temporal errors. This is not surprising considering that they have at their disposal a rich set of data to build useful models. It is thus important to now ask the following: At what cost do we obtain these improved error rates? To do this, we define an overhead metric to capture the amount of measurement used by these new methods, and we then look at the average error achieved for a given level of overhead.

Each time a flow monitor on one link is turned on, it is used for 24 hours. We thus define a metric whose units are *link-days* since it is intended to capture how many links were used (over a multi-week period) and for how long. We measure the overhead via a ratio intended to capture the number of link-days used divided by the total number of link-days that would be used if flow monitors were left on all the time. Units such as link-days can be thought of as describing a spatio-temporal metric. More precisely, we define our overhead metric as follows. Let $D(l)$ be the number of 24-hour cycles that link l used NetFlow during our three-week period. Hence $D(l)$ gives the total number of days that NetFlow was used on link l. The total overhead incurred by a single method is given by

$$OH = \frac{\sum_{l=1}^{L} D(l)}{21 * L},$$

(5.19)

where L is the number of links in the network and 21 is the number of total days of our measured validation data. Both the numerator and denominator are capturing a network-wide property. If the flow monitors were turned on all the time and network-wide, then this overhead metric would be one; this corresponds to the case when a TM is fully measured and no inference is used. If flow monitoring is never used, the metric is zero; this corresponds to the case of the tomogravity method. Our intent with this metric is to explore the error performance of methods whose overhead lies somewhere between the two extremes of zero and one.

We ran each of the three flow-based methods multiple times. In each case we changed the threshold that checks for diversion, so as to encourage more or less adaptation and recalibration, thus leading to the inclusion of different amounts of flow measurements. For each such case we computed both the spatial and temporal relative L2 norm error. The spatial errors are given in Figure 5.11 and the temporal errors are displayed in Figure 5.12. In these two figures we also include the two extreme points mentioned above, displaying the error of the tomogravity method using zero flow overhead, and the point (100, 0) for the case when a TM is obtained solely by directly measuring it all the time. We assume the estimation error is zero (ignoring measurement errors) since flow

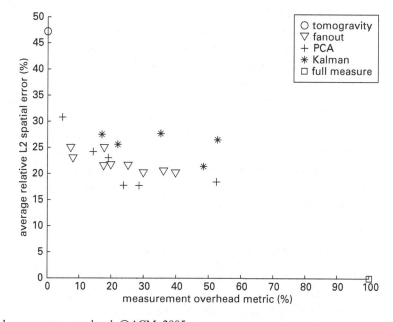

Fig. 5.11.   Spatial errors versus overhead. ©ACM, 2005.

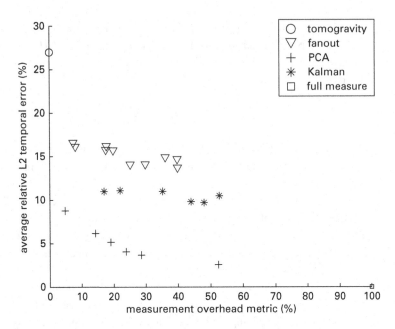

monitors are on all the time. Although the error rate is zero, the measurement overhead is at its maximum. This overhead does not even include the large communication cost, as listed in Table 5.1 at the beginning of this chapter, of a full measurement approach. Through these plots we can see the balance that can be achieved using hybrid solutions that combine flow measurements with inference to improve inherent limitations in inference-based solutions.

Two quite interesting observations are evident from these plots. First, for a measurement overhead of anywhere between 10 and 20%, we can roughly cut the errors achieved by the tomogravity method in half. This is true for both the spatial and temporal error metrics. This is a large gain in error reduction for a small measurement overhead. As compared to the full measurement approach, we see that, with 90% fewer measurements, the third-generation methods can produce highly accurate estimates. Second, beyond approximately a 20% overhead, there is little further gain in including additional measurements. In other words, a large amount of overhead is required to further reduce errors. This hints that once we exceed the benefits of 20% flow measurement overhead, we are probably reaching the limits of what inference can do. Thus to reduce errors further, one may have no choice but to rely on full measurement approaches.

5.4.4   Handling dynamic changes in the TM elements

The third-generation methods have been designed to adapt to changes in the TM by recalibrating their models. The implicit assumption in this approach is that TMs are very dynamic, and thus an adaptive approach is required. This is indeed true; this dynamic nature can take a variety of forms and be due to a variety of reasons. For example, in

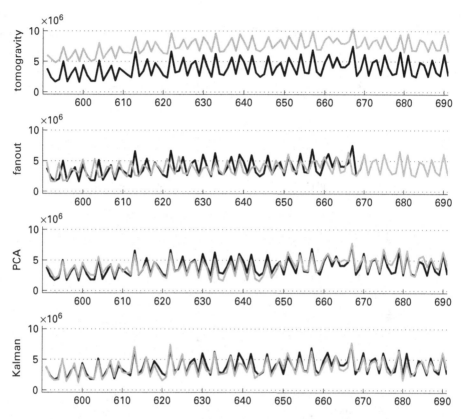

Fig. 5.13. Example of oscillating flow in bps. Original in black; estimates in gray. Variations within same order of magnitude. ©ACM, 2005.

our measurement data we found some flows that exhibit rapid oscillations for extended periods of time. Other flows are seen to undergo a long-lived, but not permanent, change for a period of a day or so. This can happen if a set of ingress links fail and take a day to repair. The addition of new customers in a single PoP will lead to a permanent increase in the traffic demand emanating from that PoP. We ask the question as to how our various methods handle these types of dynamic behavior.

First we look at the case of a flow experiencing rapid oscillations (see Figure 5.13, which we have enlarged for a period of a little less than a day). Here we see that all the methods can handle this type of behavior. We point out that, while the fluctuations are rapid, they are also contained within the same order of magnitude. We looked over the lifetime of this flow and saw that none of its oscillations lead to model recalibration for the third-generation methods. This implies that there is sufficient flexibility to capture the variability for a flow that did not change an underlying trend.

In Figure 5.14 we see the example of our second largest OD flow, which undergoes a dramatic change on day 3. This flow drops many orders of magnitude to a small, but non-zero, level. (This was most likely due to the failure of some, but not all, of the links at the router sourcing this OD flow.) This is plotted on a log scale so that we may see what happens during the severe dip in flow volume.

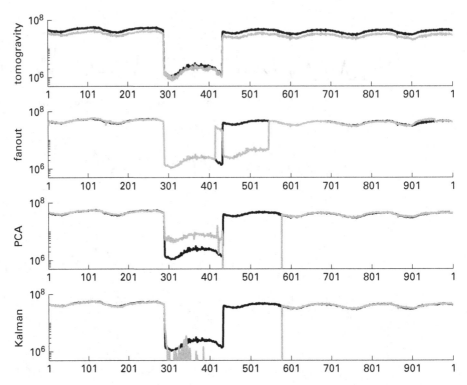

Fig. 5.14.   Dynamics in second largest flow in bps; longer-lasting change over two orders of magnitude.
©ACM, 2005.

The tomogravity method tracks this change best. This is because it starts completely anew each timeslot. In this case, the gravity model is recalculated every 10 minutes based on the newest SNMP data from access and peering links. In this case, the memoryless property of the tomogravity method serves it well.

We can see the adaptive behavior of the fanout method well here. For some other reason, the fanout method initiates a recalibration just a little before time unit 300. During the day of problems with this flow, the fanout method is using the flow measurement data to populate the TM, and hence we see no errors. However, the fanout method is learning atypical behavior, which we see repeated from time unit 410 to time unit 560 (roughly). During the fourth day the fanout method realizes that it needs another immediate recalibration, and thus we see it generates excellent estimates after time 560.

Looking back to Figure 5.6 we see that both the PCA and Kalman methods experienced larger errors during day 3 (starting at time slot 432 on that figure). We can now understand the source of this error; it was due to the large change in behavior of a dominant flow. The x-axis in Figure 5.14 starts at a different time than the one in Figure 5.6 because we have enlarged a portion of the plot for ease of viewing. We see (Figure 5.14) that the PCA and Kalman methods do adapt to this change, but less well than the tomogravity method. Because these methods realize they make errors during this period, they initiate a recalibration around time 430 (after having waited the requisite 24 hours). At time 570, they return to normal behavior.

We thus see a tradeoff between third-generation methods and the tomogravity method. The tomogravity method adapts fast to changes, but experiences larger errors, than the other methods. Of course, the rate of adaptivity has been defined by the method; tomogravity updates its model in every 10-minute timeslot because the data it uses are typically available at that level of granularity. Because the data used by the third-generation methods are more costly to obtain, they wait (defined at 24 hours) before adapting to ensure that a dynamic change is ongoing and that the method does not react to a change that will dissipate quickly.

Finally we seek to understand the issue of bias. In many figures we have observed the bias that the tomogravity exhibits (Figures 5.9, 5.10, 5.13 and 5.14). In these four examples the tomogravity method accurately tracks the shape of each flow, but exhibits a consistent over-estimate (e.g. fourth largest flow) or under-estimate (e.g. eighth largest flow). A consistent difference between estimated and true values is termed *bias*.

It is well known that biased estimators are not necessarily bad. An unbiased estimator is one whose expectation is equal to the true value of the parameter being estimated. However, even if an estimator is unbiased, it may have high variance and thereby often fail to provide estimates that are close to the parameter's true value. On the other hand, sometimes biased estimators can have smaller variance and thereby typically yield estimates closer to the true value than those of unbiased estimators. Thus any consideration of an estimator's bias should also take into account its variance.

The sample bias of an estimator for OD flow n is given by

$$bias(n) = \frac{1}{T} \sum_{t=1}^{T} (\hat{x}_t(n) - x_t(n)). \tag{5.20}$$

We choose to look at absolute bias rather than relative bias because this reflects the error a network operator would see when employing the methods.

In fact, we observe that the various methods perform very differently with respect to bias. Bias is plotted in Figure 5.15 for the four methods. On the x-axis are OD flows, sorted from largest mean to smallest. The figure shows that the most consistently unbiased method is the fanout method. Both the PCA and Kalman methods show a negative bias (i.e. an underestimation) for the largest flows, with the amount of bias increasing with flow size. However, the tomogravity method is quite different from the rest. It shows a much larger bias in general, with both positive and negative biases prominent. Furthermore, this bias is not consistently related to flow size, as in the PCA and Kalman methods. Rather, the per-flow bias can be very large, even for quite small flows.

As mentioned above, the accuracy of an estimator is a function of both its bias and variance. To assess this we define the sample standard deviation of the estimator for flow n as follows:

$$ErrStd(n) = \left(\frac{1}{T-1} \sum_{t=1}^{T} (err_t(n) - bias(n))^2 \right), \tag{5.21}$$

where $err_t(n) = \hat{x}_t(n) - x_t(n)$. Note that this metric is in the same units as sample bias, so we can directly compare the two when assessing their impact on accuracy.

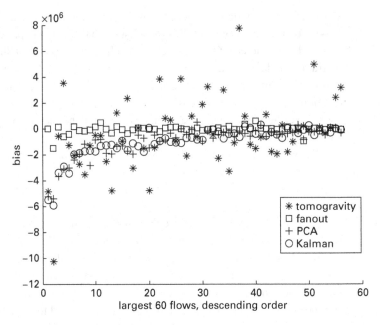

Figure 5.16 plots the sample bias for each flow against its sample standard deviation, for the four methods. This plot confirms the striking difference between the tomogravity method and the other methods. It shows that, while the tomogravity method has much lower variance than other methods, it exhibits much higher bias in general. In contrast, the Kalman and PCA methods show relatively high variance, with much lower bias in general, while the fanout method maintains relatively low bias and variance. Thus, the tomogravity method achieves a different tradeoff between bias and error variance than do the third-generation methods.

Methods with low bias and high variance tend to estimate the flow mean well over long time intervals, while giving estimates for individual timepoints with less accuracy. Methods with high bias and low variance will tend to track changes between individual timepoints well (i.e. accurately estimating short timescale variations), while long timescale averages (the mean) may be consistently wrong. These two different conditions are visible in Figure 5.10. The tomogravity method tracks the precise shape of the flow quite well, but its value is consistently offset. The Kalman and PCA methods tend to be closer to the true value on average, but their precise shape does not match that of the data as accurately.

This difference in the relative contribution of bias and variance can have implications for applications of these methods. Consider the use of TM estimation in anomaly detection at a coarser granularity (i.e. that of minutes). In this case, it is the variations from typical conditions that are important. Accurate estimation of the mean is much less important that correct tracking of the shape of the flow. Thus a method like tomogravity seems better suited in this case. On the other hand, consider an application such as traffic engineering (adjusting link weights to balance traffic) or network capacity planning.

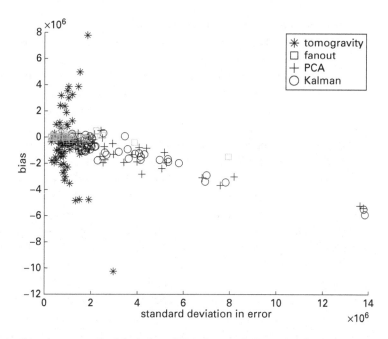

Fig. 5.16.   Estimation bias versus standard deviation of TM element (in bps). ©ACM, 2005.

In this case, accurate estimation of flow mean is paramount. Further, accurate tracking of short timescale variations is less crucial since such decisions take place on relatively long timescales. For such applications, the PCA and/or Kalman methods may offer a better choice.

5.5   Lessons learned

In this chapter we looked into the problem of inferring the amount of traffic flowing between any pair of network elements, i.e. links, routers or PoPs, in an ISP network. We refer to the above with the general term of *traffic matrix* (TM). Although this problem is fully solved when widely collecting and processing NetFlow records, it is not realistic to assume that ISPs do have such functionality fully deployed in their networks. As a consequence, either they are lacking parts of the TM or, worse, missing the entire TM. Due to the important role played by the TM in many network management and traffic engineering tasks, we dedicated this chapter to analyzing and comparing the most promising, recently proposed, techniques in the literature that approach this problem, based on different intuitions, models and correlations (spatial, temporal and spatial-temporal). We referred to those methods as first-, second- and third-generation techniques.

Based on their evaluation on the same *complete* TM data set, we showed that all of the methods handle well the short-lived changes in the TM that remain within the same order of magnitude. For longer-lived or permanent changes that involve a change in the order of magnitude of a flow (or flows), all the methods adapt reasonably well, but tomogravity adapts best. Tomogravity can adapt the fastest because its model discounts temporal history. This is a tradeoff that comes at the price of the bias.

There is a striking difference between the tomogravity and the fanout, PCA and Kalman methods. Tomogravity has lower error variance than the other methods, but exhibits a much higher bias in general. The third-generation methods are able to overcome bias and better capture the long-term behavior of an OD flow, since they incorporate the *actual* correlation data (either spatial or spatio-temporal) they have from measurements, as opposed to an *assumed* correlation structure in the gravity model.

A key feature of the PCA, Kalman and fanout methods is their reliance on actual flow measurements to build models for OD flows. These methods therefore comprise the next generation of hybrid measure-and-infer strategies to obtain TMs. In fact, we find that these methods outperform the earlier generation – route-change and tomogravity – in terms of both spatial and temporal errors. Furthermore, these new-generation methods can handle smaller flows much better, a problem that has so far plagued the known methods that rely on SNMP data alone.

An important and surprising finding is that with only 10–20% additional measurements, we can achieve half the temporal and spatial errors reported by earlier methods. Moreover, we can achieve accurate estimates using 90% less measurement than a full brute force measurement approach. The measurement, communication and computation overheads of the full measurement approach will be far greater than of the hybrid measure-and-infer strategies we have assessed here. Our computations are simple to execute and may be more efficient than a direct computation. As flow monitors become accessible in more networks, network operators will have new options to compute traffic matrices. We envision that techniques such as the fanout, Kalman and PCA methods will evolve into feasible and accurate alternatives that yield better performance than methods relying solely on inference, and at smaller costs than brute force measurement.

Part II

Network design and traffic engineering

6 Principles of network design and traffic engineering

Since the late 1990s there has been significant interest and attention from the research community devoted to understanding the key drivers of how ISP networks are designed, built and operated. While recent work by empiricists and theoreticians has emphasized certain statistical and mathematical properties of network structures and their behaviors, this part of the book presents in great detail an optimization-based perspective that focuses on the objectives, constraints and other drivers of engineering design that will help the community gain a better insight into this fascinating world and enable the design of more "realistic" models.

In this chapter we introduce the area of IP network design and the factors commonly used to drive such a process. Our discussion revolves around IP-over-WDM networks, and we define the network design problem as the end-to-end process aimed at identifying the "right" IP topology, the associated routing strategy and its mapping over the physical infrastructure in order to guarantee the efficient utilization of network resources, a high degree of resilience to failures and the satisfaction of SLAs.

We start by providing a high-level overview of the IP-over-WDM technology. We highlight the properties of the physical and IP layers (the IP layer is also known as the logical layer), we discuss their relationship, and introduce the terminology that will be extensively used in the following chapters. Then, we introduce the processes encountered in IP network design and their driving factors. We conclude the chapter by defining a multi-step methodology aimed at solving the problem of network design in an incremental fashion. Each step will be extensively studied in the following chapters.

6.1 Overview of IP-over-WDM technology

IP-over-WDM technology is aimed at combining the best features of optics and electronics. This type of architecture has been called "almost-all-optical" because traffic is carried from source to destination without electronic switching "as far as possible," but electronic switching is to be performed at some places.

IP-over-WDM envisions the IP infrastructure layed directly over the optical infrastructure without any inter-layer technology such as ATM, Frame Relay, SONET, etc. This way, the IP layer can make complete use of the entire bandwidth of each optical fiber, avoiding any overhead introduced by intermediate technologies. In some cases, the fiber is used as a simple alternative to copper wire. This means that only a single wavelength is used to carry information over a fiber and the fiber acts as the point-to-point link of a given bandwidth. Due to the intrinsic limitation of the hardware equipment and the speeds at which they can push bits into the fiber, most of the bandwidth is left unused. With WDM, i.e. wavelength division multiplexing, the upper layer can suddenly utilize bandwidths comparable to that of entire fibers. Multiple signals can travel together on different wavelengths. This operation is enabled by the usage of optical multiplexer and demultiplexer units that "bundle" and "unbundle" different signals. Each signal uses a different wavelength when traversing the fiber to avoid collision with other signals and consumes only a portion of the total bandwidth (OC-3, OC-12, OC-48, etc.).

The two layers constituting the IP-over-WDM infrastructure are named the *physical* and *logical* layers. The physical layer is composed of optical routing nodes, called optical cross-connect routers (OXCs), that are connected to each other by point-to-point optical fibers. The OXCs and their physical connections constitute the *physical topology*. Each of the OXCs has an access node connected to it anywhere electronic switching is required. The access node is called an IP router. At the logical layer, IP routers are connected via IP links, and two IP routers directly connected with an IP link are said to be logically adjacent. Note that the two routers might be geographically far away from each other from a physical perspective, but adjacent from a logical perspective. The set of IP routers and IP links constitute the *logical topology* (also known as *IP topology* or *virtual topology*). An example of an IP-over-WDM architecture is shown in Figure 6.1. Two IP links have been set up to allow communication via a clear channel between nodes. The IP links have been routed in the physical infrastructure using dotted lines and share a common fiber. On this fiber, the two IP links must use different wavelengths to avoid collision. Note that one pair of routers, i.e. (R1, R2), is directly connected by a fiber link, while the second pair of routers, i.e. (R1, R3), is not. Although R1 and R3 are logically adjacent, they are geographically distant, i.e. the packets have to cross two fibers when traveling from the source to the destination.

Packets traveling on any IP link are carried optically from end-to-end. Electronic switching is required anywhere an IP link is terminated and a new IP link is established. In order to understand what happens at each node, look at the typical node architecture shown in Figure 6.2. The bottom part of such a node deals with optical signals, i.e. OXC, while the upper part deals with electronic signals, i.e. IP router. The OXC is preceded by a wavelength demultiplexer and followed by a wavelength multiplexer. The demultiplexer is in charge of separating the incoming signal at the input ports into individual signals traveling each on a different wavelength. The multiplexer bundles together different signals into one and sends it out on the output fiber.

Wavelengths on some of the input ports carry signals that are destined for an IP node directly connected to the OXC (i.e. termination of the IP link), so the signal has to be extracted from the optical medium. The cross-connect terminates that particular wavelength, converts the data into electronic form, and delivers it to the IP layer for processing (see, for example, IP link #1 when reaching the router R3 or the IP link #2 when reaching router R2 in Figure 6.1). The number of IP links that can be initiated

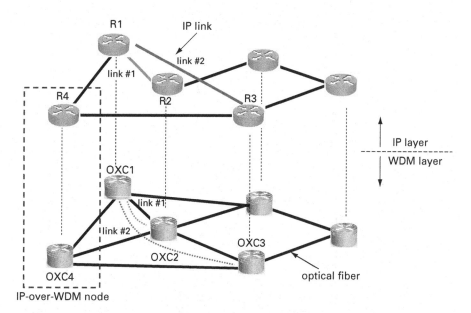

Fig. 6.1.   An example of an IP-over-WDM network. Physical nodes (OXC in the figure) are inter-connected by point-to-point fiber links constituting the physical topology. Logical nodes (IP routers R in the figure) are inter-connected by point-to-point IP links constituting the IP topology. Each IP link is assigned a wavelength and carried over a chain of physical fibers. Two IP links sharing the same optical fiber must have different wavelengths (link #1 and link #2 are represented by dotted lines sharing the fiber connecting OXC1 and OXC2.)

or terminated at each node is named *node degree* and is usually limited by both the IP technology, i.e. known as the "efficient frontier" of the router,[1] and the optical technology, i.e. number of transmitters and receivers on board of the OXC. Usually the *node degree* of each node is assumed to be equal for incoming and outgoing connectivity and across the entire logical topology.

There may also be signals on some wavelengths that need to be forwarded to other nodes (see IP link #2 when traversing OXC2 in Figure 6.1). If the OXC does not support wavelength conversion, it takes the signal on the input port traveling on a wavelength and optically switches the signal to a selected output port without modifying the wavelength being assigned. In this case, the OXC is acting as a wavelength router and routes IP links over the physical infrastructure in the same way an IP router routes packets over the logical topology. The physical layer is said to be constrained by *wavelength continuity*, indicating that an IP link must use the same wavelength on all the fibers it crosses from its origin node to its termination node.

---

[1] The efficient frontier of a router is defined as the possible bandwidth-degree connections it can sustain. That is, a router can have a few high-bandwidth connections or many low-bandwidth connections (or some combination in between). In essence, this means that the router must obey a form of *flow conservation* in the traffic it can handle. While it is always possible to configure the router so that it falls below the efficient frontier (thereby under-utilizing the router capacity), it is not possible to exceed this frontier (for example by having an increasing number of high-bandwidth connections).

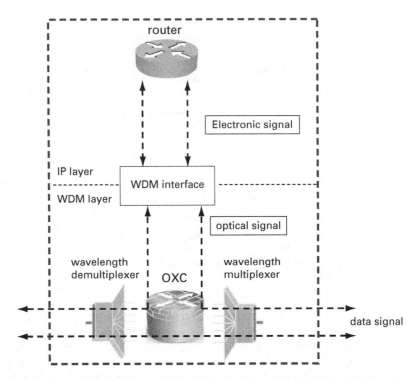

Fig. 6.2.   Architecture of an IP-over-WDM node: OXC functionality resides at the physical layer; IP routing functionality resides at the IP layer.

If the OXC is equipped with wavelength conversion functionality, then IP links might change their wavelength at each of the traversing OXCs. In-transit IP links can enter an OXC on one wavelength and leave it on a different wavelength. Although wavelength conversion allows for a better usage of the fiber infrastructure, the management of those OXCs introduces some extra level of complexity in the design process due to issues related to the tuning delay of wavelength converters and their placement in the physical infrastructure.

Note: In the remainder of this book we assume the physical topology to be an input of our design process. Furthermore, we will assume that each OXC has exactly one access IP router connected to it.

6.2   Network design: principles and processes

Understanding the large-scale structural properties of ISPs is critical for network managers, software and hardware engineers and telecommunication policy makers alike. On a practical level, models of network design factor prominently in the design and evaluation of network protocols, since it is understood that, although network topology should not affect the *correctness* of a protocol, it can have a dramatic impact on its *performance*. Accordingly, the ability to shape network traffic for the purposes of improved application performance often depends on the location and interconnection of network resources. In addition, a detailed understanding of network design is fundamental for developing improved resource provisioning, as most network design problems assume a detailed description of existing/available network components.

The design of the network should be viewed as a "one-time" task. Given the inherent cost in such a process, and the difficulty in making significant changes retrospectively, network design should aim to define the critical aspects of an IP network that should not be deviated from in a fundamental manner. Some small providers underestimate the importance of choosing the right topology and failure recovery schemes and choose to combine it in with network support and traffic engineering techniques as ongoing processes. Instead, the network design should be thought of as a process of refinement and minor modification. One cannot afford to redesign the entire network in response to growth in applications, users, or the number of sites on the network.

6.2.1   Edge versus core: from routers to PoPs

The operation of an ISP at a national scale requires the installation, management and maintenance of communication links that span great distances. At the national level, the cables are usually fiber-optic and the equipment consists of transmitters/receivers at the end points and signal repeaters along the way. While a significant portion of the link cost is often associated with obtaining the "right of way" to install the network cables, there is generally an even greater one associated with the installation and maintenance of the equipment used to send traffic across these cables. Both the installation and maintenance costs tend to increase with link distance. Thus, one of the biggest infrastructure costs faced by a network provider is that associated with the deployment and maintenance of its links. National ISPs are one type of network providers for which link costs are significant. The challenge is in providing network connectivity to millions of users spread over large geographic distances at an acceptable cost, a task made somewhat easier by the fact that most users tend to be concentrated in metropolitan areas. Such a trend leads to a natural separation of the connectivity problem into providing connectivity within a metropolitan region and providing connectivity between metropolitan regions. In its simplest form, it states that the only type of design that makes sense from an economic perspective is one that aggregates as much traffic on the fewest number of long distance links. As a consequence, the design process can be separated into the *edge* and *core* network design.

The edge network design typically occurrs at the metropolitan area, where the challenge is to provide connectivity to local customers who are dispersed over some regional geographical area. Edge design is often driven by the customers supported by the ISP. For example, in the current environment there is tremendous variability in the connection speeds used by customers to connect to the Internet, such as dial-up connections (generally 56 Kbps), broadband access (256 Kbps–6 Mbps) and large-bandwidth connections (1 Gbps and above). This connectivity is rooted at the ISP's PoP, which serves as the focal point for local traffic aggregation and dissemination between the ISP and its customers.

On the other hand, the core design is focused on providing internal connectivity within the ISP between its different PoPs, which are typically separated by larger geographic distances. Core design is often driven by different objectives aimed at achieving the most reliable and timely delivery of information from any source to any destination while using the least amount of network resources.

Thus, the separation of the network design into edge and core design is important because the resulting optimization formulations often have different objectives and constraints operating at each level of abstraction. In addition, edge design problems are inherently local, in the sense that changes to the inputs or incremental growth in the edge network have only local effects on the topology or its traffic. In contrast, changes to inputs of the core design problem or core incremental growth often have global implications on the structure and/or behavior of the network as a whole.

Note: In the following sections, we focus on the core design, although similar methodologies can be applied to solve the edge design. Accordingly, a node at the logical layer has to be interpreted as a PoP.

6.2.2   Survivability and recovery schemes

Any operational network must guarantee high reliability in the presence of failures either inherent to the physical layer (equipment failures, fiber cuts, etc.) or to the logical layer (line card failures, software problems, etc.). This property is also known as *survivability*. Generally, network survivability is quantified in terms of the ability of the network to maintain end-to-end paths in the presence of node or link losses.

Survivability to physical failures

Despite its name, the logical topology does not necessarily reflect the structure of its underlying physical layer. Depending on the technologies in use, two nodes that are "connected" by a single hop at the IP layer may or may not be physically connected to one another. Looking at just the IP layer, the reader can have the illusion of direct connectivity, even though the routers in question may be separated at the physical level by many intermediate networking devices or even an entire network potentially spanning hundreds of miles. As a consequence, failures at the physical layer are more common than we think. The analysis on physical failures for Sprint's optical backbone has highlighted that failures happen on a daily basis, and that as many as 95% of these failures last longer than 120 minutes. Moreover, it has been noted that more than 85% of these failures are isolated and do not show any temporal correlation among themselves [134]. We refer to those physical failures as *long-lived* failures.

Historically, layer 2 technologies, such as SONET, have been considered to protect the network from failures happening at the physical layer. However, in the context of IP-over-WDM, the SONET layer is not available. In this case, the IP layer is asked to restore the connectivity in case of failure. When a network element fails in the physical infrastructure, the OXCs report such a failure to the IP routers that update the routing tables with alternative logical paths. This approach only succeeds if the remaining set of logical links still forms a connected topology, i.e. any node can always find a logical path to reach any other node in the network. In order to increase the resilience of the topology to optical failures, a common practice of network operators is to establish multiple IP links between logically neighboring nodes. Some of these links are used in normal conditions and are called *primary links*; other links are inactive during normal conditions and are used only if one of the primary links is involved in a failure. These links are called *backup links*. When this happens, traffic is locally rerouted at the logical layer between the primary and the backup links according to the routing configuration. This way, the network avoids distributing a potential large volume of traffic from the failed primary links to other fully functional primary links that might get congested due to the unexpected volume of traffic being shifted over them. The failure is resolved locally by using pre-allocated network resources at the logical layer (i.e. backup links) and the network recovers in a more predictable fashion. This aspect of the design process implies the necessity that primary and backup links share no common fiber, or at least minimize the number of shared resources. The deployment of such a restoration scheme further complicates the design process. New requirements need to be added such that primary and backup links mapped over the physical infrastructure provide similar delay and cause minimal SLA impact upon a physical failure.

Survivability to logical failures

Similar considerations hold true for failures happening at the IP layer. These failures are usually less disruptive than optical failures since they involve fewer links. They are further handled completely by the IP layer using the routing protocols, and no interaction between the two layers is required. When a logical link fails, IS–IS/OSPF routing diverts the traffic from the broken link to alternate paths, increasing the load on one or more other links. Well designed network topologies with a high degree of failure resilience (i.e. use of primary and backup links between each pair of adjacent nodes) will absorb the effects of those failures by locally rerouting the traffic to pre-allocated resources. Network topologies counting on only a single link connectivity between pairs of adjacent nodes are very likely to experience performance degradation. For these networks, the most obvious way of restoring the network, to meet its original traffic engineering objectives, is to perform a network-wide recomputation and reassignment of link weights. However, changing link weights during a failure may not be practical for two reasons. First, the new weights will have to be flooded to every router in the network, and every router will have to recompute its minimum cost path to every other router. This can lead to considerable instability in the network, aggravating the situation already created by the link failure. The second reason is related to the short-lived nature of most of the link failures. In Figure 6.3 we present our analysis on inter-PoP link failures over a four-month period in the Sprint IP backbone. The reader can see that 80% of the failures last less than 10 minutes and 50% of the failures last less than 1 minute. We refer to failures that last less than 10 minutes as short-lived failures. These failures can create rapid congestion that is harmful to the network and leave a human operator with insufficient time to reassign link weights before the failed link is restored. In Figure 6.4 we report the frequency of single-link and multiple-link failures. The reader can observe that more than 70% of the short failures are single-link failures.

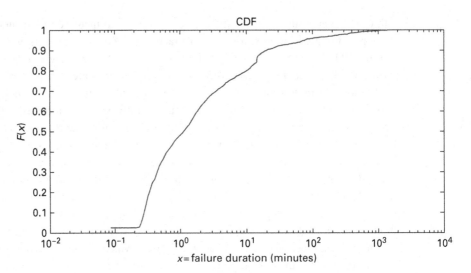

Fig. 6.3.   Failure duration distribution in Sprint's network, January to April, 2002.

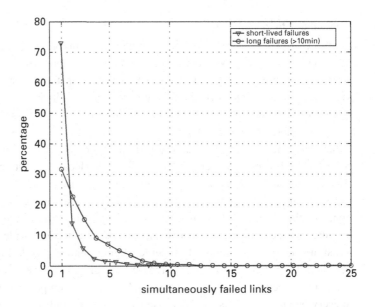

Fig. 6.4.   Distribution of simultaneous failures in Sprint's network, January to April, 2002. The
distribution is shown separately for short-lived failures (<10 minutes) and longer failures
(>10 minutes).

In order to overcome the above issues, a common practice of network operators is to
identify the one set of link weights that guarantees excellent performance under normal
network conditions and upon logical failures.

*Note: In the following sections, we consider resilience to physical and logical failures
as important requirements to guide the design process towards a robust solution.*

6.2.3   Technology limitations and business drivers

The general goal of designing a good network topology is to ensure that traffic can be delivered from any source to any destination with minimum network resource usage and under any circumstances, i.e. reliability upon network failures. Note that such a goal could be easily achieved with a topology that allows each node to be directly connected to any other node in a fully meshed manner. This states that a network with N nodes would feature a topology composed of $N(N-1)$ directed links. Such a design would imply a greater investment in terms of bandwidth to carry the same amount of traffic, and would lead to a low link utilization across the entire network. The topology, at the same time, would be characterized by outstanding resilience to failures, since packets can be routed across many different logical paths (assuming the mapping of IP links onto fibers leads to sufficient separation of physical paths across the network). Finally, the amount of electro-optical conversion needed would be minimal, and the network would be able to guarantee the fast delivery of packets to their destination; a direct link between any pair of IP routers means that the packets can be carried from the source to the destination purely in the optical domain.

Unfortunately, the limitations forced by the physical and logical layers and their inter-relationship make such a topology infeasible.

Physical layer

The physical infrastructure imposes three major constraints on the design process. First, it defines an upper bound to the node-degree, i.e. the number of IP links originated and/or terminated at each node. Since the node-degree is usually much smaller than the number of nodes in the network, it is clear that the fully meshed topology represents an unrealistic solution for real-sized networks. Second, if the OXC nodes are not equipped with wavelength conversion functionality, each IP link is constrained to use the same wavelength from its origin to its destination across all fibers along its physical path. Assigning a wavelength to each IP link and routing each wavelength over the fiber infrastructure is a problem known in the literature as the *Routing and Wavelength Assignment* (RWA) problem and proven to be NP-Hard [123]. Third, each fiber can carry a maximum number of wavelengths depending on the type of technology used. Commercial backbones, such as the Sprint's optical backbone, typically have a heterogeneous fiber infrastructure, with old technology fibers allowing no more than eight wavelengths, and newer fiber technology that allows up to 80 wavelengths. The diversity of fibers introduces another level of complexity in the design process, since the coexistence of new and old fibers requires the exposition of such information to the topology design process.

Topology and routing protocol

A good candidate topology would satisfy the above constraints while providing excellent network performance, such as minimum network congestion, targetting to minimize the maximum link load across all links constituting the topology, or to minimize the average packet delay. In order to evaluate a topology we must route packets over the topology in accordance with a defined routing protocol. IGP protocols such as IS–IS and OSPF assign a weight to each IP link and route packets from the source to the destination by choosing the logical path of minimum cost. The cost of a logical path is defined as the sum of the weights of all links constituting the path. Ideally, for each topology, we would like to route the traffic by optimizing its routing. This would guarantee high accuracy during the evaluation of a candidate topology. Unfortunately, the routing optimization is an NP-Complete problem and cannot be solved to optimality, or close to optimality, for each topology being analyzed. Network designers therefore avoid the routing optimization during the topology design. Instead, a common practice is to route the traffic according to a generic shortest-hop path routing protocol, i.e. path with minimum number of hops between the source and the destination. The optimization of the routing is considered only after the topology has been found. Due to the fact that IS–IS and OSPF provide a more general way to route packets compared with the simple shortest-hop path algorithm, the performance of the topology can be further improved during the routing optimization process.

Note: In the following sections, we consider the above technology limitations as part of our design process and adopt common practices when dealing with the routing of packets over the logical topology.

6.3   Network design process: sketching out our methodology

In the ideal case, network architects should solve all the above design issues with their annotated requirements in one single step. Unfortunately, the high complexity associated with each specific task makes this approach infeasible. In order to keep the complexity of the aforementioned optimization-based framework at a reasonable level, we separate and prioritize each task as part of a multi-step methodology while keeping under consideration how the various objectives, constraints and tradeoffs fit together. Each step is aimed at solving a specific problem in an "optimal" way by relaxing previous constraints and adding new ones. The output of each step is then used as the input of the next step, leading incrementally to the final solution. The overall methodology is presented in the context of backbone networks and briefly summarized in the following four steps.

- **Step 1:** Topology design resilient to long-lived failures. This step aims at identifying the network topology that is survivable to single-fiber cuts (i.e. long-lived failures) and shows good performance in terms of network congestion. Accordingly, we restrict our search for logical topologies to the ones for which at least one mapping schema is possible. Note that ignoring the mapping problem when designing the logical topology might lead to topologies for which no mapping exists (i.e. some nodes might be completely isolated from the network in case of failure).
- **Step 2:** Increase topology resiliency using multiple parallel links. This step aims to increase the failure resilience property of the topology found in step 1, by expanding each IP link between neighboring PoPs into multiple IP links, i.e. primary and backup. Such a provision allows the topology to recover from failures relying on

upper layer mechanisms that can reroute traffic from primary to backup links (i.e. network resources being reserved in advance to handle failure episodes). Accordingly, the number of fibers shared by primary and backup links must be minimized.

- **Step 3:** Performance enhancement and resiliency to short-lived failures via routing optimization. This step aims to improve the network performance of the topology found in step 1 by optimizing the way we route packets. During step 1 we assumed packets to be routed from their sources to their destinations using a simple shortest-hop path algorithm. IP routing protocols, such as IS–IS and OSPF, allow packets to be routed using the more general concept of shortest-cost path. Each IP link is assigned a weight (integer number between 1 and 255), and the cost of the path is defined as the sum of the weights of all IP links constituting that path. During this step we optimize the routing by playing with the link weights and then we find the set of link weights that guarantees the best performance. We approach this problem by optimizing the network performance during the normal operation, i.e. network congestion and SLA requirements. Furthermore, we extend our formulation by considering the impact that short-lived failures might have on the network due to the effect of causing global rerouting of traffic. We remark that modeling short-lived failures as part of this framework might not be necessary if multiple IP links between neighboring PoPs are considered as part of the logical topology. We conclude this step by highlighting how ignoring the interaction between IGP and BGP routing protocols can lead to an erroneous optimization and misleading results. We will extend the methodology presented in this step to consider both protocols and their dynamics;
- **Step 4:** Capacity planning. This step aims to engineer the network with the proper amount of bandwidth so as to absorb growth in traffic and applications over time. This step is not usually considered to be part of the pure design process, but rather a fundamental step for keeping the network up and running over time. After step 3, the IP topology should be up and running; failures and network performance should be optimized to serve the current demand of traffic; SLA requirements should be satisfied; and customers should experience good service performance. But, with the increase in numbers of customers and new bandwidth demanding applications, the operator will soon face the problem of when and where to add new capacity in order to provide customers with the same network performance and SLA guarantees. The most common approach one could follow is to over-provision the bandwidth in order to accommodate aggregate traffic demands. When traffic demands are relatively low, this approach is often sufficient to ensure that performance remains satisfactory for all services. It can be justified by noting that the network should be simple to operate and manage, so that savings in operating costs cancel out increased capital costs due to excess capacity. However, as the numbers of customers and the traffic demand increase, ad-hoc over-provisioning may lead to large capital investment and seriously limit the network and business growth. Therefore, a methodical approach for capacity planning bears a lot of value to the overall network design process.

Details of each specific problem will be introduced, extensively analyzed and efficiently solved in Chapters 7–10.

7 Topology design resilient to long-lived failures

Box 7.1. Network design process: Step 1

Goal Identify a logical topology (i.e. a set of IP links through which packets are routed from source to destination) and a "mapping" (i.e. a set of physical routes for each IP link) such that an appropriate objective function depending on all network states (i.e. no failure and all physical failures) is optimized.

Objective function Minimization of network congestion level, defined as maximum amount of traffic flowing on any IP link, belonging to the logical topology under any network state.

Inputs (i) Physical topology (which must be at least 2-connected), comprising nodes equipped with a limited integer number of tunable transmitters and receivers (i.e. node-degree) and with wavelength conversion functionality, connected by optical fibers that support a limited number of wavelengths; (ii) peak time traffic matrix, which defines the peak volumes of traffic exchanged between any pair of IP nodes; (iii) routing protocol, assumed to be shortest-hop-path-based; (iv) model of long-lived failures, assumed to be fiber related only and isolated in time, i.e. single-fiber failure model.

In recent years, the logical topology design problem in WDM networks was extensively studied, considering a number of different setups and technology constraints mostly imposed by the physical infrastructure over which the logical topology is layered. Examples of these constraints are the maximum number of IP links originated or terminated at each node and the maximum number of allowed wavelengths per fiber (see Box 7.1, Inputs: (i)). The design process is commonly driven by objective functions such as the minimization of network congestion (i.e. maximum link utilization experienced in the network) or the minimization of the average packet delay. Given the simplicity of modeling the network congestion as a linear function, it is often preferred as an objective function while the average packet delay is expressed in the framework as a set of tight constraints that need to be satisfied.

In order to evaluate the network congestion for each candidate solution, packets must be routed over the logical topology according to the Interior Gateway Routing protocol (IGP), such as OSPF or IS–IS. Thus the logical design process is highly influenced by both the distribution of traffic patterns, i.e. the traffic matrix (TM), and the routing protocol being used. Logical topologies that guarantee excellent network performance for specific instances of the TMs and a routing protocol may perform very poorly in other contexts. Thus, it is critical to consider an accurate model for describing the traffic exchanged by nodes in the network. Due to the fact that Internet traffic varies over time (seasonality and diurnal trends) and the TM might be prone to errors, it is common to consider the *peak traffic volumes* and to scale it up with a *ceiling factor* to avoid problems at small timescales and forecasted errors (Box 7.1, Inputs: (ii)). The same holds true for the routing protocol. In the ideal case, we would like to model perfectly the routing protocol that we will be using at the IP layer. Unfortunately, it is very hard to model the details of those protocols, such as their load balancing mechanisms. Thus, a common practice is to model the routing protocol at a very high level considering a simple shortest-hop path (Box 7.1, Inputs: (iii)).

When considering the above objective and constraints, it was shown that finding the optimal logical topology is a NP-Hard problem and thus computationally intractable for large-size networks [38, 213]. Therefore, several heuristic approaches have been proposed in the literature (see, for instance, refs. [148], [121] and [166]).

Unfortunately, most of the work done in this area has ignored the presence of physical failures and their potential impact on the logical topology. This is motivated by the fact that, in the past, carriers used to implement a multi-layer recovery scheme to defend their infrastructures. Each layer was equipped with its own protection/restoration schemes and reacted to its own layer equipment failures. SONET was used to offer protection and fast restoration of service at the WDM layer. Protection paths were pre-computed, and wavelengths were reserved in advance, at the time of connection setup. Physical failures were completely transparent to the IP layer, and the restoration was provided in less than 50 ms. The dynamic capabilities of IGP were used to react to IP link failures. When a failure happened at the IP layer, the IGP detected the failure and automatically recomputed alternative routes around the failed link.

Today, most ISPs are deciding to remove SONET gradually due to the high cost of optical equipment and the huge amount of redundant capacity needed to reroute traffic in case of physical failures but never used in the normal operation state [85]. SONET framing is being kept only for failure-detection purposes, and SONET protection is allowed only in highly dense areas with high failure probability. In this context (i.e. pure IP-over-WDM network with no SONET layer available), the restoration is obtained by exploiting only the dynamic capabilities of the IP routing. When a failure happens in the optical network, the IP routing algorithm is able to update its tables and restore disrupted paths, if the set of non-disrupted IP links still forms a connected topology. In order to achieve a good degree of fault resilience, it is fundamental *to map* (i.e. to route) each IP link onto the physical topology in such a way that, given any physical failure, the set of non-disrupted IP links still forms a connected network. Thus, an optimization of the physical mapping of IP links is desirable [61, 62].

In this chapter we generalize the approach proposed in refs. [26] and [61] and partially in ref. [62]. By considering the resilience properties of the topology directly during the design of the logical topology, and thus extending the optimization of the network resilience properties also to the space of logical topologies, we can drastically reduce the *level of physical protection* required to achieve a desired level of physical fault tolerance. This implies less wasted bandwidth and fewer reserved wavelengths in the normal operation state, but slower recovery after failure due to the execution of signaling and management procedures. Thus, a fundamental input of our design process is a model that accurately captures the characteristics of physical failures. By using the results shown in ref. [134], we consider a simple single-fiber failure model in our design process (Box 7.1, Inputs: (iii)).

Now that we have defined inputs, constraints and the overall goal, we conclude by emphasizing the importance of choosing carefully an objective function that will drive the optimization so as to obtain the best tradeoff between network performance in normal conditions (i.e. historical goal) and fault-resilience properties (i.e. novelty introduced in this chapter). Since the network performance depends on the network failure states, the objective function must combine the network performance levels under different network failure states. We select the minimization of the network congestion across all possible network states as our objective function (Box 7.1, Objective function).

The rest of the chapter is organized as follows. In Section 7.1 we introduce the problem and formalize the approach aimed at finding logical topologies with a good degree of fault resilience. We first present an integer linear programming formulation by relaxing the shortest-path routing requirement. Then we consider the more realistic case where shortest-path routing is assumed, which leads to an integer non-linear formulation. In Section 7.2 we present a Tabu Search methodology to find good solutions while limiting computational effort. Details of the method are provided in Sections 7.3 and 7.4, while Section 7.5 is devoted to the analysis of its theoretical complexity. Section 7.6 contains the results of several sets of experiments, and Section 7.7 concludes the chapter.

7.1   Fault-tolerant logical topology design problem

We refer to the problem of computing topologies that are resilient to long-lived failures as FLTDP (i.e., Fault-Tolerant Logical Topology Design Problem) that can be simply stated as follows.

Given:

(i) an existing physical topology (which must be at least 2-connected), comprising nodes equipped with a limited integer number of tunable transmitters and receivers and with wavelength conversion functionality, connected by optical fibers that support a limited number of wavelengths;

(ii) a TM whose elements represent the maximum traffic volumes exchanged by sources and destinations;[1]

---

[1] In this chapter we assume traffic to be stationary. In addition, we assume that each traffic element represents the peak volume of traffic exchanged by the corresponding source–destination pair. However, extensions of our approach are possible which consider either the effects of the traffic non-stationarity or the effects of traffic fluctuations around the average value.

(iii) a multi-hop IP routing strategy for packets;

(iv) a set of single physical link failures.

Find:

A logical topology (i.e. a set of IP links through which packets are routed from source to destination) and a "mapping" (i.e. a set of physical routes for each IP link), such that an appropriate objective function depending on all network states (i.e. no failure and all single link failures) is optimized.

The objective function must be carefully selected in order to obtain the best trade-off between network performance in normal conditions and fault-resilience properties (see refs. [173] and [174]). Since the network performance depends on the network failure states, the objective function must combine the network performance levels under different network failure states. We select as the objective of the optimization process the minimization of the network congestion level, defined as the maximum amount of traffic flowing on any IP link, belonging to the logical topology under any failure state.

7.1.1 Problem formulation

In this section we report two variants of the FLTDP formulation. For both, we consider that wavelength converters are available at each node. In the first case, the paths taken by the IP packets are not restricted to be the shortest. This leads to an ILP (Integer Linear Programming) formulation. When the shortest-path requirement is added, however, the formulation becomes non-linear, which greatly increases the complexity of its resolution. This variant is presented at the end of the section. Unfortunately, since all Tier-1 ISPs use routing protocols based on shortest paths, the more realistic formulation would be the non-linear one. However, we believe that the ILP model represents a powerful tool to find a theoretical lower bound to test the accuracy of the heuristic approaches presented for the solution of the non-linear formulation.

Notation

In this section, we introduce the notation used to formulate our problem. We introduce a notational typology for multi-layered networks. In this context, the superscript indicates the layer, starting with the lowest layer, zero, that represents the physical network. Let $G^0 = (V, E^0)$ be the unidirectional graph representing the physical topology. It is composed of a set of OXC nodes V interconnected by optical fibers represented by set E^0. Let $|V| = N$ be the cardinality of set V and $|E^0| = M$ that of set E^0. Let R_i and T_i be the numbers of receivers and transmitters at physical node $i \in V$. Let S_k be the network state, where S_0 represents the no-failure state, while S_v for $v \geq 1$ is the state of failure of optical fiber $v \in E^0$. Let S be the set of all operational states, whose cardinality is $|S| = M + 1$. Let $\mathcal{E}$ be the set of all possible IP links in any logical topology. Let $G^1(S_0) = (V, E^1(S_0))$ be the directed graph representing the logical topology in the no-failure state. It is composed of IP routers V interconnected by IP links $E^1(S_0) \subseteq \mathcal{E}$. Note that in order to simplify the notation, we assume that there is a router associated with each OXC, and, by abuse of notation, we equate the set of routers with the set of OXC. However, our formulation can be easily extended to the more general case.

Let $G^1(S_v) = (V, E^1(S_v))$ denote the logical topology in the network state S_v, obtained from $G^1(S_0)$ by dropping all the IP links $u \in E^1(S_0)$ crossing the optical fiber $v \in E^0$. Let $\Lambda = (\lambda_{sd})$ indicate the peak-time TM, where each entry λ_{sd}, in arbitrary units, represents the peak-time traffic flow between the source s and the destination d.

Decision variables

Three types of binary variables are introduced into the formulation: X_u, Y_{uv} and $t_u^{sd}(S_v)$, which correspond, respectively, to logical topology, mapping and routing.

The logical topology variables $X_u \in \{0, 1\}$ describe the IP links included in the logical topology $G^1(S_0)$:

$$X_u = \begin{cases} 1, & \text{if IP link } u \in \mathcal{E} \\ & \text{belongs to the logical topology } G^1(S_0) \\ 0, & \text{otherwise.} \end{cases}$$

Then we can state that the logical topology $G^1(S_0) = (V, E^1(S_0))$ comprises the IP links $E^1(S_0) = \{u : X_u = 1, \quad u \in \mathcal{E}\}$.

The mapping variables $Y_{uv} \in \{0, 1\}$ contain the routing information of IP links belonging to the logical topology $G^1(S_0)$ over the physical topology G^0:

$$Y_{uv} = \begin{cases} 1, & \text{if IP link } u \in \mathcal{E} \text{ crosses} \\ & \text{the optical fiber } v \in E^0 \\ 0, & \text{otherwise.} \end{cases}$$

The variables $t_u^{sd}(S_v)$ contain the information related to the routing of packets on the logical topology $G^1(S_v)$:

$$t_u^{sd}(S_v) = \begin{cases} \lambda_{sd}, & \text{if traffic } s \to d \text{ crosses} \\ & \text{IP link } u \in \mathcal{E} \text{ in state } S_v \\ 0, & \text{otherwise.} \end{cases}$$

We note that traffic splitting is not allowed in our model (i.e. all the traffic originated in s and destined to d is forced to follow the same route). The model can be easily extended to consider traffic splitting by relaxing the variables $t_u^{sd}(S_v)$ to be continuous.

Constraints

Let $\Gamma^+(i)$ be the set of IP links outgoing from node $i \in V$ and let $\Gamma^-(i)$ be the set of IP links incoming to node $i \in V$. Let $\Theta^+(i)$ be the set of physical links outgoing from node $i \in V$ and let $\Theta^-(i)$ be the set of physical links incoming to node $i \in V$. Finally, let $O(u)$ be the origin node and let $D(u)$ be the destination node of IP link $u \in \mathcal{E}$. We can then write the model constraints.

- Connectivity:

$$\sum_{u \in \Gamma^+(i)} X_u \leq T_i \qquad \forall i \in V, \tag{7.1}$$

$$\sum_{u \in \Gamma^-(i)} X_u \leq R_i \qquad \forall i \in V, \tag{7.2}$$

where inequality (7.1) indicates that the number of IP links outgoing from each node cannot be larger than the number of transmitters in the node, for each logical topology in the no-failure state ($G^1(S_0)$); inequality (7.2) indicates that the number of IP links incoming to each node cannot be larger than the number of receivers in the node, for each logical topology in the no-failure state ($G^1(S_0)$).

- Routing:

$$\sum_{u\in\Gamma^+(i)} t_u^{sd}(S_v) - \sum_{u\in\Gamma^-(i)} t_u^{sd}(S_v) = \begin{cases} \lambda_{sd} & \text{if } s = i \\ -\lambda_{sd} & \text{if } d = i \\ 0 & \text{otherwise,} \end{cases}$$
$$\forall s, d, i \in V, \forall S_v \in \mathcal{S}, \tag{7.3}$$

$$t_u^{sd}(S_v) \leq X_u \lambda_{sd},$$
$$\forall s, d \in V, \forall S_v \in \mathcal{S}, \forall u \in \mathcal{E}, \tag{7.4}$$

where equation (7.3) represents the routing continuity constraints for packet routes on the logical topology $G^1(S_v)$. It states that, for each network operational state, an available (working) path on the logical topology must exist for each source–destination pair; equation (7.4) instead states that traffic can be routed only on IP links belonging to the logical topology.

- Mapping:

$$Y_{uv} \leq X_u \qquad \forall u \in \mathcal{E}, v \in E^0, \tag{7.5}$$

$$\sum_{v\in\Theta^+(i)} Y_{uv} - \sum_{v\in\Theta^-(i)} Y_{uv} = \begin{cases} 1 & \text{if } O(u) = i \\ -1 & \text{if } D(u) = i \\ 0 & \text{otherwise,} \end{cases}$$
$$\forall u \in \mathcal{E}, \forall i \in V, \tag{7.6}$$

$$\sum_{s,d\in V} t_u^{sd}(S_v) \leq \left(\sum_{s,d\in V} \lambda_{sd}\right)(1 - Y_{uv}),$$
$$\forall u \in \mathcal{E}, \forall v \in E^0, \forall S_v \in \mathcal{S}/\{S_0\}, \tag{7.7}$$

where inequality (7.5) ensures that only the IP links in the considered logical topology are mapped; equation (7.6) represents the routing continuity constraint for IP links on the physical topology G^0; inequality (7.7) imposes that all the IP links that cross the physical link v are not available in state S_v.

- Limit on the number of wavelengths: let W_v be the number of wavelengths supported on each fiber. The set of IP links $v \in E^1(S_0)$ must satisfy the following constraint:

$$\sum_{u\in\mathcal{E}} Y_{uv} \leq W_v \qquad \forall v \in E^0, \tag{7.8}$$

which indicates that the number of IP links that cross each optical fiber has to be smaller than the wavelength number.

Objective function

The objective function must be carefully selected in order to obtain the best tradeoff between network performance in normal conditions and fault resilience (see refs. [173] and [174]). Since the network performance depends on the network failure states, the objective function must combine the network performance levels under different network failure states. We selected as the objective of the optimization process the minimization of the network congestion level, defined as the maximum amount of traffic flowing on any IP link, belonging to the logical topology under any failure state:

$$\min H,$$

where

$$H \geq \left\lceil \sum_{s,d \in V} t_u^{sd}(S_v) \right\rceil \qquad \forall S_v \in \mathcal{S}, \forall u \in \mathcal{E}. \tag{7.9}$$

Observations

Note that in the above formulation the routing of packets on the logical topology is unspecified, thus the minimization of the network congestion level is jointly performed on all admissible logical topologies and routing configurations.

Also note that, under the assumption that at least one topology exists, the above Integer Linear Programming (ILP) model provides a logical topology that can tolerate any single physical link failure. Indeed, the resulting logical topology is connected, under any single link failure (equation (7.3)). If no topology exists, the ILP model produces an infeasible solution warning message.

IP link capacity constraints are ignored in the above formulation for the sake of model simplicity; we note, however, that the minimization of the network congestion level corresponds to the minimization of the IP link capacity needed to guarantee an efficient transport of the offered traffic. Thus, the minimization of the network congestion level leads to the minimization of the capacity needed to guarantee good performance.

Extension to shortest-path routing

In order to restrict the optimization to act only on the set of the admissible logical topologies with shortest-path routing, we need to introduce some extra variables and constraints.

Let us introduce an extra set of variables $\tau_u^{sd}(S_v) \in \{-1, 0, 1\}$ that represent a possible alternative routing with respect to the routing specified by $t_u^{sd}(S_v)$ on the logical topology:

$$\tau_u^{sd}(S_v) = \begin{cases} 1, & \text{if traffic } s \to d \text{ is rerouted on} \\ & \text{IP link } u \in \mathcal{E} \text{ in state } S_v \\ -1, & \text{if traffic } s \to d \text{ is no longer routed on} \\ & \text{IP link } u \in \mathcal{E} \text{ in state } S_v \\ 0, & \text{otherwise.} \end{cases}$$

We note that $\tau_u^{sd}(S_v)$ must satisfy the following constraints:

$$\sum_{u \in \Gamma^+(i)} \tau_u^{sd}(S_v) - \sum_{u \in \Gamma^-(i)} \tau_u^{sd}(S_v) = 0,$$

$$\forall s, d, i \in V, \forall S_v \in \mathcal{S}, \quad (7.10)$$

$$\tau_u^{sd}(S_v) \leq X_u - Y_{uv} - t_u^{sd}(S_v)/\lambda_{sd},$$

$$\forall s, d \in V, \forall u \in \mathcal{E}, \forall v \in E^0, \forall S_v \in \mathcal{S}/\{S_0\}, \quad (7.11)$$

$$\tau_u^{sd}(S_0) \leq X_u - t_u^{sd}(S_0)/\lambda_{sd},$$

$$\forall s, d \in V, \forall u \in \mathcal{E}, \quad (7.12)$$

$$-\tau_u^{sd}(S_v)\lambda_{sd} \leq t_u^{sd}(S_v),$$

$$\forall s, d \in V, \forall u \in \mathcal{E}, \forall S_v \in \mathcal{S}, \quad (7.13)$$

where equation (7.10) represents the routing continuity constraint for the rerouting on logical topology $G^1(S_v)$. Equations (7.11) and (7.12), instead, state that traffic can be rerouted only on a path consisting of working IP links belonging to the logical topology; equation (7.13) finally states that, after rerouting, routes defined by $t_u^{sd}(S_v)$ may be no longer valid.

Finally, we define G as follows:

$$G = - \sum_{\substack{S_v \in \mathcal{S} \\ u \in \mathcal{E} \\ s, d \in V}} \tau_u^{sd}(S_v).$$

Note that G represents the difference between the total path length before rerouting and after rerouting. It is possible to find a set of $\tau_u^{sd}(S_v)$ such that G assumes positive values whenever the set of $t_u^{sd}(S_v)$ does not describe a shortest-path routing. In conclusion, if and only if the set $t_u^{sd}(S_v)$ defines a shortest-path routing, we find $\max_{\tau_u^{sd}} G = 0$; thus, selecting

$$\min_{t_u^{sd}}[\epsilon H + \max_{\tau_u^{sd}} G]$$

as the objective function, where H is defined in equation (7.9) and $\epsilon < 1/\sum_{sd} \lambda_{sd}$, we obtain the result of restricting the optimization to the set of logical topologies implementing a shortest-path routing. Indeed, we observe that, by construction, $0 < \epsilon H < 1$, while $\max G$ can assume only non-negative integer values. Thus, $\epsilon H + \max_{\tau_u^{sd}} G > 1$, whenever variables $t_u^{sd}(S_v)$ do not define a shortest-path routing; on the other hand, $\epsilon H + \max_{\tau_u^{sd}} G < 1$ if variables $t_u^{sd}(S_v)$ define a shortest-path routing. As a consequence, we can state that the optimal solution of the previous problem is the logical topology which minimizes the network congestion level under a shortest-path routing. Note, however, that the resulting objective function in this specific case is non-linear. Thus, the formulation falls in the class of integer non-linear programming problems, and no general methodologies and tools are available for an optimal solution of this formulation.

7.2   Solution strategy

The FLTDP is NP-Hard, since it is a generalization of the traditional Logical Topology Design Problem (LTDP) that was proved to be NP-Hard. Even for moderate-size networks, an optimal solution of the FLTDP appears to be quite problematic due to the large number of variables and constraints involved in the formulation. Thus, the development of heuristic solution methodologies is required. The heuristic approach to the FLTDP proposed in refs. [26] and [140] consists in decomposing the whole problem into two independent subproblems: the LTDP, in which the logical topology optimization is performed on the basis of the congestion level in the full operational state (S_0), thus ignoring the resilience property of the solution; and the Fault-Tolerant Mapping (FMP) Problem, according to which the mapping of the logical topology onto the physical topology aims to achieve good resilience properties. While the LTD problem has been widely investigated in the literature, and many algorithms have been proposed [121, 148, 166, 213], the FM problem has been considered only recently. In ref. [26] this problem was found NP-Complete and an heuristic approach based on the application of the Tabu Search optimization algorithm has been proposed, while in ref. [140] an ILP formulation of the problem is provided and solved for instances of moderate size (e.g. physical topology with 14 nodes and 21 links and logical topologies with 14 routers and node-degree equal to 3, 4 and 5) applying the CPLEX optimization tool [1].

In this chapter, we adopt a different strategy for the solution of the FLTDP. We apply Tabu Search for the optimization of the logical topology, considering the case of no failure and all possible cases of a single physical link failure in the network. For each considered logical topology, IP links are routed over the physical topology and the number of wavelengths to be used on each fiber is computed. Both IP link routing and wavelength assignment are obtained through a new heuristic algorithm, called Greedy Disjoint Alternate Path (GDAP), described in Section 7.3. Finally, the traffic routing on the IP links forming the logical topology is taken to be shortest path.

7.3   Mapping between physical and logical topology: GDAP

The definition of algorithms that optimally map the IP links on the physical topology is an important subproblem of the FLTDP. This problem is related to equations (7.5) and (7.6) in Section 7.1.1. The mapping problem can be stated as follows: given a logical topology, find a routing for each IP link of the logical topology over the physical topology, such that the negative effects of a single optical link failure are minimized.

Since the mapping problem is only a part of FLTDP, the utilization of a computationally expensive algorithm to solve the mapping could have a disruptive impact on the CPU time necessary for the solution of the entire problem. Thus, for the solution of the mapping problem, we present a simple greedy algorithm, the Greedy Disjoint Alternate

Path (GDAP) algorithm, whose computational complexity is small. A brief description of the GDAP algorithm follows.

GDAP algorithm

Let $OR(i)$ and $IR(i)$ be the sets of already routed IP links, respectively outgoing from and incoming to node i, and let $ON(i)$ and $IN(i)$ be the sets of outgoing and incoming IP links not yet routed. Let V_{ij} be the IP link belonging to the logical topology, with end points i and j. Let $\mathcal{O}$ denote a set of nodes. Initialize $\mathcal{O}$ to the set of all nodes in the network.

> **Step 0** Route all IP links V_{ij} whose end points are adjacent in the physical topology. Insert V_{ij} in $(OR(i), IR(j))$ and remove V_{ij} from $(ON(i), IN(j))$.
>
> **Step 1** If $\mathcal{O} = \emptyset$, STOP; otherwise randomly select a node $i \in \mathcal{O}$ and remove i from $\mathcal{O}$.
>
> **Step 2** If $ON(i) = \emptyset$, GOTO Step 3; otherwise randomly select each $V_{ik} \in ON(i)$ and find the shortest path for V_{ik} which is physically disjoint from the routes on which the $V_{ij} \in OR(i)$ and the $V_{jk} \in IR(k)$ have already been routed. If no such physical path exists, V_{ik} is routed on the shortest path. If in addition the shortest path is not available, due to the lack of free wavelengths, IP link V_{ij} is not mapped.
>
> **Step 3** If $IN(i) = \emptyset$, GOTO Step 1, randomly select each $V_{ki} \in IN(i)$ and try to find a route for V_{ki} which is physically disjoint from the routes on which the $V_{ji} \in IR(i)$ and the $V_{kj} \in OR(k)$ have already been routed. If a physically disjoint route for V_{ki} has not been found, V_{ki} is routed on the shortest path. If in addition the shortest path is not available, due to the lack of free wavelengths, IP link V_{ji} is not mapped.

An example of mapping produced by GDAP is shown in Figure 7.1. The IP links that are mapped first over the physical topology are those whose end points are two adjacent physical nodes (see, for example, the IP links $1 \rightarrow 2$, $2 \rightarrow 3$, $3 \rightarrow 2$, etc.). Then, starting from node 1, Steps 2 and 3 of GDAP are iteratively and sequentially applied to all nodes of the network. Focusing on node 1, GDAP maps the outgoing remaining IP link $1 \rightarrow 6$ over the optical fibers $(1, 3)$ and $(3, 6)$. Note that all the possible routes for IP link $1 \rightarrow 6$ must comprise fiber $(1, 3)$, since IP link $1 \rightarrow 2$ is already routed on fiber $(1, 2)$. Concerning the incoming IP links of node 1, GDAP maps IP link $4 \rightarrow 1$ over the optical fibers $(4, 2)$ and $(2, 1)$. It is interesting to look at the mapping for IP link $6 \rightarrow 1$. It must cross optical fiber $(6, 3)$, since $(6, 5)$ has been already used by IP link $6 \rightarrow 5$. Then the only possible physical path for IP link $6 \rightarrow 1$ is represented by the sequence of optical fibers $(6, 3)$ and $(3, 1)$. Note that if no disjoint path for $6 \rightarrow 1$ were possible, the algorithm would have selected one shortest path.

It is worth noting that GDAP computes both the number of IP links crossing each fiber and the number of wavelengths used per fiber; if the maximum number of wavelengths over a fiber is limited, it may be impossible to map some IP links over the physical topology.

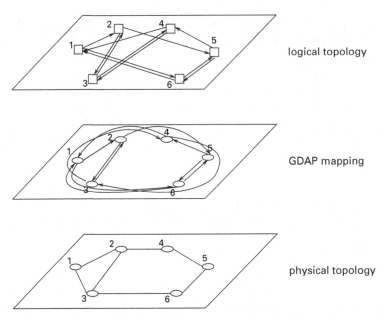

Fig. 7.1.   Example of GDAP mapping with $W_v = 3$ $\forall v \in E^0$ and $T_i = R_i = 2$ $\forall i \in V$.

7.4   Tabu Search for the FLTDP: TabuFLTDP

7.4.1   General description of Tabu Search

The heuristic we present for the solution of the FLTDP relies on the application of the Tabu Search (TS) methodology [86] and is called TabuFLTDP. The TS algorithm can be seen as an evolution of the classical local optimum solution search algorithm called the Steepest Descent (SD) algorithm; however, thanks to the TS mechanism that allows worsening solutions to be also accepted, contrary to SD, TS is less likely to be subject to local minima entrapments. TS is based on a partial exploration of the space of admissible solutions, finalized to the discovery of a good solution. The exploration starts from an initial solution that is generally obtained with a greedy algorithm, and, when a stop criterion is satisfied, the algorithm returns the best visited solution. For each admissible solution, a class of neighbor solutions is defined. A neighbor solution is defined as a solution that can be obtained from the current solution by applying an appropriate transformation, called a *move*. The set of all admissible moves uniquely defines the *neighborhood* of each solution.

At each iteration of the TS algorithm, all solutions in the neighborhood of the current one are evaluated, and the best is selected as the new current solution. Note that, in order to explore the solution space efficiently, the definition of neighborhood may change during the solution space exploration; in this way it is possible to achieve an *intensification* or a *diversification* of the search in different solution regions.

A special rule, the *Tabu list*, is introduced in order to prevent the algorithm from deterministically cycling among already visited solutions. The Tabu list stores the last accepted moves; while a move is stored in the Tabu list, it cannot be used to generate a new move. The choice of the Tabu list size is very important in the optimization procedure: too small a size could cause the cyclic repetition of the same solutions, while too large a size can severely limit the number of applicable moves, thus preventing a good exploration of the solution space.

7.4.2 Fundamental aspects of TabuFLTDP

In order to put in place a Tabu procedure, we must define the following elements:

- the choice of an initial solution;
- the definition of the moves and the neighborhood;
- the evaluation of the visited solutions;
- the stopping criterion.

Initial solution

As an initial solution we selected the result of the D-MLTDA heuristic introduced in ref. [121]. This heuristic initially considers a fully connected logical topology and sequentially removes a set of least-loaded IP links from the logical topology, until the node-degree constraints are satisfied. We invite the reader to look at ref. [121] for the details of the D-MLTDA algorithm.

Moves and neighborhood generation

Let T represent a given feasible topology and let $\mathcal{N}(T)$ be the neighborhood of such a topology when the Tabu moves are applied. A new solution $T' \in \mathcal{N}(T)$ is found by searching for cycles of a given length and erasing the right number of IP links to keep the degree constraint feasible. In a more detailed manner, let us denote by l the fixed length of the cycle, l being an even number such that $l \leq 8$. Let us assume that $n_1, n_2, \ldots, n_l$ are the nodes to be visited in the cycle, starting at node n_1. From a given node n_i the next node to be visited, n_{i+1}, is found as follows.

- If i is an odd number, choose an incoming IP link and travel in the opposite direction. The resulting node is n_{i+1}.
- If i is even, choose any node that has not yet been visited in the cycle as n_{i+1}.

Once the cycle has been defined, the new degree of each node is assessed. The superfluous IP links are removed from those nodes presenting a degree larger than their original value. An example of the procedure is given in Figure 7.2 for a cycle of length 6. We found that the visited nodes in the cycle are: $n_1 = 1, n_2 = 4, n_3 = 6, n_4 = 5, n_5 = 3, n_6 = 2$. IP links $2 \rightarrow 3, 4 \rightarrow 1$ and $5 \rightarrow 6$ are removed from the topology and replaced by IP links $5 \rightarrow 3$ and $4 \rightarrow 6$ to get to the new topology.

This procedure guarantees that the degree constraints are not violated, thus generating a valid move. Note that, with this perturbation, it is very easy and fast to implement a *diversification* and/or *intensification* criterion by exploring a region of the solution space with small cycles and to move to another region of the solution space with large cycles.

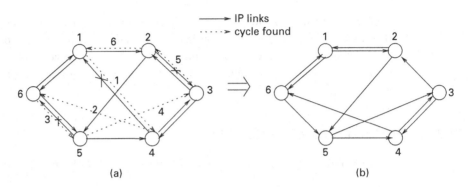

Fig. 7.2.   Perturbation using a cycle $C(1, 6)$. (a) Old solution; (b) new solution.

Then the neighborhood of the current solution T (i.e. $\mathcal{N}(T)$) is generated by considering sequentially all network nodes and applying all possible cycles of length l.

Solution evaluation
Each solution in the neighborhood is evaluated by routing the traffic on the topology for all network states (i.e. no-failure and all single-link-failure states), and computing the network congestion level in each state. The solution with the minimum congestion level is selected as the new current solution, and the IP links selected during its generation are stored in the Tabu list.

Stopping criterion
The search procedure is stopped when a given number of iterations is reached. The number of iterations should be chosen relative to the size of the network and such that a good tradeoff between computational time and quality (distance from the optimal solution) of the solution reached is achieved.

7.4.3   TabuFLTDP pseudo-code

In this subsection, we present the pseudo-code of the Tabu procedure. First, let us define some useful notation.

- $F_{eval}(T)$ is the evaluation function required to compute the merit coefficient of logical topology T. It returns M, the network congestion level. Note that the evaluation of M requires the execution of the routing algorithm on the logical topology and of the GDAP mapping algorithm.
- *BuildInitialSolution* is used to build an initial logical topology applying D-MLTDA.
- *BuildCycle(T, l)* is used to build the neighbor solution of the current solution T, using cycles of length l. When the *diversification* criterion has to be used, the cycle is longer than in normal TabuFLTDP. We denote by q the length of the normal cycle and by p the length of the cycle used for the *diversification* criterion, where $p \gg q$.

- *BuildNeighborhood*(T) is a procedure to build the neighborhood of the current solution T, by iteratively applying the *BuildCycle(T, l)* procedure.
- *BestNeighSol(T)* is a procedure that evaluates each solution in the neighborhood of T and returns the best solution. The evaluation is based on $F_{eval}(T)$;
- *TabuList* is a fixed size Tabu list to store the latest moves.
- $\mathcal{N}(T)$ is the neighborhood of logical topology T, built applying procedure *BuildNeighborhood(T)* and using only cycles not belonging to *TabuList*.
- T, T^* and T^{**} represent, respectively, the current logical topology, the best logical topology in $\mathcal{N}(T)$ and the best solution found by FLTDP.
- M, M^* and M^{**} represent, respectively, the merit associated with the logical topologies T, T^* and T^{**}.
- *IterationsNumber* is the number of iterations.
- *LimitDiv* is the number of consecutive iterations without improvements, expressed by the variable *counterDiv*, after which the *diversification criterion* is applied. When this happens, only *one cycle* of length p is generated; the cycle is such to change several IP links at the same time and then visit a different area of the solution space. After the new solution is generated, the procedure works as before, using cycles of length q.
- *IterBest* is the iteration at which T^{**} is found.

The pseudo-code for TabuFLTDP is given in Figure 7.3.

The implementation of TabuFLTDP can be carried out using any custom programming language such as C or C++ (see refs. [110] and [190], respectively).

Algorithm: TABUFLTDP(T, *IterationsNumber*, *LimitDiv*, p, q)

```
T = BuildInitialSolution;
IterBest = 1;
T** = T;
M** = M;
TabuList = {}; %the TabuList is empty
counterDiv = 1;
for counter ← 1 to IterationsNumber
    if counterDiv = LimiDiv
        T = BuildCycle(T, p);
        counterDiv = 1;
    N(T) = BuildNeighborhood(T);
    T* = BestNeighSol;
    TabuList = UpdateTabuList(TabuList);
    if M* ≤ M**
        T** = T*;
        M** = M*;
        IterBest = counter;
        counterDiv = 1;
    else if counterDiv + +;
    T = T*;
return (T, IterBest)
```

Fig. 7.3.   Pseudo-code for searching the best logical topology using TabuFLTDP.

7.5   Complexity

We now discuss the complexity of the presented heuristics. Let $\Delta = T_i = R_i \ \forall i \in V$ denote the identical in/out degrees for each node in the logical topology. We further suppose $\Delta \leq N$. For each analyzed logical topology, we route (i) its IP links over the physical topology with the GDAP algorithm and (ii) the traffic over the logical topology.

The GDAP algorithm has complexity $O(N\Delta(M + \Delta N \log(N\Delta)))$, since at most $O(N\Delta)$ iterations are executed, while at each iteration at most $O(M + \Delta N \log(N\Delta))$ operations are required; $O(M)$ operations, indeed, are necessary to update the cost of the links of the physical topology and $O(N \log(\Delta N))$ operations are necessary to run the Dijkstra algorithm. Since $M \leq N^2$, the GDAP complexity is upper bounded by $O(N^3\Delta + \Delta^2 N^2 \log(N\Delta))$.

To evaluate solutions, it is necessary to route the traffic. The routing algorithm requires $O(N^2 \log(N\Delta))$ operations (Dijkstra algorithm).

The D-MLTDA heuristic to evaluate the initial solution has complexity $O(N^4)$, since at most $O(N)$ iterations are needed to complete the algorithm, and at each iteration it is necessary to route traffic and execute a 1-minimum Weight Matching (1-mWM) algorithm, whose complexity is $O(N^3)$.

Let us now focus on the complexity of the TS algorithm. At each iteration, the evaluation of all solutions in the neighborhood is necessary; this requires $O(\Delta^2 N^2 (N^3 \Delta + (\Delta^2 N^2 + MN^2) \log(N\Delta)))$ operations, since $O(\Delta^2 N^2)$ neighbors are evaluated (assuming perturbations are generated using cycles of length 4), and the evaluation of each solution requires the execution of the GDAP algorithm and the execution of the routing algorithms for each failure state (i.e. $M + 1$ times). If the number of iterations is I, the resulting complexity is $O(I\Delta^2 N^2 (N^3 \Delta + (\Delta^2 N^2 + MN^2) \log(N\Delta)))$. Thus, the computational complexity of TS is upper-bounded by $O(I\Delta^2 N^6 (\log(N\Delta)))$.

7.6   Numerical results

In this section we present numerical results obtained with the described approach (called joint optimization), and we compare them with those obtained by performing a conventional optimization of the logical topology. The conventional optimization approach incorporates the optimal mapping of the IP links on the physical topology according to the algorithm proposed in ref. [140], and is extended in order to deal with a unidirectional logical topology (this approach is called disjoint optimization).

Optimal results are reported for the medium-sized (ten-node) topologies plotted in Figure 7.4, since we were unable to run the optimal mapping for larger networks (such as the networks shown in Figure 7.5). Larger instances were heuristically obtained using the presented approach. The network in Figure 7.4(a) was obtained by removing some nodes and links from the NSF-net topology (shown in Figure 7.6), while the network of Figure 7.4(b) has the structure of an Italian backbone IP network.

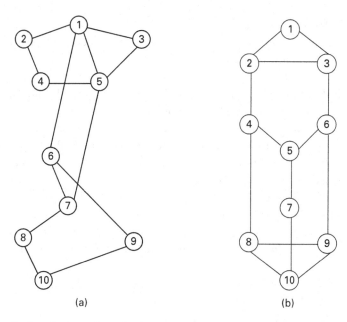

Fig. 7.4.   Physical topologies of two medium-sized networks. (a) Network 1 is composed of 10 nodes and 14 links. (b) Network 2 is composed of 10 nodes and 13 links.

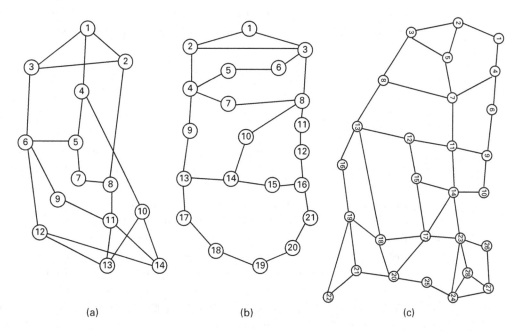

(a)   (b)   (c)

Fig. 7.5.   Physical topologies of three large-sized networks. (a) NSFNET: 14 nodes and 21 links; (b) ARPANET: 21 nodes and 26 links, (c) USA Long Distance: 28 nodes and 45 links.

We consider randomly generated traffic patterns. As an example, the bandwidth required for every source–destination pair is randomly extracted from an exponential distribution with mean $\mu = 1$.

While TabuFLTDP does not consider IP link capacities, in the presentation of numerical results in this section we will consider that each IP link has a fixed capacity, and can thus carry a limited amount of traffic.

As already noted, considering capacities is not essential in the optimization, where topologies are ranked according to their maximum flow on IP links, thus implicitly minimizing the IP link capacity required to carry a given traffic pattern. On the other hand, including IP link capacities in the optimization process would lead to a significant increase of complexity. Considering (variable) IP link capacities in the presentation of results is important, since it allows us to estimate the actual amount of traffic that cannot be carried over the logical topologies produced by the different optimization algorithms.

After an initial calibration, the Tabu parameters used in our experiments were set as follows:

- *TabuList*: a Tabu list of fixed size equal to 7 is used.
- *Cycle size*: during the exploration of the solution space, cycles of length 4 are used. In some cases, however, different perturbation rules are used to implement the *diversification criterion*. In particular, to ease the exit from local minima regions, after 50 iterations without improvement a cycle of size 6 is used.
- *Stopping criterion*: the procedure is stopped after a fixed number of iterations. The number of iterations is set to 300, since this value seems to provide a good tradeoff between the conflicting requirements of limiting the CPU time and obtaining good results.

In Tables 7.1, 7.2, 7.3 and 7.4 we report results obtained with the disjoint- and the joint-optimization techniques, for different logical network configurations on the physical topology plotted in Figure 7.4(a). In each column, the results for a particular IP link capacity value LC are portrayed.

Three important network performance indices are reported: the network congestion level for the no-failure state (C_{S_0}), the network congestion level (C_S), i.e. the maximum network congestion level over all the states S_v, and the maximum amount of traffic (TL) that is lost in the network due to a single link failure. The latter is expressed as a percentage of the total offered traffic. Traffic losses are encountered whenever the flow on an IP link exceeds the IP link capacity. For the three measures, we report the mean and the worst values obtained over ten randomly generated traffic instances.

Tables 7.1, 7.2, 7.3 and 7.4 present results on four different values of nodal in/out degree ($\Delta = 2, 3, 4, 5$). It can be observed that the joint-optimization approach in these cases outperforms that of the disjoint optimization, especially in terms of the maximum traffic lost because of failures. For example, in Table 7.1 we see that, with disjoint optimization, the maximum lost traffic is still non-null when the link capacity LC is 50, while, under joint optimization, almost null losses are observed when the IP link capacity is 30.

Table 7.1. Comparison between FLTDP and the disjoint optimization of LTDP and optimal mapping (Opt-MP) with $\Delta = 2$ for Network 1, Figure 7.4(a)

Parameters	Disjoint optimization (LTDP+Opt-MP)							
	$LC=20$	$LC=25$	$LC=30$	$LC=35$	$LC=40$	$LC=45$	$LC=50$	$LC=55$
C_{S_0} mean	13.02	13.02	13.02	13.02	13.02	13.02	13.02	13.02
C_S mean	20.00	25.00	30.00	35.00	38.98	40.48	41.04	41.35
TL mean (%)	27.19	20.30	13.58	7.48	2.75	1.01	0.36	0.00
C_{S_0} worst	14.72	14.72	14.72	14.72	14.72	14.72	14.72	14.72
C_S worst	20.00	25.00	30.00	35.00	40.00	45.00	50.00	52.81
TL worst (%)	39.39	32.25	26.45	20.65	14.85	9.05	3.25	0.00

Parameters	Joint optimization (FLTDP)							
	$LC=20$	$LC=25$	$LC=30$	$LC=35$	$LC=40$	$LC=45$	$LC=50$	$LC=55$
C_{S_0} mean	15.96	15.96	15.96	15.96	15.96	15.96	15.96	15.96
C_S mean	20.00	23.44	24.43	24.54	24.54	24.54	24.54	24.54
TL mean (%)	6.58	1.36	0.13	0.00	0.00	0.00	0.00	0.00
C_{S_0} worst	19.84	19.84	19.84	19.84	19.84	19.84	19.84	19.84
C_S worst	20.00	25.00	30.00	30.97	30.97	30.97	30.97	30.97
TL worst (%)	18.76	7.74	1.14	0.00	0.00	0.00	0.00	0.00

Table 7.2. Comparison between FLTDP and the disjoint optimization of the LTDP and optimal mapping (Opt-MP) with $\Delta = 3$ for Network 1, Figure 7.4(a)

Parameters	Disjoint optimization (LTDP+Opt-MP)				
	$LC=15$	$LC=20$	$LC=25$	$LC=30$	$LC=35$
C_{S_0} mean	7.51	7.51	7.51	7.51	7.51
C_S mean	15.00	20.00	24.46	26.85	27.7
TL mean (%)	15.98	9.26	3.76	0.98	0.00
C_{S_0} worst	9.18	9.18	9.18	9.18	9.18
C_S worst	15.00	20.00	25.00	30.00	33.55
TL worst (%)	22.16	15.72	10.56	4.75	0.00

Parameters	Joint optimization (FLTDP)				
	$LC=15$	$LC=20$	$LC=25$	$LC=30$	$LC=35$
C_{S_0} mean	9.11	9.11	9.11	9.11	9.11
C_S mean	12.51	13.11	13.11	13.11	13.11
TL mean (%)	0.15	0.00	0.00	0.00	0.00
C_{S_0} worst	11.29	11.29	11.29	11.29	11.29
C_S worst	15.00	15.66	15.66	15.66	15.66
TL worst (%)	0.76	0.00	0.00	0.00	0.00

Table 7.3. Comparison between FLTDP and the disjoint optimization of LTDP and optimal mapping (Opt-MP) with $\Delta = 4$ for Network 1, Figure 7.4(a)

	Disjoint optimization (LTDP+Opt-MP)						
Parameters	$LC = 10$	$LC = 15$	$LC = 20$	$LC = 25$	$LC = 30$	$LC = 35$	$LC = 40$
C_{S_0} mean	6.32	6.32	6.32	6.32	6.32	6.32	6.32
C_S mean	10.00	14.42	17.36	19.96	21.73	22.82	23.31
TL mean (%)	17.42	10.36	6.90	3.85	1.83	0.57	0.00
C_{S_0} worst	7.91	7.91	7.91	7.91	7.91	7.91	7.91
C_S worst	10.00	15.00	20.00	25.00	30.00	35.00	39.41
TL worst (%)	34.11	28.31	22.51	14.41	9.41	5.11	0.00

	Joint optimization (FLTDP)						
Parameters	$LC = 10$	$LC = 15$	$LC = 20$	$LC = 25$	$LC = 30$	$LC = 35$	$LC = 40$
C_{S_0} mean	7.05	7.05	7.05	7.05	7.05	7.05	7.05
C_S mean	8.90	9.01	9.01	9.01	9.01	9.01	9.01
TL mean (%)	0.22	0.00	0.00	0.00	0.00	0.00	0.00
C_{S_0} worst	8.63	8.63	8.63	8.63	8.63	8.63	8.63
C_S worst	10.00	10.98	10.98	10.98	10.98	10.98	10.98
TL worst (%)	2.04	0.00	0.00	0.00	0.00	0.00	0.00

Table 7.4. Comparison between FLTDP and the disjoint optimization of LTDP and optimal mapping (Opt-MP) with $\Delta = 5$ for Network 1, Figure 7.4(a)

	Disjoint optimization (LTDP+Opt-MP)				
Parameters	$LC = 10$	$LC = 15$	$LC = 20$	$LC = 25$	$LC = 30$
C_{S_0} mean	6.07	6.07	6.07	6.07	6.07
C_S mean	10	13.88	15.76	16.43	16.76
TL mean (%)	8.98	3.61	1.14	0.38	0.00
C_{S_0} worst	8.12	8.12	8.12	8.12	8.12
C_S worst	10.00	15.00	20.00	25.00	27.99
TL worst (%)	20.87	15.07	9.27	3.47	0.00

	Joint optimization (FLTDP)				
Parameters	$LC = 10$	$LC = 15$	$LC = 20$	$LC = 25$	$LC = 30$
C_{S_0} mean	6.27	6.27	6.27	6.27	6.27
C_S mean	7.62	7.62	7.62	7.62	7.62
TL mean (%)	0.00	0.00	0.00	0.00	0.00
C_{S_0} worst	7.48	7.48	7.48	7.48	7.48
C_S worst	9.74	9.74	9.74	9.74	9.74
TL worst (%)	0.00	0.00	0.00	0.00	0.00

Table 7.5. Comparison between FLTDP and the disjoint optimization of LTDP and optimal mapping (Opt-MP) with $\Delta = 5$ for Network 2, Figure 7.4(a)

Parameters	Disjoint optimization (LTDP+Opt-MP)						
	$LC = 20$	$LC = 25$	$LC = 30$	$LC = 35$	$LC = 40$	$LC = 45$	$LC = 50$
C_{S_0} mean	12.71	12.71	12.71	12.71	12.71	12.71	12.71
C_S mean	20	25	30	35	38.95	41.54	42.28
TL mean (%)	24.77	18.68	12.94	7.52	3.44	0.75	0.00
C_{S_0} worst	14.23	14.23	14.23	14.23	14.23	14.23	14.23
C_S worst	20.00	25.00	30.00	35.00	40.00	45.00	47.94
TL worst (%)	35.90	27.03	19.57	13.36	8.20	3.04	0.00
Parameters	Joint optimization (FLTDP)						
	$LC = 20$	$LC = 25$	$LC = 30$	$LC = 35$	$LC = 40$	$LC = 45$	$LC = 50$
C_{S_0} mean	14.95	14.95	14.95	14.95	14.95	14.95	14.95
C_S mean	16.08	23.51	24.76	25.81	25.81	25.81	25.81
TL mean (%)	7.27	2.92	1.07	0.00	0.00	0.00	0.00
C_{S_0} worst	18.07	18.07	18.07	18.07	18.07	18.07	18.07
C_S worst	20.00	25.00	30.00	34.17	34.17	34.17	34.17
TL worst (%)	17.30	11.68	4.31	0.00	0.00	0.00	0.00

The difference between the two optimization procedures increases when the logical topology degree increases. Table 7.4 shows that with joint optimization no losses are observed for configurations in which the IP link capacity is 10, while under disjoint optimization, losses are still registered when $LC = 25$.

Differences become even larger when the physical topology plotted in Figure 7.4(b) is considered. In this case, trying to build a logical topology with nodal-degree equal to 2, fails 10 times out of 20 tested traffic instances. The reason behind such a behavior is that there is no such mapping (in those ten cases) that guarantees that the logical topology remains connected under any single-link-failure scenario. This means that some source–destination pairs will not be able to communicate in the presence of failures, regardless of the capacity allocation across the network. In those cases, of course, the solution provided by disjoint-optimization algorithms leads to unacceptable performance in terms of failure resilience. Table 7.5 reports results restricted to the ten cases in which the disjoint optimization does not fail.

Table 7.6 instead reports a comparison between joint optimization and disjoint optimization in terms of the average required number of wavelengths. Results refer to four topologies with different numbers of nodes and links: Network 1, shown in Figure7.4(a) (10 nodes, 14 links), the NSFNET topology, shown in Figure 7.5(a) (14 nodes, 21 links), the ARPANET topology, shown in Figure 7.5(b) (21 nodes, 26 links) and the USA Long Distance topology, shown in Figure 7.5(c) (28 nodes, 45 links). We observe that whereas for the ten-node topology the logical topology resulting from disjoint optimization was obtained by applying the optimal mapping algorithm proposed in ref. [140], for larger

Table 7.6. Number of wavelengths required to map the logical topology

Mean wavelength-number	Disjoint optimization (LTDP+Opt-MP)		
	$\Delta = 2$	$\Delta = 3$	$\Delta = 4$
Network 1: 10 nodes, 14 links	3.2	5.0	6.1
NSFNET: 14 nodes, 21 links	3.9	5.1	6.7
ARPANET: 21 nodes, 26 links	6.4	9.2	12.6
USA Long Distance: 28 nodes, 45 links	7.2	11.4	14.5
Mean wavelength-number	Joint optimization (FLTDP)		
	$\Delta = 2$	$\Delta = 3$	$\Delta = 4$
Network 1: 10 nodes, 14 links	2.5	3.8	5.1
NSFNET: 14 nodes, 21 links	2.3	4.4	6.3
ARPANET: 21 nodes, 26 links	5.4	8.4	11.7
USA Long Distance: 28 nodes, 45 links	7.0	9.2	12.1

Table 7.7. CPU times for one iteration of the Tabu Search algorithm

CPU time	$\Delta = 2$			$\Delta = 3$			$\Delta = 4$		
	1 it. (s)	M O-It	W O-It	1 it. (s)	M O-It	W O-It	1 it. (s)	M O-It	W O-It
Network 1	0.78	19.88	85	1.38	31.55	91	1.49	65.67	96
NSFNET	6.69	62.22	98	7.12	24.44	41	9.66	64.67	94
ARPANET	60.46	36.22	67	47.48	43.22	91	46.98	40.78	90
USA Long Distance	246.78	36.1	60	177.20	68	114	160.43	63.5	81

networks the results were obtained by performing the heuristic GDAP mapping algorithm over the outcome of the logical topology optimization procedure, because the algorithm of ref. [140] is too complex for networks of this size. Results show that, in terms of required number of wavelengths, the application of the joint-optimization algorithm appears to be advantageous, yielding average savings of about 20%.

Table 7.7 reports the CPU time needed to run an iteration of the joint-optimization Tabu Search algorithm. All results were obtained over a 800 MHz Pentium III PC running Linux 6.2.

Table 7.7 also shows the iteration number at which the optimal solution was found; the average value (the Mean Optimum Iteration, M O-It) over ten instances and the worst case (Worst Optimum Iteration, W O-It) value are reported. In all cases, 300 iterations were run before stopping the algorithm. We note that, only in one instance, were more than 100 iterations (114) necessary to find the optimum value.

Finally, Figure 7.6 reports the average number of iterations required to find a solution that differs by a given percentage from the optimum. It is worth noting that a solution

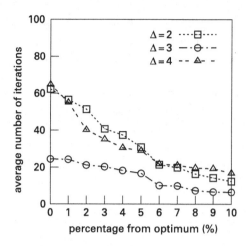

Fig. 7.6.   Average number of iterations versus the percentage distance from the optimal solution.

that is few percentage points worse than the best can be obtained in a significantly smaller number of iterations.

We remark to the reader that the results presented in this chapter serve merely as an illustration. Different results are likely to be obtained when considering different models of TMs and different topologies.

7.7   Lessons learned

In this chapter we presented a new methodology for the design of logical topologies resilient to long-lived failures.

Our approach to protection and restoration generalizes the concepts first proposed in refs. [26], [61] and [62], and relies on the exploitation of the intrinsic dynamic capabilities of IP routing, thus leading to cost-effective fault-tolerant logical topologies. It further differs from those proposed in refs. [26], [61] and [62], in that it considers the resilience properties of the topology during the logical topology-optimization process, thus extending the optimization of the network resilience performance also into the space of logical topologies. The intuition behind our joint optimization of the logical topology and the mapping of the topology over the physical infrastructure resides on the fact that a logical topology characterized by great performance might not be optimally mappable over the physical infrastructure and may thus drammatically suffer in case of optical failures.

By designing and applying a search heuristic called TabuFLTDP aimed at exploring the space of the logical topologies, and applying a simple heuristic called GDAP for the mapping, we have found that the proposed joint-optimization approach largely outperforms those previously proposed (i.e. optimizing the topology and the mapping into two separate steps), in terms of both traffic lost because of failures (yielding average bandwidth savings of about 2.5 times for comparable losses) and the required number of wavelengths (yielding an average wavelength reduction of about 20%).

In Chapter 8 we focus our attention on enhancing the resilience of the topology found in this chapter by introducing the concept of parallel links between adjacent nodes and solving the mapping problem more efficiently (here solved in a very simple way). Furthermore, we will remove the constraint of the presence of wavelength converters in the physical infrastructure, with the objective of showing the reader how both the mathematical formulation of the problem and the presented heuristics can be easily adapted to this case.

8 Achieving topology resilience using multiple-parallel links

---

> **Box 8.1.** Network design process: Step 2
>
> **Goal** Identify a mapping between the logical and physical topologies (i.e. a set of physical routes for each IP link) and an assignment of wavelengths for each IP link, so as to ensure that any long-lived failure does not disconnect any pair of nodes (i.e. PoPs). Our secondary goal is to drive the mapping to a point where each long-lived failure brings down the smallest number of inter-PoP links as possible.
>
> **Objective function** Minimization of the number of optical fibers shared by the parallel logical links between each pair of adjacent nodes. This objective is formalized with the definition of a sophisticated new metric called *global jointness*.
>
> **Inputs** (i) Physical topology (which must be at least 2-connected), comprising nodes (OXCs) which do not support wavelength conversion, interconnected by optical fibers that support a limited number of wavelengths and are of limited capacity; (ii) logical topology, comprising nodes interconnected to adjacent nodes via multiple parallel links that must have similar delay characteristics; (iii) model of long-lived failures, assumed to be fiber related only and isolated in time, i.e. single-fiber failure model; (iv) priorities across PoP pairs, parallel IP links between adjacent nodes and SLA metrics.

8.1 Introduction

As already discussed in Chapter 2, an IP backbone network is made up of a set of Points-of-Presence (PoPs) or nodes interconnected by logical links, as illustrated in Figure 8.1. Each PoP is itself a mini-network composed of a small number of core routers and a large number of access routers. The core routers are typicallty fully (or nearly fully) meshed. In addition, each access router is typically attached to a minimum of two core routers. Customers connect to access routers (not represented in Figure 8.1). PoPs that are directly connected by one or more logical links are said to be neighboring or adjacent. Typically, neighboring PoPs are connected by many parallel logical links

Portions reprinted, with permission, from Giroire, F., Nucci, A., Taft, N. and Diot, C. (2003). Increasing the robustness of IP backbones in the absence of optical level protection. *Proceedings IEEE Infocom*, San Francisco, CA, March/April, 2003.

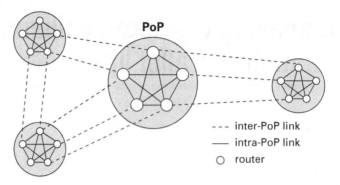

Fig. 8.1.   PoP interconnection with multiple links.

terminating at different core routers in each PoP in such a way that a single *router* failure cannot bring down a customer or compromise the connectivity between a pair of neighboring PoPs. The number of parallel logical links is different for each PoP pair, and their speed varies from OC-48 (2.5 Gbps) to OC-192 (10 Gbps). For example, in the Sprint backbone, the number of logical links between any pair of adjacent PoPs varies between 2 and 12 (the average being 3). Some of these links are used under normal conditions and are called *primary links*; other links are inactive during normal conditions and are only used if one of the primary links is affected by a failure. These links are called *backup links*. When this happens, traffic is locally rerouted at the logical layer between the primary and the backup links according to the routing protocol. This way the network avoids distributing new routing tables to all nodes. The failure is resolved locally and the network reacts faster to avoid potential disruption.

The IP or logical topology is layered directly over the physical infrastructure, composed of optical fibers grouped into fiber conduits that connect optical cross-connects distributed across large geographical areas. Investigating the characteristics of Sprint's optical infrastructure, we observe that such physical topologies are characterized by the almost complete absence of wavelength converters and by great diversity in fiber quality (fibers can support between 8 and 80 wavelengths). Therefore a shortage of wavelengths in such networks is not unusual. In this chapter, we assume that the physical topology is provided as an input to our problem (Box 8.1, Input: (i)). Similarly, we assume that the logical topology is available as well (Box 8.1, Input: (ii)).[1]

In this chapter, one of our goals is to identify a mapping between the logical topology and the physical topology that ensures that any long-lived failure does not eliminate the connectedness of any pair of nodes. Such a task necessitates access to an accurate model reflecting the characteristics of those failures. In accordance with what was presented in ref. [134] and adopted in Chapter 7, we use a simple single-fiber failure model (Box 8.1, Input: (iii)).

---

[1] If the logical topology is not available, the network designer can apply a methodology similar to the one presented in Chapter 7. The obtained logical topology can then be enriched with parallel links between adjacent nodes for resilience to failures. We advise that links carrying the majority of traffic should be expanded into more parallel links than links carrying smaller volumes of traffic.

The general problem of mapping the logical topology over a physical infrastructure with constraints, requirements and goals described above, has already been studied in the literature [61, 62, 152]. The mapping problem with wavelength conversion functionality at the physical layer is known to be NP-Complete [26]. An ILP formulation is provided in ref. [140], and the problem is optimally solved for moderate size networks by applying a branch and cut algorithm [71]. The same problem, but with wavelength continuity, is known as the Wavelength Assignment Problem (WAP) [27, 35, 115]. It is similar to the path coloring problem in standard graphs, which is, in turn, equivalent to the general vertex coloring problem [55]. It has been proved to be NP-Complete [73, 94], and numerous heuristics have been proposed for different types of topologies [27, 74, 115].

Unfortunately, all previous work ignores constraints and requirements present in realistic operational environments when approaching the mapping problem. Examples are the existence of priorities across PoP pairs, the presence of multiple IP links between adjacent PoPs and SLA metrics. These requirements must be considered at every point in time (Box 8.1, Input: (iv)). Carriers usually have a strong desire to prioritize some PoP pairs versus others. For example, logical links connecting PoPs that carry the largest amount of traffic must be endowed with greater fault resilience than logical links carrying small amounts of traffic. On the Sprint network, the *priority PoP pairs* are those transcontinental links that connect two major cities such as New York and San Francisco. In addition to the previous constraints, SLAs must be met at any time for any PoP pair in the network. Maximum PoP-to-PoP delay is an important SLA parameter. Its value is defined by each ISP. In the continental USA, the maximum delay is typically between 50 ms and 80 ms. The delay between any PoP pair must be below the value defined in the SLA. In addition to the maximum delay, we must also restrict the relative delay on the alternative inter-PoP paths. Many applications cannot tolerate a major change in delay in the event of a failure. For example, a VoIP application would suffer dramatically if rerouted on a link that caused the end-to-end delay to increase by 50 ms.

The introduction of these new constraints and requirements complicates the problem even further. Given a particular topology of the physical network, it is not always possible to find *simultaneously* completely disjoint physical links and to maintain the delay below the SLA for all logical links between a given PoP pair. For example, there may not necessarily exist two short delay paths that are also completely disjoint. In order to find completely disjoint paths, sometimes one has to use a long circuitous route for the second path that substantially increases the delay. Finding the right balance between delay and disjointness is a very important task. For example, in this chapter we focus on finding maximally disjoint paths when completely disjoint paths cannot be found. It is not easy to manage this tradeoff when solving the mapping problem manually for networks the size of Sprint's backbone. As a consequence, it is important to define performance metrics that lead the optimization process toward a specific set of goals. In this chapter our primary goal in terms of robustness is to ensure that a single-fiber failure does not eliminate the connectivity between a pair of PoPs. Our secondary goal is to drive the mapping to a point where a single-fiber cut brings down the smallest number of inter-PoP links possible (Box 8.1, Goal). To achieve these goals we consider the notion of *link priority* and introduce the metric of *jointness* (Box 8.1, Objective function).

We will examine the impact of these strategies in the Sprint IP backbone and especially the tradeoff between delay and jointness. The consideration of different strategies allows a wider diversity of path selection, which helps meet a larger number of requirements simultaneously.

The chapter is organized as follows. In Section 8.1, we start by describing the problem and the associated requirements and constraints. Then we introduce some concepts used to drive our optimization process and examine their impact to the overall problem. The problem is then formalized in Section 8.2, where we provide an ILP model for the problem and present a new heuristic algorithm that achieves a good balance between optimality and computational complexity. We briefly compare the ILP model to the heuristic on a medium-size network. In Section 8.3, we use the heuristic algorithm to study the mapping problem on the Sprint IP backbone network. Findings are summarized in Section 8.4.

8.2   The problem: constraints, requirements and objective function

In this section we define the constraints and requirements that must be included in any practical solution to the mapping problem. In Section 8.1.2 we introduce the criteria that will be used in our methodology to prioritize one candidate solution to the problem versus one other. These priorities will be included in the definition of our objective function.

8.2.1   Impact of constraints and requirements

Given the context described in Section 8.1, we study the mapping problem when the following requirements are considered:

- parallel logical links must be mapped on to physical links that are as disjoint as possible, i.e. maximally disjoint;
- the worst case delay between any PoP pair must be less than the corresponding SLA requirement;
- the parallel links between two PoPs must be mapped onto physical links of similar delay, so that the differences in delay are limited;
- the solution must take into consideration the availability of wavelengths.

Network protection and disjointness
In order to maximize fault resilience, parallel logical links need to be mapped onto the fiber network in such a way that either a fiber conduit or an optical equipment failure does not cause all the parallel logical links between a pair of PoPs to go down simultaneously. Thus the parallel logical links should be mapped onto physically disjoint fibers whenever possible.

Finding completely disjoint fiber paths for logical links is often difficult, if not impossible. This is because there is a limited set of conduits containing fibers in the ground and because these fibers have been layed out according to terrain constraints (mountains, bridges, etc.) or next to already existing facilities, such as train tracks or pipelines. When completely disjoint paths cannot be found, our strategy is to search for maximally disjoint paths. It is well known that the problem of finding maximally disjoint paths is hard; it is particularly challenging in the case of a real backbone such as Sprint's because the multiplicity of parallel links between PoPs is not merely 2, but can be as large as 12 (although is more commonly between 4 and 7). Hence the physical topology may limit the number of options for alternative disjoint paths, but the logical topology demands large numbers of disjoint options. Our approach is intended to minimize the number of logical links that are disrupted over all possible physical failures.

Therefore, the objective of our mapping function is to *minimize the jointness* of the parallel logical links between each pair of adjacent PoPs. Minimizing the jointness is equivalent to maximizing the disjointness. To do this we first define a Local Jointness metric that is assigned to a pair of PoPs. Later, we define a network-wide jointness metric, called Global Jointness.

Consider two neighboring PoPs, s and t. The parallel links between s and t use a set of fiber segments $\{(i, j)\}$ that start at node i and terminate at node j. Each fiber segment will be assigned one jointness value for each pair of adjacent PoPs using that segment. (Thus each fiber gets a set of values, one for each PoP pair traversing it.) For a given PoP pair, the fiber segment is assigned a jointness value equal to the number of parallel logical links sharing this fiber segment minus one. Therefore, the *jointness of a fiber segment* used by a single link between s and t is zero. The local *jointness of a PoP pair* (s, t) is defined as the sum of the jointness of each fiber segment $\{(i, j)\}$ used by any of its parallel logical links. Note that a local jointness of zero for PoP pair (s, t) means that all the parallel logical links between s and t use fully disjoint physical paths.

We illustrate this definition using the example in Figure 8.2. We want to map three logical links between these two PoPs onto the physical network represented in the figure. The solid lines indicate fibers separated by optical cross-connects. The dashed lines represent the candidate physical paths for the three logical links. In Figure 8.2(a) (case 1), the three parallel links share a single fiber segment, and thus the jointness of PoP pair (A, B) is 2. In Figure 8.2(b) (case 2), there are two fibers that each have a

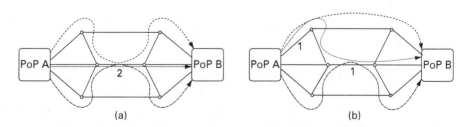

Fig. 8.2.   Example of jointness and priorities. (a) Case 1; (b) Case 2.

jointness of 1 (since two paths share each link), and thus the jointness of the PoP pair (A, B) is also 2.

Our goal in defining a jointness metric is to have a quantity to minimize in our optimization. The lower the jointness metric, the less fiber sharing there is. As we reduce the jointness, we are essentially decreasing the likelihood that a single-fiber failure will affect a large number of parallel logical links of the same PoP pair. Thus minimizing the jointness metric pushes us in the direction of improving robustness.

We point out that our jointness metric has the following limitation. As we saw in our example, two different scenarios can give rise to the same jointness value for a PoP pair. Therefore the jointness metric does not distinguish between the two scenarios in Figure 8.2 in terms of robustness. Since different mapping scenarios can lead to the same jointness value, the jointness metric is not unique in the sense that it cannot uniquely differentiate all possible mapping scenarios. In this example, we would typically consider case 2 more robust than case 1. In case 1 a single-fiber failure will bring down all the logical links between PoPs A and B, whereas in case 2 the two PoPs will remain connected under any single-fiber failure scenario. Recall that our definition of robustness was to ensure that no single-fiber failure can completely disconnect a pair of adjacent PoPs. Hence, although such a metric pushes us in the right direction for achieving robustness, it is not sufficient in and of itself to guarantee we do as best as possible for our definition of robustness. We will show later how we use priority information to add further robustness differentiation to our jointness metric.

We define *global jointness* as the sum of the local jointness over all neighboring PoP pairs in the backbone. The global jointness is a useful way to compare various mappings. We can also evaluate the impact of delay SLAs on fiber sharing using this jointness metric.

Delay constraints

A SLA is a contract between an ISP and its customer. This contract may specify a maximum end-to-end delay between any arbitrary pair or PoPs (not just neighboring PoPs) that must be satisfied at any moment in time, both under normal operation and during failures. We introduce this constraint into the problem as the *maximum delay* constraint. We assume that the delay comes primarily from propagation delay [156]. The delay between a pair of PoPs is defined as the worse case total transmission time, among all possible routes, between these two PoPs. We have to consider all the possible routes, since any one of them could be used as the alternative route in the event of a failure.

The physical layout of fibers in today's networks tends to yield the following situation: two PoPs that are geographically close often have one route between them that is short (in terms of distance and hence propagation time), while all other routes are much longer (on the order of five to ten times longer). If there are many parallel links to be mapped for a given PoP pair, this makes it difficult to minimize jointness without increasing the length of alternative fiber paths. As mentioned earlier, it is not acceptable for SLAs to be broken when routes change. Furthermore, ISPs cannot allow delay-sensitive applications to experience a degradation in delay if low delay is critical.

We thus introduce a second delay constraint, called the *relative delay* constraint, that limits the allowable difference in delay between two paths.

In order to control the relative delay constraint, we define the notion of a *default path*. For each pair of neighboring PoPs, we choose one of its paths to be the default path. We require that the delay of each of the parallel logical links, for a given PoP pair, be no more than $u\%$ longer than the default path delay. Conceptually the default path is a reference path used to control the delay differences between alternate paths. Because the default path is an artifact of our method, it may or may not be used itself. In Section 8.2.2 we will define three different strategies for computing the default path.

Wavelength limitation

In WDM networks, each fiber has a fixed number of wavelengths. While performing the mapping, we need to assign wavelengths and verify that a sufficient number of wavelengths exist for this mapping. In the case of no wavelength conversion, we have to make sure that the same wavelength is available on all the fiber segments involved in the physical paths. The limitation on the number of available wavelengths significantly complicates the problem. A solution that is optimal from a jointness standpoint might not be feasible from the wavelength allocation standpoint. In other words, assigning one wavelength to a logical link of PoP pair (A, B) can reduce the possibilities of fiber path choice for PoP pair (C, D) and increase the jointness for all other PoP pairs. Therefore, our approach needs to take wavelength limitations into consideration in the computation of jointness.

8.2.2   Approach

In this section we explain the objective function we use both for the ILP model and the heuristic algorithm (the same objective function is used for both). Our objective is to minimize the global jointness while simultaneously meeting the maximum and relative delay requirements. However, finding an optimal mapping with regard to all of the constraints introduced is a complex problem because the search space is still large. Before stating our objective, we introduce two types of priorities that help us manage the distribution of resources across PoPs and that help us to improve robustness further. These priorities also limit the search space in a way that makes a lot of practical sense.

Sometimes the mapping of one PoP pair can compromise the mapping of another. In particular, if there is a shortage of wavelengths, then the order in which PoP pairs are mapped can be critical. Those PoP pairs mapped first may use up some wavelengths that are then no longer available to other PoP pairs. This can limit the choices of alternate paths for the latter PoP pairs. We allow a set of PoP pairs to be considered as *priority PoP pairs* and map their logical links first. These priority PoP pairs should be granted the minimum local jointness possible, even if it means that the non-priority PoP pairs end up with a larger local jointness than they would receive if no priorities existed at all. Priority PoP pairs have a natural justification in any network topology. They correspond to the inter-PoP logical links that are the most important to protect because they have a special status in the network (e.g. they carry more traffic, or they connect major geographical locations). In the Sprint backbone, transcontinental east–west links are usually considered to be high-priority PoP pairs.

Recall that in our discussion of the jointness metric, we mentioned that sometimes different mappings can give rise to the same value of the jointness metric, but not have the same robustness. To increase the robustness of the mapping we produce, we introduce the notion of *priority logical links*. Among all the parallel links that must be mapped for a given PoP pair, we want at least two of them to be completely disjoint (if possible). The number of links we choose to put in the priority group is two, because of our definition of robustness. Note that we do not assign ahead of time a priority to a link; the links in this group can be by any two of the logical links. We want any two for which complete disjointness can be found. Thus, instead of mapping all parallel logical links for each PoP pair simultaneously, we initially focus on finding two logical links that can be mapped to completely disjoint paths. If we can find such paths, then the remaining parallel links are mapped afterwards. For the remaining parallel paths, we try to find physical paths that minimize the local jointness for that PoP pair (given the mapping of the first two paths). If we cannot find two such paths, then all the links are mapped together – again trying to minimize the local jointness. With this second priority notion, we increase the chances of each PoP pair to have at least two completely disjoint fiber paths. This makes the PoP pair more robust because then there is no single-fiber failure that can completely disconnect the PoP pair. Recall our example in Figure 8.2. The two priority logical links would have a jointness of 0 in case 2 and a jointness of 1 in case 1. With this notion of priority we would choose the solution in case 2 rather than case 1 because case 2 includes two completely disjoint paths whereas case 1 does not. Priority links thus help to differentiate the robustness of two mappings of equal local jointness.

Objective function

We integrate these priorities into our objective of minimizing global jointness as follows. Using our priorities, we define a *mapping sequence*. The goal is to minimize the global jointness while respecting this sequence.

- Step 1. Map the priority logical links for the priority PoP pairs.
- Step 2. Map the remaining logical links of the priority PoP pairs.
- Step 3. Map the priority logical links for the remaining non-priority PoP pairs.
- Step 4. Map the remaining links (non-priority links of non-priority PoP pairs).

Delay requirements

In addition to jointness minimization, we must guarantee the following two delay constraints:

- the delay between any PoP pair in the network must be bound by the maximum delay value found in the SLA (known as the maximum delay constraint);
- the delay difference between all parallel links for any given neighboring PoP pair must be within $u\%$ of the default path (known as the relative delay constraint).

The relative delay requirement appears as a constraint in our optimization formulation and in our algorithm. Instead of adding the maximum delay constraint as an input to the objective function, we compute the maximum delay after the mapping has been performed, i.e. as an output of our solutions. We can analyze the tradeoff between jointness and maximum delay by varying the value of u in the set of constraints.

We consider the following three strategies for selecting the default path.

- SP: Shortest Path. The default path is the shortest physical path between a given neighboring PoP pair. "Shortest" here refers to the path with the shortest propagation delay.
- SSP: Second Shortest Path. The default path is the second shortest path that exists between a given pair of neighboring PoPs.
- SDP: Smallest Disjoint Path. For each pair of neighboring PoPs we can always find two completely disjoint paths if we temporarily remove the constraints on relative delay and wavelength availability. This is true because the minimum cut of Sprint's network is two. Given these two disjoint paths, we select the longer of the two as our *default path*.

We will examine the impact of these strategies in our network and especially the tradeoff between delay and jointness. The consideration of different strategies allows a wider diversity of path selection that helps meet a larger number of requirements simultaneously.

8.3 Formalization of the problem

8.3.1 Problem definition

In this formalization, we represent a PoP by a single node, where this node has all of the inter-PoP links for the whole PoP attached to it. Nothing is lost in this topology representation since our immediate goal is to map the inter-PoP links and not the intra-PoP links. (Of course the same technique could be applied to intra-PoP links as well.)

Given:

(i) an existing physical topology (which must be at least 2-connected), comprising nodes equipped with a limited integer number of tunable transmitters and receivers and no wavelenth conversion functionality, connected by optical fibers that support a limited number of wavelengths and with limited capacity; and

(ii) an existing IP topology (which must be at least 2-connected), comprising IP PoPs interconnected by IP links.

Find: maximally disjoint physical paths for the parallel logical links of all pairs of neighboring PoPs, such that they satisfy the relative delay constraint. Furthermore, find an assignment of wavelengths for each logical link.

Note that the search for disjoint paths and the wavelength assignment must be conducted in parallel because the wavelength assignment has a direct impact on the feasibility of physical paths.

The maximum delay over all PoP pairs is an output of the solution (and our algorithm in the case of the heuristic). As explained in Section 8.1, the maximum delay can be controlled by tuning the parameter u. Therefore, the maximum delay is computed in a post-computation step, after a mapping solution has been found.

8.3.2   ILP model

We formulate the mapping problem as an ILP model whose objective is to minimize the global jointness of the network. We compute first all the default path lengths between each pair of neighboring PoPs, as defined in Section 8.1.2.

Notation

Let $\mathcal{E} = \{(i, j)\}$ denote the set of fibers and let $\mathcal{S} = \{(s, t)\}$ denote the set of neighboring PoP pairs. We use n^{st} for the number of inter-PoPs links between the two PoPs s and t. Let $\mathcal{S}_{\text{priority}} \subset \mathcal{S}$ represent the subset of the priority PoP pairs.

We let w_{ij} represent the number of wavelengths for fiber (i, j) and w_{max} is the number of wavelengths available on the fiber with the most wavelengths. It will be used as a bound for the channel index in the constraints. We introduce $^{(x)}a_{ij} \in \{0, 1\}$ for all $(i, j) \in \mathcal{E}$ and $x \in \{1, 2, ..., w_{\text{max}}\}$ such that $^{(x)}a_{ij} = 1$ if the wavelength x belongs to fiber (i, j).

The notation pertaining to delays is as follows. Let $l_{ij} \geq 0$ be the length of the physical link (i, j) for all $(i, j) \in \mathcal{E}$. The values are in the millisecond range. Let d^{st} for all $(s, t) \in \mathcal{S}$ be the delay between PoPs s and t using the *default path*. The maximum delay difference among all parallel links between each pair of neighboring PoPs is specified via the parameter u.

Decision variables

To compute the routing we define $\pi_{ij}^{st}(m)$ for all $(i, j) \in \mathcal{E}, (s, t) \in \mathcal{S}, m \in \{1, 2, ..., n^{st}\}$. We have $\pi_{ij}^{st}(m) = 1$ if the mth logical link of the PoP pair (s, t) traverses the fiber (i, j).

We now define the decision variables used to handle wavelengths. We use $^{(x)}\lambda^{st}(m)$, defined for all $(s, t) \in \mathcal{S}, m \in \{1, \dots, n^{st}\}$, and $x \in n_{\text{max}}^{st}$, where $^{(x)}\lambda^{st}(m) = 1$, if the mth logical link of (s, t) uses the wavelength x. We also define $^{(x)}\lambda_{ij}^{st}(m) \in \{0, 1\}$, where $^{(x)}\lambda_{ij}^{st}(m) = 1$, if the mth logical link of (s, t) traverses fibers (i, j) and uses wavelength x.

The decision variables for handling the SLA are as follows. Let $\Lambda^{st}(m)$ be the total length of the mth logical link of (s, t) for all $(s, t) \in \mathcal{S}$ and $m \in \{1, 2, ..., n^{st}\}$, defined by $\Lambda^{st}(m) = \sum_{(i, j) \in \mathcal{E}} (\pi_{ij}^{st}(m) * l_{ij})$. Let $\Lambda_{\text{max}}^{st}$ be a length longer than the longest logical link of (s, t).

The jointness is computed in the model with two variables q and q', where q represents the jointness for all logical links and q' denotes the jointness for the two *priority logical links*. These two variables allow us to analyze separately the local jointness for only two priority logical links (for all neighboring pairs) and for all logical links in the network. We define $q_{ij}^{st} \geq \sum_{m=1}^{n^{st}}(\pi_{ij}^{st}(m) + \pi_{ji}^{st}(m)) - 1$ for all $(i, j) \in \mathcal{E}$ and $(s, t) \in \mathcal{S}$. It is the number of paths of (s, t) minus one that use the fiber (i, j). We define $q_{ij}'^{st} \geq \sum_{m=1}^{2}(\pi_{ij}^{st}(m) + \pi_{ji}^{st}(m)) - 1$ for all $(i, j) \in \mathcal{E}$ and $(s, t) \in \mathcal{S}$. If the two paths use the fiber (i, j), $q_{ij}'^{st}$ is equal to one; otherwise it is null.

Constraints

- The flow continuity constraints for the physical paths of the inter-PoP links of the pair of PoPs (s, t) are given by

$$
\sum_{j \in V:(i,j) \in \mathcal{E}} \pi_{ij}^{st}(m) - \sum_{j \in V:(j,i) \in \mathcal{E}} \pi_{ji}^{st}(m)
$$
$$
= \begin{cases} 1 & \text{if} \quad i = s \\ -1 & \text{if} \quad i = t \\ 0 & \text{otherwise} \end{cases} \tag{8.1}
$$
$$
\forall i \in V, \forall ((s, t), m) \in \mathcal{S} \times \{1, \ldots, n^{st}\}.
$$

Equation (8.1) defines the physical path associated with each logical link.
- Wavelength assignment: $\forall ((s, t), m) \in \mathcal{S} \times \{1, \ldots, n^{st}\}$,

$$
\sum_{1 \leq x \leq w_{\max}} {}^{(x)}\lambda^{st}(m) = 1. \tag{8.2}
$$

Equation (8.2) does the wavelength assignments for all the paths.
- The following equation ensures that the physical paths use only fibers where wavelengths are available: $\forall (i, j) \in \mathcal{E}, \forall 1 \leq x \leq w_{ij}, \forall ((s, t), m) \in \mathcal{S} \times \{1, \ldots, n^{st}\}$,

$$
\pi_{ij}^{st}(m) \leq (1 - {}^{(x)}\lambda^{st}(m)) * B + {}^{(x)}a_{ij}. \tag{8.3}
$$

If the mth path of the pair (s, t) uses the wavelength x, then the constraint reduces to $\pi_{ij}^{st}(m) \leq {}^{(x)}a_{ij}$ since the term $(1 - {}^{(x)}\lambda^{st}(m)) = 0$. Furthermore, if the fiber (i, j) does not support the wavelength x, i.e. ${}^{(x)}a_{ij} = 0$, the variable $\pi_{ij}^{st}(m)$ is forced to be equal to zero. B is a large arbitrary number and its use is explained at the end of this section.
- Equation (8.4) ensures that one wavelength can only be used once per fiber:

$$
\sum_{(s,t) \in \mathcal{S}} \left(\sum_{m=1}^{n^{st}} \left({}^{(x)}\lambda_{ij}^{st}(m) + {}^{(x)}\lambda_{ji}^{st}(m) \right) \right) \leq 1 \tag{8.4}
$$
$$
\forall (i, j) \in \mathcal{E} : i < j, \forall 1 \leq x \leq w_{ij}.
$$

For each fiber (i, j) and each wavelength x, only one ${}^{(x)}\lambda_{ij}^{st}(m)$ or ${}^{(x)}\lambda_{ji}^{st}(m)$ can be used, for all the logical links of all the paths.

- Constraints on $^{(x)}\lambda_{ij}^{st}(m)$: $\forall((i,j),x) \in \mathcal{E} * \{1,\dots,w_{ij}\} : i < j, \forall((s,t),m) \in \mathcal{S} \times \{1..n^{st}\}$,

$$^{(x)}\lambda_{ij}^{st}(m) \geq \lambda_{ij}^{st}(m) + \pi_{ij}^{st}(m) + \pi_{ji}^{st}(m) - 1, \tag{8.5}$$

$$^{(x)}\lambda_{ij}^{st}(m) \leq {}^{(x)}\lambda^{st}(m), \tag{8.6}$$

$$^{(x)}\lambda_{ij}^{st}(m) \leq \pi_{ij}^{st}(m) + \pi_{ji}^{st}(m). \tag{8.7}$$

Equations (8.5), (8.6) and (8.7) ensure that $^{(x)}\lambda_{ij}^{st}(m) = 1$ if both $^{(x)}\lambda^{st}(m) = 1$ and $\pi_{ij}^{st}(m) = 1$, and $^{(x)}\lambda_{ij}^{st}(m) = 0$ otherwise.
- We incorporate our constraint on the relative path lengths as follows: $\forall((s,t),m) \in \mathcal{S} \times \{1,\dots,n^{st}\}$,

$$\Lambda_{\max}^{st} - \Lambda^{st}(m) \geq 0, \tag{8.8}$$

$$\Lambda_{\max}^{st} \leq d^{st} * (1+u). \tag{8.9}$$

Equation (8.8) forces $\Lambda_{\max}^{st}$ to be longer than all the physical paths between PoPs (s,t). The minimization process will search for solutions less than this largest value. Equation (8.9) requires this largest value to be within $u\%$ of the delay of the default path length for each (s,t).

Avoiding loops

The flow continuity constraints (equation (8.1)) are insufficient to guarantee that our physical paths avoid loops. To solve this problem, we add new constraints as proposed in ref. [153]. The principle is to make sure that a path uses only fibers that are part of a subset of the physical topology called a covering tree.

Objective function

The objective function is to minimize

$$B^3 * \sum_{(i,j)\in\mathcal{E}} \sum_{(s,t)\in\mathcal{S}_{\text{priority}}} q_{ij}^{\prime st}$$

$$+ B^2 * \sum_{(i,j)\in\mathcal{E}} \sum_{(s,t)\in\mathcal{S}_{\text{priority}}} q_{ij}^{st}$$

$$+ B * \sum_{(i,j)\in\mathcal{E}} \sum_{(s,t)\in\mathcal{S}} q_{ij}^{\prime st} \tag{8.10}$$

$$+ \sum_{(i,j)\in\mathcal{E}} \sum_{(s,t)\in\mathcal{S}} q_{ij}^{st}$$

The four components of the objective function correspond to the four steps outlined in the mapping sequence in Section 8.1.2. For each component, we are trying to minimize the corresponding jointness; B is a large number that needs to be much larger than the sum of all the jointness parameters. In this objective, the jointness of the links included in step 1 of our mapping sequence is multiplied by B^3, step 2 is multiplied by B^2, and so on. By multiplying the first term by B^3, we guarantee that the first term of the objective function is minimized first. Thus step 1 (step 2) has the highest importance (second highest importance) within this objective function, respectively. Whenever there is a tie (i.e. two solutions produce the same jointness for the first term), then the following term is used to break the tie. The rest of the objective function is structured the same way. As pointed out in Chapter 7, ILP models can be solved using standard software packages such as that in ref. [1].

8.3.3 Tabu search for the mapping problem: TabuMap

The heuristic we present for identifying the ideal mapping relies on the application of the Tabu Search (TS) methodology [86], extensively described in Chapter 7. We refer to our specific implementation as *TabuMap*.

Before describing the TabuMap heuristic, we point out an important issue. During the search for an optimal solution, we allow TabuMap to investigate solutions outside the space of admissible solutions, e.g. solutions that violate one or more constraints. By non-admissible solutions, we mean solutions that require more wavelengths on some fibers than provided by the WDM topology. All solutions, even non-admissible ones, always satisfy the SLA requirements. For some scenarios (e.g. when a fiber has a few wavelengths) even finding a single admissible solution can be hard because of the wavelength assignment problem. To ensure we continue to move forward in our exploration, we allow the heuristic to temporarily go outside the space of admissible solutions. We operate a strategic oscillation (see ref. [87]) between the space of admissible solutions and the space of non-admissible solutions. When inside the space of admissible solutions, we try to improve the current solution; when outside this space, we try to come back inside by applying a special kind of move (described below). In the following we do not report the pseudocode of TabuMap since it is similar to the one shown in Figure 7.3. The analysis of its computational complexity can be obtained in a similar fashion as described in Section 7.5.

8.3.4 Fundamental aspects of TabuMap

We now describe the seven components of the TabuFLTDP heuristic we have designed to solve the mapping problem.

Pre-computation step
Before running the TabuMap heuristic, we need to pre-compute the following information:

- for each pair of neighboring PoPs, we compute the default length path according to the three strategies described in Section 8.1.2;
- for each pair of neighboring PoPs, we build the set of physical paths satisfying the relative delay constraint – this set is then sorted according to the length of each physical link, from shortest to longest;
- we build the IP routes for all arbitrary PoP pairs according to the IS–IS routing protocol.

Initial solution

The choice of the initial solution is very important since it can significantly reduce the convergence time. For each logical link, we choose the shortest physical path between neighboring PoPs to be the initial mapping. Typically, this solution is outside the space of admissible solutions. But it is optimal in terms of delay.

Moves and neighborhood generation

Because we visit admissible and non-admissible solutions during the exploration, we define two different kinds of moves. When the search is focused on the space of admissible solutions, the selected move will find a solution without considering the wavelength constraint; when the search takes place outside the space of admissible solutions the move will try to minimize the number of logical links that share fibers on which there is a shortage of wavelengths.

- Admissible space. Given a currently admissible solution, the next solution is generated according to the following three steps: (i) randomly select an adjacent PoP pair; (ii) randomly select one of the pair's parallel links not present in the Tabu List; and (iii) change the physical path of this link by picking a new path satisfying the maximum length constraint. All other physical paths associated with all the other logical links during past moves remain unchanged.
- Non-admissible space. If the current solution does not meet the wavelength constraints, a special move is applied to force the solution to become admissible by looking at the fibers on which the shortage of wavelengths was experienced. The new solution is built as follows: (i) randomly select a fiber experiencing a shortage of wavelengths; (ii) randomly select a logical link that uses this fiber; and (iii) change the physical path of this link to a set of fibers that does not experience wavelength shortage. All other physical paths associated with all the other logical links during past moves remain unchanged.

A new solution is consequently built by applying one of the moves defined above to a random subset of all the physical paths chosen from the previous step. The cardinality of this subset defines the size of the neighborhood investigated by TabuMap.

Wavelength Assignment Problem (WAP)

The WAP is NP-Complete. Since this problem must be solved for each solution visited during the exploration, we need a heuristic that is simple enough to reach a good tradeoff between running time and quality of the solution. The principle of our algorithm is to assign the wavelength, with the smallest channel index available, each time a new physical path is mapped.

Tabu list

We use a static Tabu list and store the most recent moves made. A move is not allowed to be reselected while in the list.

Diversification

If we do not see any improvements in the solution after a given number of iterations (say 100), then we employ a *diversification* strategy. We carry out a unique perturbation by (i) selecting randomly a neighboring PoP pair and (ii) changing all the physical paths of its parallel logical links. The selection of the paths is a random process. After having applied this perturbation, the traditional move defined previously is applied.

Stop criterion

The search procedure is stopped after a fixed number of iterations. This number is defined based on the size of the network studied and on a good tradeoff between computational time and quality of the solution. Our experience has shown that setting this parameter to 3500 is sufficient for the Sprint network.

8.3.5 Validation

We compare the performance of TabuMap to that of the ILP model. We demonstrate that the heuristic works well for medium- size networks, so that we can be confident that it works well for large networks. A medium-size topology is defined as the maximum topology that we can solve with the ILP model. We consider a physical topology with 15 nodes and 23 fibers, which is a simplified version of the Sprint backbone. We generate random logical topologies made of six to eight PoPs and six to ten neighboring PoP pairs with two to four parallel logical links. For each of these topologies, we ran several simulations with different numbers of wavelengths on the fibers. On average, after 1000 iterations of the heuristic, the results for the global jointness differed by less than 3%.

TabuMap can be implemented using any custom programming language such as C or C++ (see refs. [110] and [187], respectively).

8.4 Results

8.4.1 Topologies and metrics

We use the TabuMap heuristic algorithm to map the logical topology onto the physical topology of the Sprint US continental IP backbone. Simplified views of the IP topology and the physical topology are shown, respectively, in Figure 8.3 and Figure 8.4.[2] The WDM layer is composed of 51 OXC and 77 WDM fibers that have between 16 and 48 channels. The delay associated with each fiber is proportional to its length and equal to the amount of time required for light to cover such a distance. The logical topology consists of 101 logical links and 16 PoPs (excluding Pearl Harbor), with 35 neighboring PoP pairs. Each neighboring PoP pair has a minimum of two parallel logical links, a

---

[2] Because topology information can be sensitive, we do not show exact maps, but we do use the exact ones in the computation of the mapping problem.

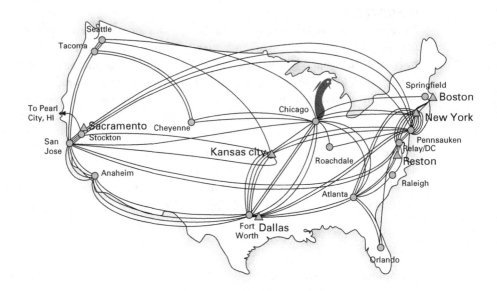

Fig. 8.3.   North American Sprint IP backbone network topology (third quarter, 2001): 16 PoPs with 35 bidirectional inter-PoP aggregated links. Each aggregated link comprises many multiple links ranging from two to six.

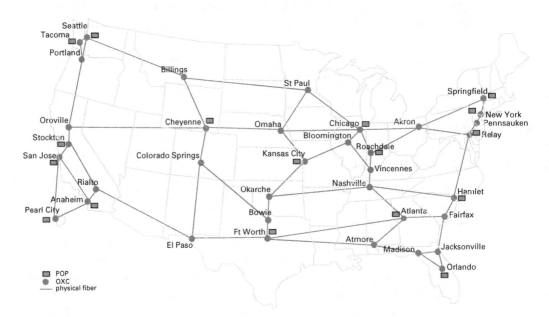

Fig. 8.4.   North American Sprint fiber-optic backbone topology (third quarter, 2001, simplified version): 51 OXCs and 77 WDM fibers.

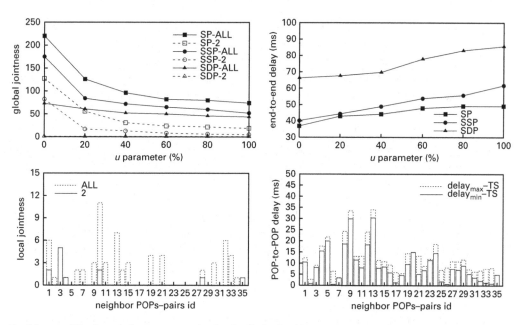

Fig. 8.5.   The four performance metrics for the Sprint backbone: (a) global jointness; (b) end-to-end
delay; (c) local jointness (SSP, $u = 50\%$); and (d) PoP-to-PoP delay (SSP, $u = 50\%$).

maximum of six and an average of three. We have six priority PoP pairs (indexed by the
numbers 6, 9, 13, 15, 33 and 34 in the plots in Figures 8.5 and 8.6). There are a total
of 136 arbitrary PoP pairs, each of whose IP routes are between one and four hops in
length.

We use four metrics to evaluate our approach. The first two are the global jointness
and the local jointness, which is defined in Section 8.1. The other two metrics are the
maximum delay and the PoP-to-PoP delay, defined in earlier chapters.

8.4.2   Jointness and maximum delay

Performance metrics are plotted in Figure 8.5. The terms SP, SSP and SDP refer to the
default path computation strategies defined in Section 8.1.2. The suffix -2 means that
the jointness is computed for the priority parallel logical links only. Recall that we use
two priority links per pair of adjacent PoPs. The suffix -ALL is used when the jointness
is computed on all logical links.

Figure 8.5(a) illustrates the tradeoff between the jointness and the relative delay
increase expressed by u. We see how the global jointness decreases as the relative delay
increases. Larger values of u allow for larger sets of acceptable paths, which in turn
make it easier to find disjoint paths. If we restrict u to be small, then the set of candidate paths meeting the relative delay requirement is small, and this limits our choices
in trying to find paths that do not overlap. All curves flatten out when u reaches 40%,
leading to very similar jointness values. This means that, if we allow paths to have a
difference in delay of 40%, then the relative delay constraint is sufficiently loose that it

no longer has any impact on the jointness of paths. For all except one strategy (SDP-2), the jointness is significant for u below 20%. The minimum achievable jointness in the Sprint backbone, given the current physical topology, is approximately 50.

The choice of the default path has a great impact on the jointness of the solution. The Shortest Path (SP) strategy produces the largest amount of jointness. This result confirms our observations on network topologies: there typically exists only a single short physical path between two PoPs, and all alternate paths are much longer. The large jointness of the SP strategy can thus be explained as follows. If there is only one short path, whereas all others are rather long, and the relative delay constraint u is small, then the algorithm will have to pick two long paths to satisfy the u requirement. Since this leads to two long paths, they are less likely to be disjoint.

The Shortest Disjoint Path (SDP) strategy yields the minimal jointness. The fact that SDP-2 exhibits a global jointness of zero for any value of u is good news for network designers. This means that, with this strategy, we can find completely disjoint paths for the two priority links of all PoP pairs. The same plot also tells us that it is impossible to find a physically disjoint path for all logical links in the backbone, whatever the value of u.

These strategies also need to be assessed in terms of the maximum delay they yield. Figure 8.5(b) plots the maximum delay as a function of u for each of the three strategies. The maximum delays increase as the relative delay u increases. The SP strategy that was the worst in terms of jointness performs best in terms of maximum delay. Similarly, the SDP strategy that was best in terms of jointness performs the worst in terms of maximum delay. This illustrates the tradeoff between jointness and maximum delay. Moreover, we also see that the only strategy (SDP) that provides totally disjoint solutions for at least two logical links per PoP pair (an SDP-2 jointness of zero) will often fail to meet the SLA requirement (for SLAs below 60 or 70 ms, depending upon u). Therefore, we learn that if a large ISP such as Sprint wants to have two disjoint logical links between each adjacent PoP pair, they must set their SLA as high as 65 ms. Note that u does not matter here as the jointness is zero for any value of u for the SDP-2 strategy. In addition, if Sprint wants to set their SLAs as low as 50 ms, then they must accept some path overlap among priority links. With an SLA of 50 ms, the optimal strategy is the Second Shortest Path (SSP), which has a global jointness of roughly 10 for the priority links (i.e. global jointness for SSP-2)

In Figures 8.5(c) and (d), we use the SSP strategy and $u = 50\%$. We have chosen this value of u because it corresponds to a maximum delay of 50 ms and to a global jointness for the priority links of priority PoPs that is fairly close to zero for the SSP strategy. (In the figure, $GJ = 14$ for priority links of priority PoPs and $GJ = 71$ for all logical links in the whole network, where GJ denotes the value of the global jointness). Figure 8.5(c) shows the local jointness achieved for each neighboring PoP pair. Only six of these PoP pairs cannot find two completely disjoint physical paths. On the other hand, 13 PoP pairs have completely disjoint paths for all their parallel logical links. The priority PoP pairs are not among these 13 PoP pairs; however, these priority PoP pairs have at least two disjoint parallel logical links (LJ-2 $= 0$). This is an important result, which says that, on the current Sprint network, it is impossible to protect fully all logical

Table 8.1. Impact of the notion of priority PoP pairs on the metric of global jointness

	Priority links of priority PoPs	All links of priority PoPs	Priority links of all PoPs	All links of all PoPs
Mapping with priority PoPs	0	20	33	103
Mapping without priority PoPs	2	25	30	108

links between all priority PoP pairs with a SLA of 50 ms. We will discuss this issue in Section 8.4.3.

Figure 8.5(d) shows the delay experienced by the longest and the shortest logical links among all parallel logical links for each adjacent PoP pair. The difference between the maximum and the minimum delay corresponds to the relative delay parameter. For most pairs, the relative delay is small. The maximum relative delay is 7 ms. It is important to note that, despite the SDP strategy, the maximum relative delay is observed for short physical links. PoP pairs that are far apart geographically always experience a small relative delay, on the order of 3 ms.

8.4.3 Impact of priorities

In this section we examine the impact of having a priority for PoP pairs. First we carry out a mapping, according to the sequence stated in Section 8.1.2, that includes all the priorities. Then we carry out a second mapping in which we drop the notion of priority PoP pairs (while still retaining the notion of priority links). This is easy to do with our objective function (equation (8.10)) because we simply drop the first two terms while retaining the latter two. For both of these scenarios we calculate the resulting jointness on four sets of logical links: the priority links of the priority PoP pairs, all links of the priority PoP pairs, priority links of all PoP pairs, and all links among all PoP pairs. We calculate the jointess of a few logical links by summing the jointness value of each fiber segment belonging to those logical links. Similarly, we calculate the jointness of a subset of the PoP pairs by summing the jointess for those PoP pairs included in the subset. Note that even if we do not include the priority PoP pairs in our second mapping, we can still calculate the resulting jointness for those PoPs. This way we can see what happens to those particular PoPs when their priority is removed. Again, we use the SSP default path strategy and a relative delay requirement of $u = 50\%$. Table 8.1 shows the global jointness for the four sets of links under both mapping scenarios.

As we have seen already, with PoP priorities we cannot guarantee complete disjointness for all the parallel links of the priority PoPs (case $GJ = 20$) but we can achieve complete disjointness for the priority links of the priority PoPs (case $GJ = 0$). If we remove the priority of these special PoP pairs, then we are no longer guaranteed that all the priority PoP pairs are completely robust. In this case, at least one PoP pair, and possibly two, have not achieved complete robustness.

It is interesting to note that, by eliminating PoP priorities, the overall jointness measure on all links increases (from 20 to 25 for priority PoPs, and from 103 to 108 when measured over all PoPs). All four sets of links do better in terms of jointness with priority PoPs except for one, namely the priority links of non-priority PoPs. These links have better jointness without PoP priorities. Thus, the tradeoff in having PoP pair priorities is that three of the link groups do better with priorities while one group does worse. In the scenario studied here, the cost of having priorities is a 10% increase (GJ goes from 30 to 33) in the jointness of this group of logical links. This example illustrates how our priority mechanism manages the global set of network resources across all logical links.

8.4.4   Improving the network design

The previous results show that it is critical to understand how to improve the robustness of a network. In this section, we use our mapping algorithm to analyse where new fibers or new wavelengths should be added to increase the robustness of the IP topology.

Figure 8.6 uses a gray-scale to indicate the fraction of logical links that use a given fiber segment for each adjacent PoP pair. A black square means that 100% of the parallel

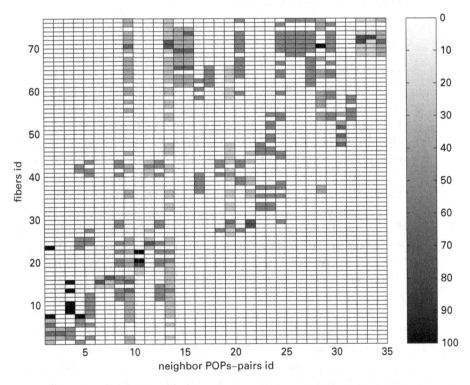

Fig. 8.6.   Fiber network upgrade. The x-axis reports all pairs of adjacent PoPs ordered by an id; the y-axis reports all fiber segments constituting the physical infrastructure ordered by an id. A black square means that 100% of the parallel logical links associated to a pair of adjacent PoPs use a specific fiber segment (i.e. these PoPs will be disconnected if that fiber fails). A white square means that the PoP pair does not use that fiber segment at all.

logical links use the fiber segment. A white square means the PoP pair does not use that fiber segment at all.

Ten out of the 77 fibers can cause a pair of adjacent PoPs to lose connectivity completely in case of failure (e.g. the black boxes in the grid). From the logical link standpoint, five out of 35 PoP pairs can lose connectivity because of a single fiber-segment failure (e.g. PoP pairs that feature at least one black box). Note that none of the priority PoP pairs are concerned. The rest of the network is well protected since, in 66% of the fiber cuts, none of the other PoP pairs would lose more than 50% of their parallel logical links.

Using such a visualization tool, a carrier can quickly identify the high risk fiber segments whose failure could bring down the entire direct connectivity between a pair of PoPs. The PoP pair would have to communicate via another intermediate PoP in such an event. The identify of the PoP pairs subject to complete disconnection due to a single fiber-segment failure can also be readily found from this visualization. For example, PoP pair number 3 can lose all its parallel logical links from the failure of any one of the five fiber segments marked as black; similarly PoP pair 10 can lose all its logical links from the failure of any one of the two fibers marked with black. By chance, these seven high-risk fibers are not the same and these two PoP pairs are not located in the same geographical area.

Viewing this from the physical topology standpoint, we can see, for example, that the failure of fiber segment 7 disconnects two PoP pairs. Similarly for fiber segment 72. The locations of these two fiber segments can thus be considered as high risk, critical areas in the USA where large problems can occur. Adding fibers in these areas (along similar close by routes) would increase the disjointness without paying a large price in terms of delay.

We used our tool to compute the optimal mapping after having added two fiber segments to the physical topology in order to improve the robustness around fiber segments 7 and 72. With these fiber additions we were able to reduce the number of fiber cuts that would completely disconnect adjacent PoP pairs from ten to four. We also brought the number of adjacent PoP pairs impacted by these fiber cuts from five down to three. Hence the addition of a few new fibers in well chosen locations can substantially improve the protection of the logical topology while still meeting the SLA.

It is usually more cost effective to improve network robustness by upgrading existing fibers via additional wavelengths than by deploying new fibers in the ground. To know where it is most useful to upgrade fiber segments, we disable the wavelength availability constraint before running the mapping algorithm. By comparing this visualization graph of the mapping with and without WAP, we can identify those black boxes that turn to lighter shades of gray when WAP is removed. These boxes will identify the fiber segments that should be upgraded. In the Sprint backbone, we can decrease the general jointness for the priority links by 66% (from 33 to 12) and the jointness of all the paths by more than 25% (from 103 to 71) by upgrading only six fiber segments in the country (with an average of five wavelengths each). This illustrates that a small shortage of wavelengths can have a huge influence on the robustness of the network.

8.5   Lessons learned

In this chapter we looked into the problem of how to increase the robustness of IP backbones in the absence of optical level protection. We presented a new approach that focuses on minimizing the number of physical fiber segments that are shared by all IP layer logical links between two adjacent PoPs. The problem has been solved taking into consideration operational constraints such as maximum and relative delay requirements, a limited number of wavelengths and priorities for PoP pairs and certain IP layer logical links.

The method we presented has been implemented as an ILP model and as a heuristic based on Tabu Search called TabuMap. We applied the method to the Sprint IP backbone network, and found that, if the SLA can be set as high as 65 ms, then full robustness can be achieved for all PoP pairs. If the SLA must stay below 50 ms, we showed that we can fully protect the most important PoP pairs and achieve a high level of protection on all other PoP pairs. In this case, the worst case relative delay difference between all the parallel logical links for any adjacent PoP pair was less than 7 ms.

We concluded our analysis by showing how the technique presented can be readily used to identify the vulnerable areas in which fibers or wavelengths should be added to the network in order to increase the robustness to single fiber failures. We illustrated using an example how the robustness of the Sprint backbone can be improved by the addition of a few fibers or wavelengths in the right place.

In Chapters 9 and 10 we present Step 3 of the design process, which is aimed at enhancing the performance of the logical topology (now being designed with robust resilience properties upon long-lived optical failures) and its resilience to logical failures (i.e. short-lived failures) by optimizing the IGP routing protocol configuration. The methodology is first presented for the case that ignores the interaction between IGP and BGP (Chapter 9), and is then extended to incorporate such an interaction in the problem formulation (Chapter 10).

9 Performance enhancement and resilience to short-lived failures via routing optimization

---

> **Box 9.1.** Network design process: Step 3(a)
>
> **Goal** Identify *one* set of static IGP link weights (i.e. integer numbers between 1 and 255 to be assigned to each link constituting the logical topology) that works *well* in both the normal operation state (i.e. the absence of logical failures) and during any logical failure.
>
> **Objective function** Satisfy the SLA requirements during the normal operation state and achieve the best tradeoff in terms of network congestion (i.e. defined as maximum amount of traffic flowing on any link composing the logical topology) between the normal operation state and under any logical failure.
>
> **Inputs** (i) Logical topology, comprising nodes interconnected to other nodes via single links, characterized by a given propagation delay and limited capacity; (ii) traffic matrix, which describes the peak-time volumes of traffic exchanged between any pair of IP nodes; (iii) SLA requirements for the normal operation state, defined as the average end-to-end delay between any pair of IP nodes; (iv) model of short-lived failures, assumed to last for a short time and affecting one link at the time, i.e. single-link-failure model.

Large IP networks use a link-state protocol such as IS–IS or OSPF as their Interior Gateway Protocol (IGP) for intra-domain routing (see Chapter 2). Every link in the network is assigned a weight, and the cost of a path is measured as the sum of the weights of all links along the path. Traffic is routed between any two nodes along the minimum-cost path which is computed using Dijkstra's Shortest Path First (SPF) forwarding algorithm. Thus setting the link weights is the primary traffic engineering technique for networks running IS–IS or OSPF. A wrong selection of link weights can lead to severe congestion of one or more links and thus to packet drops, long end-to-end delay and large jitter that can dramatically impact the quality of service of real-time applications like voice and video over IP. In the following we focus on IS–IS, although the same concepts can be extended to OSPF as well.

Portions reprinted, with permission, from Nucci, A., Bhattacharyya, S., Taft, N. and Diot, C. (2007). "IGP link weight assignment for operational tier-1 backbones." *IEEE Transactions on Networking*, **15**: 4, 789–802.

The most common description of the IGP link-weight selection problem requires to have access only to the IP topology, where each link comes with the associated propagation delay and capacity (Box 9.1, Input:(i)), and to an accurate description of the traffic exchanged between any two end points in the network (Box 9.1, Input:(ii)). The optimization process is focused on finding the *one* set of static IGP link weights that guarantees the minimum level of network congestion.

A common recommendation of router vendors is to set the weight of a link to the inverse of its capacity [154]. The idea is that this will attract more traffic to high-capacity links and less traffic to low-capacity links, thereby yielding a good distribution of traffic load. Another common recommendation is to assign a link a weight proportional to its physical length in order to minimize propagation delays. In practice, many backbone operators use the ad-hoc approach of observing the flow of traffic through the network and iteratively adjusting a weight whenever the load on the corresponding link is higher or lower than desired. An increase in a link weight is bound to make the paths going through it higher cost, and therefore detract traffic from the link, lowering its utilization. On the other hand, a reduction in the weight of a link will attract more traffic to it, due to the decreased cost of the paths going through.

Despite the obvious importance of the IS–IS weight selection problem, a formal approach to the problem has not been undertaken. By considering the same inputs as above, the first work to address this problem is given in ref. [79], which showed that the problem of finding *optimal* IS–IS link weights is NP-Hard with respect to many objectives. The authors propose a local search heuristic to find weights for two objective functions. The first objective function is to minimize the network congestion, and the second one is to minimize a cost function that assigns every link a cost depending on its utilization. They show, for both objective functions, that their heuristic can compute weight sets that lead to solutions within a few percent of a theoretical lower bound. This bound is computed using linear programming in much the same way as we will show later in this chapter.

Others have evaluated the use of other methods from operations research, for example implement heuristics based on local search, simulated annealing, Lagrangean relaxation and evolutionary algorithms [92, 163]. Furthermore, these authors propose a new method that uses a combination of simulated annealing and linear programming. This method is not guaranteed to find a weight set for every problem instance, but is shown to work in 95% of cases. The authors find that, in terms of the number of overloaded links and the degree of overload, all methods perform similarly well. For larger topologies the new combined method is much faster than any of the other methods.

The use of linear programming duality was proposed in ref. [201] to compute link weights. However, that method is limited to the case where unequal splitting is feasible. Neither OSPF nor IS–IS currently support unequal splitting, only even splitting on equal-cost paths.

One drawback of current approaches is that SLA bounds are not incorporated into the link-weight assignment problem. This is a crucial aspect for ISPs since delay bounds are both a vehicle for attracting customers and a performance requirement which carriers are contracted to meet. Each ISP defines its own metrics and the values for those metrics in their SLA. For the North American Sprint IP backbone the SLA is set to 50 ms, whereas for the European Sprint IP backbone, it is set to 20 ms. The average delay between any pair of nodes, referred to in this chapter as an Origin–Destination (OD) pair, must be below the value specified in the SLA. As a consequence, it is imperative to incorporate SLA bounds into the problem definition and solution (Box 9.1, Input:(iii)).

Another and more crucial aspect of the routing optimization problem is related to its dynamics upon failure. Most of the work proposed in this area views the link-weight assignment problem as a static problem largely ignoring network dynamics. In practice, one of the main challenges for network operators is to deal with link failures that occur on a daily basis in large IP backbones [100]. When a link fails, IS–IS/OSPF routing diverts the traffic from the broken link to alternate paths, increasing the load on one or more other links.[1]

The most obvious way of restoring the network to meet its original traffic engineering objectives, is to perform a network-wide recomputation and reassignment of link weights. It has been shown that changing just a few link weights is usually sufficient to rebalance the traffic [80].

Changing link weights during a failure may not, however, be practical for two reasons. First, the new weights will have to be flooded to every node in the network, and every node will have to recompute its minimum-cost path to every other node. This can lead to considerable instability in the network, aggravating the situation already created by the link failure. The second reason is related to the short-lived nature of most of the link failures. As shown in Figure 6.3, 80% of the failures last less than ten minutes and 50% of the failures last less than one minute. We refer to failures that last less than ten minutes as short-lived failures. Short failures can create rapid congestion that is harmful to the network. However, they leave a human operator with insufficient time to reassign link weights before the failed link is restored. Given that our measurements reveal 70% of the short-term failures to affect a single link inside the network (Figure 6.4), our formulation focuses on such scenarios (Box 9.1, Input:(iv)).

Accordingly, network architects must choose a performance metric able to take into account all the issues raised above and lead the optimization process toward *one* set of link weights that works well in *both* the normal operating state (i.e. the absence of failures) and during short-lived failures. By "perform well" during short failures we mean that none of the remaining active links should be overloaded, in general, nor should the new flow of traffic across the network lead to a SLA violation (Box 9.1, Objective function).

The rest of the chapter is organized as follows. Section 9.1 formally states the problem addressed in this chapter. Section 9.2 describes the model for the general routing problem that considers SLA bounds and single-backbone link-failure scenarios. Section 9.3

---

[1] As explained in Chapter 6, this problem can be alleviated by allowing multiple links between each pair of logical adjacent nodes (i.e. primary and backup). In this case, the traffic affected by the failure will be shifted from primary links to the corresponding backup links where enough capacity has been reserved in advance. We remark to the reader that the methodology we will present in this chapter can be easily adapted to ignore logical failures during the optimization process.

describes the Tabu Search heuristic for the link-weight selection problem. We explain the topologies and TMs used for the evaluation in Section 9.4. The performance of the heuristic is presented in Section 9.5, along with a discussion of the tradeoffs. Section 9.6 concludes the chapter.

9.1   Link-weight selection problem

The problem of computing IS–IS link weights can be formally stated as follows. We represent a network by an undirected graph $G(\mathcal{V}, \mathcal{L})$, where $\mathcal{V}$ corresponds to the set of nodes and $\mathcal{L}$ corresponds to the links connecting the nodes. Each edge $l \in \mathcal{L}$ has a bandwidth capacity c_l and integer IS–IS weight w_l. Let L be the cardinality of the set $\mathcal{L}$. Let D be a TM representing the traffic demands such that $d^{sd} \in D$ corresponds to the traffic demand between origin node s and destination node d. The traffic demand between an OD pair s and d is routed along the minimum-cost path from s to d, where the cost of a path is measured as the sum of the weights of the links along the path. If multiple paths exist with the same minimum cost, the traffic demand is split equally among all these paths.[2] The problem is to choose a single set of link weights $\mathcal{W} = (w_1, w_2, ..., w_L)$ so as to optimize a given objective function. The objective function is typically selected to meet the traffic engineering goals of a network. An important goal for IP networks is to distribute traffic evenly so that no link carries an excessive load. A common approach is to choose routes so as to minimize maximum load over all network links.

At the same time, carriers have to guarantee that the new routing configuration satisfies the SLA requirement, defined as the average end-to-end delay for all OD pairs traversing the backbone. Selecting link weights that minimize the maximum link load yet violate the SLA the carrier offers its customers yields a solution that cannot be deployed in an operational network. In our work, we assume that only propagation delay contributes to end-to-end delay since queuing delays have been observed to be negligible in a core network [156].

We model the occurrence of isolated short-lived failures in the network as state transitions. In the absence of any such failure, the state of the network is denoted by S_0. During the short-lived failure of link i $(i = 1, 2, ..., L)$, the state of the network is denoted by S_i. We refer to the maximum link load of the network in any given state as the *bottleneck load*. Let $\mu(S_j) \geq 0$ be the bottleneck load in state S_j; S_{wst} is defined as the state with the maximum bottleneck load over all network states, i.e. $\mu(S_{\text{wst}}) \geq \mu(S_j), \forall j \in [0, L]$. Henceforth we refer to state S_0 as the no-failure state, and the state S_{wst} as the worst-failure state.

---

[2] Technically IS–IS/OSPF do equal splitting per outgoing interface, not per path. During our research efforts, we considered both equal-cost splitting per outgoing link and per path (in the problem formulation). We found that incorporating the splitting per outgoing link adds considerable complexity into the problem, but yields little difference in final performance. We thus selected the per-path formulation as a good approximation.

To handle both the short-lived failure events and the SLA requirement in the problem, we present an objective function that penalizes the SLA violation for the no-failure state S_0 and limits the maximum link overload due to *any* single-link failure event. Selecting a weight set that is optimal for the network in a failure state may result in suboptimal performance in the absence of failures. We want to meet three objectives: (i) satisfying the SLA constraint and (ii) preventing link overloads during a failure, while simultaneously (iii) minimizing performance degradation in the absence of failures. We thus choose the following objective function:

$$\mathcal{F} = (1 - W)\mu(S_0) + W\mu(S_{\text{wst}}) + B\Delta(S_0), \qquad (9.1)$$

where $W \in [0, 1]$ is a tradeoff parameter that helps in balancing the two goals minimizing performance degradation in the absence of failures while maximizing the gain for any short-lived failure. Setting W to zero corresponds to finding link weights without considering failures. Setting $W = 1$ results in link weights that minimize the maximum load under any failure, but may be suboptimal in terms of performance when the network operates in the no-failure state.

The $\Delta(S_0)$ represents the excess SLA (the amount exceeding the requirement) for the no-failure state S_0. To compute this, we linearly sum the excess SLA for all routes. If the routing configuration selected meets the bound, then $\Delta(S_0) = 0$. However, if, for example, the SLA is set to be 50 ms for all routes and we generate all routes below 50 ms except for two routes with 53 ms SLA, then $\Delta(S_0) = 6$ ms. Let B denote a large scalar number used to give high priority to the SLA constraint. Only solutions characterized by the same $\Delta(S_0)$ value are further differentiated by the first two terms of $\mathcal{F}$.

9.2   ILP for general routing problem

In this section, we formulate the *general routing problem* with single-link failures as an ILP problem. The goal is to derive a lower bound on the value of the objective function used in our heuristic, equation (9.1).

Since the IS–IS link-weight selection problem is NP-Hard, it is computationally intractable to obtain the optimal solution. On the other hand, the general routing problem is computationally tractable. We therefore model the general routing problem to obtain a lower bound on the performance of our heuristic.

Unlike the link-weight selection problem, the general routing problem directly selects routes between node pairs (instead of doing it indirectly via link weights). Also, route selection is not restricted to minimum-cost routes based on link weights. Instead, traffic can be arbitrarily routed along several paths between two nodes and split in arbitrary ratios across these paths. Therefore an optimal solution to the general routing problem yields a lower bound on the objective function value for the link-weight selection problem.

Notation

We aggregate all the traffic between a given OD node pair in the network into a single traffic demand. Let $C = \{c_i\}$ be the set of all OD pairs whose cardinality is described by $C = |C|$ and let $\mathcal{L} = \{l_i\}$ be the set of bidirectional links belonging to the network topology whose cardinality is represented by $L = |\mathcal{L}|$. The traffic demand associated with OD pair $c \in C$ is represented by d^c. Let $\mathcal{S}$ be the set of all the network states, with cardinality $S = |\mathcal{S}| = L + 1$; S_0 represents the no-failure case, while S_j with $j > 0$ represents the failure case when the link l_j is broken. Let $R^c(S_j) = \{r_i^c(S_j)\}$ be the set of all the feasible routes for the OD pair $c \in C$ in the network state S_j. A route in a failure network state S_l is defined as *feasible* if it does not use the broken link l. Moreover, a route is defined feasible, for the no-failure state S_0, if and only if it does not violate the SLA constraint defined as an input to the problem. Let $u_{ij}^c(l)$ be the binary parameter that is equal to one if the ith route associated with OD pair c in the network state S_j, i.e. $r_i^c(S_j)$, uses the link l, and zero otherwise. Let $v_{ih}^c(S_0, S_j)$ be a binary parameter equal to one if $r_i^c(S_0) = r_h^c(S_j)$, and zero otherwise.

Decision variables

Let $\alpha_i^c(S_j) \in \{0, 1\}$ represent the set of variables used to define which routes are selected by OD pair c in network state S_j. The variable $\alpha_i^c(S_j) = 1$ if the OD pair c uses the ith route in network state S_j $(r_i^c(S_j))$ and equals zero otherwise. Let $d_i^c(S_j) \geq 0$ be the fraction of traffic associated with OD pair c that is sent to the ith route in the network state S_j. The variables $\beta^l(S_j) \geq 0$ represent the aggregated load carried by link $l \in \mathcal{L}$ in network state S_j. Let $\gamma_i^c(S_j) \in \{0, 1\}$ describe the feasibility of each route $r_i^c(S_0)$ in each network state S_j; $\gamma_i^c(S_j) = 1$ if OD pair $c \in C$ uses route $r_i^c(S_0)$ and this route uses the broken link l_j. When a link fails, all OD pairs going over the link must be rerouted. Several feasible routes can be used to reroute this traffic. For route $r_i^c(S_0)$, let $\theta_i^c(S_j) \geq 0$ be the traffic associated with OD pair c that is routed over failed link l_j. In other words, $\theta_i^c(S_j)$ is equal to $d_i^c(S_0)$ if $\gamma_i^c(S_j) = 1$ and zero otherwise. Finally, let $\mu(S_j) \geq 0$ be the maximum link load in network state S_j, while let $\mu(S_{\text{wst}})$ be the maximum link load over all the network states, i.e. $\mu(S_{\text{wst}}) = max_{j \in [0, L]} \mu(S_j)$. The corresponding state is denoted by S_{wst}.

Constraints

• The constraints associated with the maximum link load in each network state $S_j \in \mathcal{S}$ are as follows:

$$\mu(S_j) \geq \beta^l(S_j) \qquad \forall S_j \in \mathcal{S}, \forall l \in \mathcal{L}, \qquad (9.2)$$

$$\mu(S_{\text{wst}}) \geq \mu(S_j) \qquad \forall S_j \in \mathcal{S}, \qquad (9.3)$$

where equation (9.2) defines the maximum link load in each network state S_j, and equation (9.3) defines the maximum link load over all network states.

The variable $\beta^l(S_0)$ describes the aggregated traffic flowing on the link $l \in \mathcal{L}$ in the no-failure state S_0:

$$\beta^l(S_0) = \sum_{c \in \mathcal{C}, i \in |R^c(S_0)|} d_i^c(S_0) u_{i0}^c(l) \qquad \forall l \in \mathcal{L}. \tag{9.4}$$

- The constraints associated with routes used in S_0 and their associated traffic are given by:

$$\sum_{i \in |R^c(S_0)|} \alpha_i^c(S_0) \geq 1 \qquad \forall c \in \mathcal{C}, \tag{9.5}$$

$$d_i^c(S_0) \leq d^c \alpha_i^c(S_0) \qquad \forall c \in \mathcal{C}, \forall i \in |R^c(S_0)|, \tag{9.6}$$

$$\sum_{i \in |R^c(S_0)|} d_i^c(S_0) = d^c \qquad \forall c \in \mathcal{C}, \tag{9.7}$$

where equation (9.5) ensures that each OD pair in S_0 has to select at least one route to send its own traffic, while equation (9.6) restricts the traffic carried by each route. If $\alpha_i^c(S_0) = 1$, the total amount of traffic for OD pair c_i on this route is restricted by d^c; if $\alpha_i^c(S_0) = 0$, then no part of the traffic for OD pair c_i is sent over this route. The traffic associated with each OD pair has to be sent using one or more routes, equation (9.7).

- The constraints associated with routes affected by the failure of link $j \in [1, L]$ are given by:

$$\gamma_i^c(S_j) \leq \alpha_i^c(S_0) u_{i0}^c(l_j) \qquad \forall c \in \mathcal{C}, \forall i \in |R^c(S_0)|, \forall l_j \in \mathcal{L}, \forall S_j \in \{\mathcal{S} - S_0\}, \tag{9.8}$$

$$\gamma_i^c(S_j) \geq 2\alpha_i^c(S_0) u_{i0}^c(l_j) - 1 \qquad \forall c \in \mathcal{C}, \forall i \in |R^c(S_0)|, \forall l_j \in \mathcal{L}, \forall S_j \in \{\mathcal{S} - S_0\}, \tag{9.9}$$

where equations (9.8) and (9.9) detect which routes used in S_0 are affected by the failure of link $l_j \in \mathcal{L}$. Note that if OD pair $c \in \mathcal{C}$ uses the route $r_i^c(S_0)$ and the link l_j belongs to $r_i^c(S_0)$ $(u_c^{i0}(l_j) = 1)$, the route is not feasible for state S_j $(\gamma_i^c(S_j) = 1)$. The traffic flowing on this route has to be rerouted using other feasible routes.

- The constraints associated with a single route $r_i^c(S_j)$ affected by the failure of link $j \in [1, L]$ (i.e. $\gamma_i^c(S_j) = 1$) are given by:

$$\theta_i^c(S_j) \leq B\gamma_i^c(S_j) \qquad \forall c \in \mathcal{C}, \forall i \in |R^c(S_0)|, \forall S_j \in \{\mathcal{S} - S_0\}, \tag{9.10}$$

$$\theta_i^c(S_j) \leq d_i^c(S_0) \qquad \forall c \in \mathcal{C}, \forall i \in |R^c(S_0)|, \forall S_j \in \{\mathcal{S} - S_0\}, \tag{9.11}$$

$$\theta_i^c(S_j) \geq d_i^c(S_0) - B[1 - \gamma_i^c(S_j)] \qquad \forall c \in \mathcal{C}, \\ \forall i \in |R^c(S_0)|, \forall S_j \in \{\mathcal{S} - S_0\}. \tag{9.12}$$

These constraints evaluate the total traffic associated with each OD pair traversing routes affected by the link failure. This amount of traffic must be dropped from the routes in S_0 affected by the failure and rerouted using other *feasible* routes. The model decides whether to keep the routes in S_0 unaffected by the failure or to select new feasible ones in the current failure state. Note that if $\gamma_i^c(S_j) = 0$, then $\theta_i^c(S_j) = 0$; otherwise, $\theta_i^c(S_j) = d_i^c(S_0)$.

- The constraints associated with OD pair $c \in C$ affected by the failure of link $j \in [1, L]$ are given by

$$\sum_{i \in |R^c(S_j)|} d_i^c(S_j) = \sum_{k \in |R^c(S_0)|} \theta_k^c(S_j) \qquad \forall c \in C, \forall S_j \in \{S - S_0\}, \qquad (9.13)$$

$$\alpha_i^c(S_j) \leq B \sum_{k \in |R^c(S_0)|} \theta_k^c(S_j) \qquad \forall c \in C, \forall i \in |R^c(S_j)|, \forall S_j \in \{S - S_0\}, \qquad (9.14)$$

$$\sum_{i \in |R^c(S_0)|, h \in |R^c(S_j)|} (\alpha_i^c(S_0) + \alpha_h^c(S_j) v_{ih}^c(S_0, S_j)) \leq M \qquad \forall c \in C, \forall S_j \in \{S - S_0\}, \qquad (9.15)$$

$$d_i^c(S_j) \leq d^c \alpha_i^c(S_j) \qquad \forall c \in C, \forall i \in |R^c(S_j)|, \forall S_j \in \{S - S_0\} \qquad (9.16)$$

Equation (9.13) defines the total amount of traffic associated with OD pair c that is affected by the failure of link l_j. This amount of traffic has to be rerouted using other feasible routes in this failure state. Equations (9.13) and (9.14) define which feasible routes for the failure state S_j are to be used and how much traffic is to be assigned to each. Equation (9.15) upper bounds the maximum number of routes (M) that can be used by each OD pair in a given network state. Note that if none of the routes associated with OD pair c, $(r_i^c(S_0))$, are affected by failure S_j, then no traffic has to be rerouted (i.e. $d_i^c(S_j) = 0$) and the OD pair does not need to use new routes for this state (i.e. $\alpha_i^c(S_j) = 0$). Otherwise the traffic affected ($\sum_{i \in |R^c(S_j)|} d_i^c(S_j)$) has to be rerouted and the model has to choose new routes.

- The constraints associated with load on each link in the failure state S_j with $j > 0$ are given by

$$\beta^l(S_j) = \beta^l(S_0) - \sum_{c \in C, i \in |R^c(S_0)|} \theta_i^c(S_j) u_{i0}^c(l)$$

$$+ \sum_{c \in C, i \in |R^c(S_j)|} d_i^c(S_j) u_{ij}^c(l) \qquad \forall l \in L, \forall S_j \in \{S - S_0\}, \qquad (9.17)$$

where equation (9.17) defines the load on each link in each failure state S_j. The load of each link l is modified starting with the load in S_0. Traffic crossing link l for routes affected by failure S_j is subtracted out, and traffic crossing link l associated with the new routes in failure state S_j is added in.

The general routing problem uses the same objective function as our heuristic (equation (9.1)) without the third term $\Delta(S_0)$. Recall that in Section 9.1 we stated that the only routes considered feasible (to search over in optimization) are those that meet the SLA constraint. Thus we have essentially prefiltered routes that violate

SLA constraints and hence do not need the third term in our objective function. The objective function for this optimization problem is to minimize $\mathcal{F}$, where

$$\mathcal{F} = (1 - W)\mu(S_0) + W\mu(S_{\text{wst}}). \tag{9.18}$$

As pointed out in Chapter 7, ILP models can be solved using standard software packages such as that given in ref. [1].

9.3   Tabu Search for the IS–IS link weight selection problem: TabuISIS

The heuristic we present for the identifcation of the ideal set of IS–IS link weights relies on the application of the Tabu Search methodology [86], extensively described in the previous chapters. We refer to our specific implementation as *TabuISIS*.

9.3.1   Fundamental aspects of TabuISIS

We now describe the fundamental components of the TabuISIS heuristic we have designed to solve the IS–IS link-weight assignment problem.

Precomputation Step

The time taken to find the optimal solution grows exponentially with the size of the search space, thereby making the search procedure very slow. Before running the Tabu Search, we use an approximation to speed up the search procedure by reducing the size of the search space. Each solution in the search space consists of a set of routes for all traffic demands. We choose a criterion to filter out a subset of possible routes. This filtering is based on a hop-count threshold. Only routes whose hop-count is less than the threshold are deemed admissible and considered during the search procedure. This approximation is based on the intuition that TabuISIS will avoid very long routes anyway because they consume network resources unnecessarily and will lead to long end-to-end delays. Long delays for OD flows will, in turn, lead to SLA violations. Moreover, it seems pointless to waste time searching paths that do not have any serious potential as candidates. The hop-count threshold allows our heuristic to explore the search space efficiently with a reasonable computational overhead.

Care needs to be taken in selecting the hop-count threshold, which will be topology-dependent. On the one hand, choosing a small value for the threshold will eliminate many routes from consideration, including potentially the optimal solution. On the other hand, a large value of the threshold may result in a very large search space, thus slowing down the run-time performance of the algorithm. We explore this tradeoff in Section 9.5 and explain our choice of hop-count threshold.

Initial solution

The choice of the initial solution is important since it can significantly affect the time taken by the search procedure to converge to the final solution. We set the initial weight of each link to be the inverse to the link capacity – a common recommendation of router vendors [154].

Moves and neighborhood generation

Each move corresponds to perturbing one or more of the weights in the current weight set. The first step in a move is to run Dijkstra's SPF algorithm for the current weight set, generate all minimum-cost routes and compute the traffic load for each link. We then identify two sets of links – those whose loads are within a small percentage of the maximum load (heavily loaded) and links whose loads are within a small percentage of the minimum load (lightly loaded). A link is selected at random from the heavily loaded set and its weight is increased (in order to divert traffic to other paths and reduce its load). Then a link is selected at random from the lightly loaded link set and its weight is decreased. The goal is to attract traffic towards this link and potentially reduce the load on other, more heavily loaded, links. A new neighborhood is designed by repeating the above procedure.

Evaluation of each solution

Each iteration consists of generating a new neighborhood and selecting the best solution in the neighborhood. Every solution in the neighborhood is evaluated as follows. Minimum-cost routes are computed using the SPF algorithm and the traffic load on each link is determined for a given TM. This evaluation is performed when there is no failure in the network *and* for each possible link failure. Then the objective function in equation (9.1) is computed.

Diversification

This step is needed to prevent the search procedure from indefinitely exploring a region of the solution space with only poor quality solutions. It is applied when there is no improvement in the solution after a certain number of iterations. The diversification is a modification of the move described earlier. For a regular move, only one link, from each of the heavily and the lightly loaded link sets, is chosen at random. For a diversification move, several links are picked from each set. The weights of the selected links from the heavily loaded set are increased while the weights of the selected links from the lightly loaded links are decreased. This diverts the search procedure to a rather different region of the solution space, where it resumes the process of neighborhood generation and solution evaluation.

Tabu list

The Tabu list serves to remember the most recent moves made and consists of the links whose weights have been changed as well as the amount of increase/decrease applied to the corresponding link weight.

Stop criterion

The search procedure stops when a fixed pre-determined number of iterations is reached. The number of iterations is defined based on the size of the network, the computational time needed and the quality of solution desired.

In the following we do not report the pseudocode of TabuISIS since it is similar to the one shown in Figure 7.3. The analysis of its computational complexity can be obtained in a similar fashion as described in Section 7.5. TabuISIS can be implemented using any custom programming language such as C or C++ (see refs. [110] and [190], respectively).

9.4   Experimental configuration

To evaluate the presented algorithm and its impact on link loads and SLA guarantees, as well as its performance under failures, we first need to select topologies and TMs on which to assess our heuristic. We describe these now, as well as our choice of hop-count threshold for limiting the search.

9.4.1   North American and European Sprint IP backbones

For IP backbones, the failure of long-haul inter-PoP links between cities is significantly more critical than link failures within a PoP, since every PoP has a highly meshed topology. Accordingly, we consider two representative PoP-level topologies extracted from the real North American Sprint IP backbone, consisting of 16 nodes (each corresponding to a PoP) and 35 bidirectional links (Figure 9.1) and the real European Sprint IP

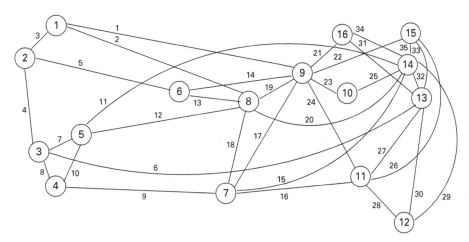

Fig. 9.1.   North American Sprint IP backbone network topology (third quarter, 2001): 16 PoPs, 35 bidirectional links (pseudo-real).

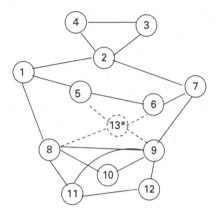

Fig. 9.2.   European Sprint IP backbone network topology: 12 PoPs, 18 bidirectional links (pseudo-real).

backbone, consisting of 12 PoPs and 18 bidirectional links (Figure 9.2).[3] We emphasize that our heuristic does not rely on the properties of any specific network topology and is therefore equally applicable to any topology or graph. According to current practices at Sprint, links are typically assigned integer weights in the range of 5 to 255 for the North American topology and in the range of 5 to 80 for the European topology. We will compute link weights in the same range using our algorithm. For ease of readability, we will sometimes refer to Sprint's North American backbone as the "NA topology," and Sprint's European backbone as the "Europe topology."

9.4.2 Traffic matrices

The second input to a link-weight selection algorithm is the traffic demands between OD node pairs, represented by a TM. Unfortunately, few measurement data have been collected on PoP-to-PoP TMs for the North American Sprint IP backbone. To evaluate our heuristic on the North American Sprint topology, we therefore use synthetic TMs, based on models derived from OD flow characteristics as observed in ref. [137]. For the European Sprint topology, we do have an exact TM at our disposal, which is described below. First, we state our two models used for synthetic TM generation.

Gravity model
This form of the TM is based on the findings in ref. [137] about the characteristics of PoP-to-PoP TMs in Sprint's IP backbone. The volume of traffic from node i to node j is selected as follows:

$$d_{i,j} = O_i e^{V_j} / \sum_{j \in [1,V]} e^{V_j} \qquad \forall i \in [1, V], \tag{9.19}$$

---

[3] We remark to the reader that both the topologies being used for the evaluation of our methodology are slightly different from the real Sprint IP backbone topologies since the level of analysis carried out may reveal insightful routing properties considered confidential by Sprint.

where O_i is the total traffic originating at node i, given by

$$O_i = \begin{cases} \text{uniform}(10,50), & \text{if } prob \in [0, 0.6) \\ \text{uniform}(80,130), & \text{if } prob \in [0.6, 0.95) \\ \text{uniform}(200,300), & \text{if } prob \in [0.95, 1] \end{cases}$$

and $prob$ is a uniform random variable between zero and one. Note that $V_j / \sum_{j\in[1,V]} e^{V_j}$ is the share of traffic originating at node i that is destined to node j. The V_j is a random number picked according to a uniform distribution between 1 and 1.5. The idea is to create three kinds of traffic demands by volume [37]. The fan-out of traffic originating at a given PoP is then determined by equation (9.19) in accordance with the observations in ref. [137].

Negative exponential

Each entry in the matrix is generated according to a negative exponential distribution with a mean value of 40 Mbps. This is one of the several other forms of TMs that we evaluated in order to validate the generality of our results. We choose this type of TM because it closely approximates the distribution of a subset of traffic demands seen in the Sprint network [37].

For Sprint's European IP backbone, we collected NetFlow data for roughly a three-week period during the summer of 2003. NetFlow was enabled on all the incoming links from gateway nodes to backbone nodes. The version of NetFlow used is called Aggregated Sampled NetFlow and deterministically samples one out of every 250 packets. Using local BGP tables and topology information, we were able to determine the exit link for each incoming flow. The resulting link-by-link TM is aggregated to form a PoP-to-PoP TM. For more details on this process, please refer to Chapter 5.

Traffic matrices exhibit strong diurnal patterns, and although they vary a lot throughout the day, we are seeking link weights that can be static for long periods of time (i.e. even though short-lived failures). In order to ensure that our link weights can perform well under any traffic conditions, we consider the worst-case traffic demand scenario. We thus extracted the peak hour (noon to 1 pm) of our TM to be used as input to our problem.

9.4.3   Reducing the size of the search space

Recall that in our precomputation step (Section 9.3) we eliminate candidate routes for each OD pair whose hop-count is above a threshold. We now describe our choice for this threshold parameter that influences the size of the search space. The choice of this threshold depends on the characteristics of the network topology. The hop-count threshold T has to be larger than the longest minimum-hop path for any OD pair across all network states S_j, $j \in [0, L]$. Otherwise, there will be no admissible route between some OD pair(s) in one or more network states. By counting the hop-counts for all routes in the North American Sprint network, we found this value to be 4; hence the hop-count threshold has to be *at least* 4.

Figure 9.3 illustrates the tradeoff between the computational overhead of the heuristic presented and the quality of the solution for the North American network with different values of T. Link weights are selected without considering failures ($W = 0$). We have verified that the behavior is similar for other values of W. Figure 9.3(a) shows the number of iterations needed by our heuristic to reach the final solution. The number of iterations grows exponentially fast – from 2000 for $T = 4$ to 16 000 for $T = 12$. This shows that the parameter T is necessary for enabling our heuristic to explore the search space efficiently with a reasonable computational overhead.

Figure 9.3(b) plots the maximum link load in network state S_0 for the link-weight set chosen by the heuristic presented. As expected, the quality of the solution improves as the threshold grows. The problem with small threshold values is that most routes are eliminated from consideration, and, having too few routes to choose from, our heuristic

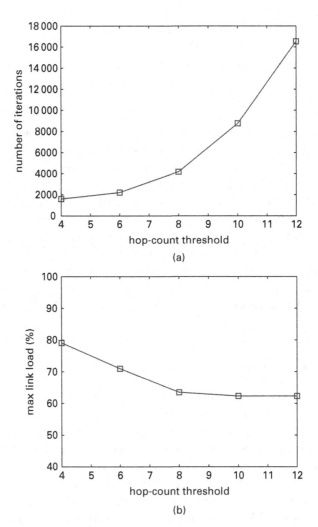

(a)

(b)

Fig. 9.3.   Effect of hop-count threshold on (a) the number of iterations needed to reach the final solution and (b) the associated maximum link load. Case $W = 0$.

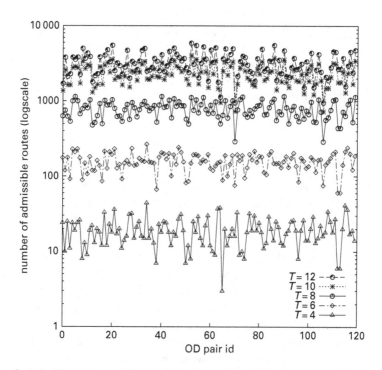

Fig. 9.4.   Number of admissible routes per OD pair for various values of T.

yields a solution whose maximum link load is high. By increasing the threshold from $T = 4$ to $T = 8$, we see a large reduction in maximum link load: from roughly 80% to 60%. Consequently, on one hand, we do not want too small a threshold as this can degrade the quality of the solution found by our heuristic. On the other hand, we see that for values of $T > 8$, there is little further gain from increasing the threshold.

Figure 9.4 explains the above observations. It shows the number of admissible routes per OD pair for different values of the hop-count threshold (T). The number of admissible routes for each OD pair is roughly 30 for $T = 4$ and roughly 200 for $T = 6$. We know from Figure 9.3 that there exist good solutions among the many solutions being eliminated by such low thresholds. For $T = 8$, the number of solutions retained increases ten-fold to about 1000. Consequently the maximum link load for the final solution found by our heuristic could be reduced to 62%. For higher values of the threshold, the number of admissible routes increases at a slower rate – from 1000 for $T = 8$ to 3000 for $T = 12$. But at this point, the search space is so big that any further increase in its size causes the computation time of our heuristic to grow exponentially. Based on these results, we pick the hop-count threshold to be 8.

9.5   Applications

An algorithm that selects link weights can also be used for other traffic engineering tasks. We now show how our heuristic can be used by operations personnel to assess the impact of design choices (such as maximum allowed link weight) and for tasks such as topology design. Topology changes lead to changes in the routing domain, which in turn lead to changes in network performance metrics. An application tool using our heuristic as its engine can rapidly assess the impact of a topology change on the performance metrics through its impact on routing. We use Sprint's networks in a case study to illustrate these applications.

9.5.1   Selecting the range of allowable link weights

The original specification of IS–IS allowed for link weights in the range $[0, 63]$ [154]. Recent modifications to the protocol have increased this range to $[0, 2^{24} - 1]$. Network operators typically choose weights in $[0, 255]$. But the tradeoffs of increasing or decreasing the range of link weights are poorly understood. For example, a carrier might want to know by how much they have to increase the maximum link weight before any gain is realized from the broader range. By allowing an operator to change the maximum weight value as an input parameter to a tool, this question can be explored.

We examine the case when the maximum weight allowed is successively increased. Figure 9.5 shows how the maximum link load varies with the allowable range of link weights in the Sprint network for the negative exponential and gravity model TMs. In each graph, the solid line shows the maximum link load in the absence of failures and the dashed line shows the maximum load under any single link failure. Interestingly, the maximum link load does not decrease uniformly with an increase in the maximum weight value allowed. Increasing the maximum weight has no impact over a large range of values and then suddenly we see a sharp change. It is surprising that this curve is similar to a decreasing step function. For example, in the case of the negative exponential TM, the sharpest reduction happens around a maximum weight of 60. After that, there is no reduction until a value of 254 is reached, implying that if a carrier wants to reduce the maximum link load beyond what is achievable with a maximum weight value of 60, they will need to increase the weights all the way to 254. We suspect that, in general, the particular value (e.g. 60) at which these sharp transitions occur will depend

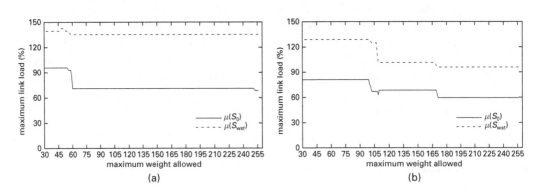

Fig. 9.5.   Variation in maximum link load with maximum link weight allowed; North American Sprint topology. (a) Negative exponential TM; (b) gravity model TM.

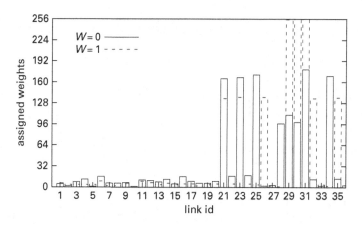

Fig. 9.6.   Link weights selected for the gravity model TM.

upon the topology and the traffic mix. We limit our conclusions here to the observation that the impact of the maximum weight value on maximum link load is that the maximum load value *decreases* in broad discrete steps for *increasing* maximum weight value. This insight can be useful for operators when they decide to alter the range of link-weight values used by their routing protocols.

9.5.2   Aiding topology design

In this section we show how a network operator can use the output of our weight selection algorithm to flag potential problems or limitations in the topology of a network. By examining the set of weights selected, as in Figure 9.6, an operator can observe that two links have weight settings that are more than three times the value of any other link and are also near the maximum allowed. This suggests that those links are protecting a path for one of their end nodes. This is because by setting the weights so large, most nodes (except the immediate neighbors of those links) are discouraged from using the link since using it would yield a high-cost path. Some paths may be "protected" or "saved" for particular users in this way if there is a small number of (or no) alternate paths. Viewing the ensemble of all link-weight settings can thus serve as a warning that there may be topological limitations in the number of paths available to a given OD pair, either during normal operation or during failure. Such a topology limitation can affect the maximum load level(s), and even the best of routing schemes cannot overcome a topology limitation.

This warning is indeed true in our topology. We can see in Figure 9.1 that links 23 and 25 are node 10's only connections to the rest of the network. When one fails, the other must absorb all of the fail-over traffic. Let us revisit Figure 9.5(b). At the maximum weight value of 255, the maximum load was 60% under normal conditions and 92% under failures. Looking inside our computations, we saw that the maximum link load over all failures occurs when link 25 fails. Indeed, when this happens link 23 becomes the most heavily loaded link. It is impossible to do anything about this because there

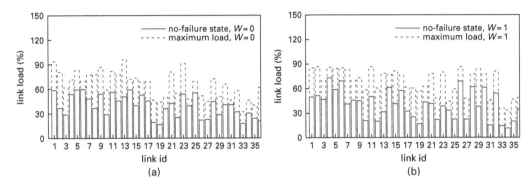

Fig. 9.7.   Impact of adding a new link in the North American Sprint topology; gravity model TM.
(a) $W = 0$, (b) $W = 1$. The solid bars show the loads in the absence of failures. The dashed
bar for each link shows the maximum load on that link for any link failure.

are no alternate paths. Hence, even though the results in Figure 9.5 were computed for
$W = 0$, in this case considering $W > 0$ would not help and thus would not alter these
maximum loads. This is an important point that illustrates that considering failures does
not help when there are no alternate paths.

A heuristic such as the one presented allows an operator to assess rapidly the
hypothesis that there is a topology limitation. We can do this simply by adding a
link into our topology, near the problematic portion of the graph, running TabuISIS
to compute the new set of link weights, and to compute the resulting worst-case loads
during failure and normal operation. A heuristic that is automated and quick allows a
network designer to consider numerous topology changes and receive fast feedback,
along with specific performance metrics quantifying the impact of such a topology
change.

As an illustration, we added a link between nodes 10 and 16 in the North American
Sprint topology. The link weights that our heuristic assigns in this new topology are
shown in Figure 9.7. We see that a second set of links (26, 29, 30, 31, 32 and 34 for $W =$
1) are now being given large weights as well. Figure 9.7 shows the distribution of link
loads for this modified topology with the gravity model TM, for two sets of link weights
(corresponding to $W = 1$ and $W = 0$). Note that if we do not consider failures ($W = 0$),
the maximum load under normal conditions (60%) and the maximum load under the
worst failure (92%) are the same as in the original topology. However, if we do consider
the failure states ($W = 1$), then the maximum load under failures can be reduced to
87%. This is possible since there are now two alternative links protecting link 25, and
our heuristic is able to select weights that balance the fail-over traffic between these two
paths.

Hence adding a link did not impact the performance in the no-failure mode, but
instead it helped make the topology more robust to failures. This implies that the
topology limitation we found was one that was dormant during normal operation but
dominant under specific failure scenarios. That our heuristic helped reveal this illustrates
a nice application of the presented approach.

Table 9.1. Comparing topology designs through their impact on routing

Metrics	Original topology	With extra PoP	Extra PoP and upgraded links
SLA	20 ms	15 ms	15 ms
Worst-case load, no-failure mode	49.4%	36%	38%
Worst-case load, failure mode	70%	70%	42.1%

9.5.3   Designing the SLA offered

One of the key questions a carrier asks when designing a service is "What is the best SLA I can offer given my network infrastructure, and what changes could change the relevant delay guarantees?" Using the European topology and TM, we show how TabuISIS can be used to answer this question.

First we consider the impact of adding a new PoP in the topology.[4] We added a new PoP (PoP id 13∗ in Figure 9.2) that is attached with four high-speed links to PoPs 2, 7, 5 and 6. The impact of this topology change on our three performance metrics is given in Table 9.1. With this additional PoP we were able to reduce the SLA by 5 ms, from 20 ms to 15 ms. The link utilization is reduced from 49.4% to 36% for the no-failure scenario, while a 15 ms SLA is simultaneously satisfied (see Table 9.1). The addition of a new PoP thus improved both the SLA and maximum load under normal operation. We can also see, however, that it did not improve the maximum load under failures.

This is most likely due to the heterogeneity of link types, and in particular their low-speed links. To test this hypothesis, we considered a second topology change, namely that of upgrading the two low-speed links by doubling their capacity. The maximum link load for any failure case dropped dramatically from 70% to 42.1%. This new routing configuration incurred a minor 2% degradation in maximum load for the no-failure scenario. Using our heuristic we are able to realize when optimization limits have been reached (e.g. due to current topology limitations) and evaluate "what if?" scenarios (e.g. alternative topologies) quickly.

9.6   Lessons learned

In this chapter, we extended the problem of link-weight assignment to include the important practical requirement coming from SLA bounds that ISPs guarantee to their customers. The presented solution further addresses a second practical constraint, that of finding weights such that the network will perform well during short-lived failures.

---

[4] It is possible to formalize another optimization problem to identify the optimal placement strategy. Our intent here is not tackle this problem, but rather to illustrate how TabuISIS can aid operators in the rapid assessment of the impact of different topology design choices.

In this way, operations personnel avoid having to change weights for short-lived failure events, which are known to occur frequently. To the best of our knowledge the presented heuristic, named TabuISIS, is the first to consider short-lived link failures and SLA constraints for link-weight selection.

The evaluation section looked at the problem from many different angles. In comparing the performance of our heuristic to an optimal lower bound computed using an ILP formulation, we found that our solution was within 0–10% of optimal, over a variety of scenarios considered. We compared our solution to the link weights currently deployed today and found that we could meet the same performance metrics achieved today (delay and load under no-failure conditions). This shows that our method yields practical solutions. We showed that we can improve upon today's solutions, which were not designed to work well under failure. The maximum load on a link during any single-link failure can be reduced by as much as 50% at the cost of a 10% degradation in maximum load during no-failure modes. We illustrated that the SLA for the North American network can be reduced from 45 to 40 ms at the cost of a small increase (6%) in the maximum link load during no-failure modes. In our case study, we used our heuristic to surmise that the inability to reduce the SLA bound further was not due to a limitation of optimization, but rather to a limitation in the topology. With the addition of a single extra node, we were then able to reduce the SLA from 20 ms to 15 ms for the European network.

In this chapter, we have studied the optimization of the IGP link weights for a local AS in perfect isolation, assuming that hot-potato routing was absent, e.g. phenomenon that occurs when there are multiple egress points to reach a destination. In this regard we have used point-to-point TMs for illustrative purposes and because this covers a large portion of an ISP's traffic. In Chapter 10, we will extend the methodology presented here to incorporate hot-potato routing and we will study the interaction between IGP and BGP routing protocols. In order to carry out this analysis, we will extend the concept of a point-to-point TM to a point-to-multipoint TM and consider the logical topology at the router level.

10 Measuring the shared fate of IGP engineering: considerations and take-away

The Internet is made of many separate routing domains called Autonomous Systems (ASs), each of which runs an IGP such as IS–IS [154] or OSPF [147]. The IGP handles routes to destinations within the AS, but does not calculate routes beyond the AS boundary. Internet Gateway Protocol engineering [79] (or traffic engineering or IGP optimization) is the tuning of local IS–IS or OSPF metrics to improve performance within the AS. Today, IGP engineering is an ad-hoc process where metric tuning is performed by each AS in isolation. That is, each AS optimizes paths within its local network for traffic traversing it without coordinating these changes with neighboring ASs. The primary assumption behind such an assertion is that there is sufficient separation between intra-domain and inter-domain routing.

Beyond the AS boundary, the choice of AS hops is determined by the BGP, [169]; BGP engineering is a less developed and less understood process compared to IGP engineering. In addition to whether there is a physical link between two ASs over which routes and traffic can flow, there are several BGP policies that determine which inter-domain paths are exposed to a neighboring AS. Business peering policies can directly translate into which routes are exported to each AS [83, 188]. After all these policies are applied, the remaining feasible paths are subjected to the "hot-potato" routing policy. Hot-potato routing occurs when there are multiple egress points to reach a destination. BGP inside that AS will pick the egress point which is the "closest," i.e. has the smallest IS–IS or OSPF cost from the traffic ingress point. The rationale behind such a choice is that the AS in question does not consume resources to route traffic destined to non-customers, offloading it to the best next provider as soon as possible; the destination's ISP should expend the greatest cost transitting such traffic.

The first problem we consider is the case when IGP metrics change, resulting in a change in the closest egress point (hot-potato routing), which in turn causes traffic to shift within the AS towards that new egress. Typically, IGP link metrics are selected to achieve a network-wide engineering objective such as the minimization of network congestion, i.e. the minimization of the maximum link utilization across all links constituting the logical topology. As explained in Chapter 9, one of the inputs to that selection process is a TM that represents the volume of traffic flowing between every

Portions reprinted, with permission, from Agarwal, S., Nucci, A. and Bhattacharyya, S. (2005). "Towards Internet-wide network management." *IEEE Infocom*, Miami, FL, March, 2005.

ingress–egress pair within the AS [137]. When link metrics are set to new values, the IGP cost to reach various egress points from an ingress point may change. This may lead BGP to recompute the best egress point for certain destination prefixes, thereby leading to traffic shifts. In other words, the final flow of traffic in the network is different from what was considered during the IGP optimization stage. In this chapter, our goal is to quantify such an effect using real data from an operational network. Certainly, this situation can be preempted by proactively considering hot-potato routing during the IGP metric selection process. However, that comes at a cost – it can increase the running time of algorithms chartered with the NP-Hard IGP optimization problem. Therefore it is important to understand and quantify the extent to which this is a real issue.

The second problem we examine is how traffic to neighboring ASs shifts due to changes in the local ISP's IGP metrics. Local IGP tuning causes egress points to change for some traffic flows, thereby changing the ingress points for neighboring ASs. This can lead to sub-optimal network performance for those ASs. Their existing IGP metrics are no longer optimal because they were tuned for the previous TM. There may be high traffic load on some links. As a result, the IGP engineering of one AS has impacted other ASs. The goal of this chapter is therefore to measure, using operational network data, how much traffic to neighboring ASs shifts between multiple peering points.

In this chapter, we use the same heuristic presented in Section 9.3 (i.e. TabuISIS), to carry out the IGP optimization and modify it to consider inter-domain changes. We use real data gathered from the operational Sprint IP network to drive this heuristic and analyze potential outcomes. These data include the entire IP router-to-router level topology, existing IS–IS metrics, average link propagation delays, link capacities, BGP routing tables and traffic flow information from the European Sprint IP network. While the results presented are specific to a single ISP, they point to a significant interaction between local IGP engineering and inter-domain routing policies.

The remainder of the chapter is organized as follows. In Section 10.1 we describe the problem further using an illustrative example and giving details of route selection. The required measurements and follow up analysis are described in Section 10.2 and Section 10.3, respectively. We present results on the scope of the problem, the impact on IGP engineering and the impact on neighboring ASs in Section 10.4. The chapter concludes with Section 10.5.

10.1   Problem description

Each AS on the Internet is a closed network of end hosts, routers and links that runs its choice of intra-domain routing protocol or IGP. The IGP of choice today is typically a link state protocol such as IS–IS or OSPF operating on the network topology annotated with link weights. The IGP determines the path a network entity such as a router inside the AS needs to send traffic across in order to reach another router in the same AS. While there may be multiple possible paths between the ingress router and the egress router, typically the path with the lowest cost (sum of link metrics) is chosen. For

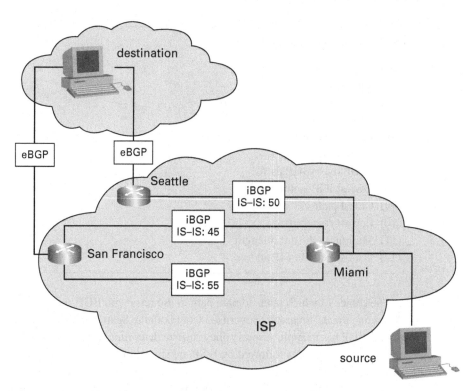

Fig. 10.1. Example of hot-potato routing. Note that iBGP stands for *internal BGP*, which is used to exchange routing information across all routers within the same AS. Conversely, eBGP stands for *external BGP*, which is used to exchange information between routers belonging to different ASs.

example, consider Figure 10.1, where a destination is multi-homed to the ISP in two locations. Traffic entering the Miami PoP (Point-of-Presence) is destined to the host denoted as destination. BGP dictates that traffic from Miami to the destination needs to exit the network through the San Francisco egress point. Of the two paths, IS–IS picks the cheaper one, with cost 45.

IGP engineering or traffic engineering is an important task in the network operation of most large ASs. It can be applied for a variety of goals. These include avoiding congestion in the network, maintaining SLA delays and reducing the detrimental impact of link failures on congestion and delay. In IGP engineering, link metrics are changed to alter the relative cost of paths. In the above example, if the path with IS–IS cost 45 experiences heavy load, IGP engineering may increase its cost to 55, to balance the traffic load between the two paths from Miami to San Francisco. In reality, the optimization process considers the entire AS topology and all the flows traversing it and does a joint optimization. The traffic load is represented as a TM showing all the ingress and egress points and the demand of traffic between them, as discussed in previous chapters.

Box 10.1. BGP route selection

(1) Largest weight
(2) Largest local preference
(3) Local paths over remote paths
(4) Shortest AS-path
(5) Lowest origin type
(6) Lowest MED
(7) eBGP over iBGP paths
(8) Lowest IGP metric
(9) Oldest path
(10) Lowest router id
(11) Minimum-cluster id length
(12) Lowest neighbor IP address

The choice of which inter-domain path to use rests with BGP, a path vector protocol. Assuming the destination is advertised from both the Seattle and San Francisco peering points, BGP has to apply several policy rules to determine which egress point traffic will go to. The policy rules followed by BGP are shown in Box 10.1. Let us examine a few of them. For more details on what the different metrics mean and how they are used, see ref. [97]. Rule (4) in Box 10.1 determines that when BGP needs to make a choice among multiple paths and the decision process has reached that rule, then the path with the smallest number of ASs will be preferred.[1] A Multi-Exit Discriminator (MED) [185] is used to control traffic on multiple links between a customer and a provider AS. If the destination advertised different MED values on the routes between the two links, the route with the lowest MED value will be preferred. However, if the destination can be reached from both Seattle and San Francisco with the same AS-path length and MED values, and there is no local policy preferring one over the other, the hot-potato policy will apply. This corresponds to rule (8) in Box 10.1. BGP will choose the egress which has the lowest IGP cost from the ingress. So, for traffic entering Miami, the cheapest exit to the destination is San Francisco, with an IS–IS cost of 45. In small ASs, with a large number of connections to other ASs through the same router, this IGP cost will be practically indistinguishable between multiple BGP routes. However, for ISPs spanning a large geographical area, there can be a significant range in the IGP cost.

In this example, if, due to IGP engineering, the cheapest IS–IS cost from Miami to San Francisco changes to 55, then BGP will change the egress to Seattle for this destination. As a result, both the performance of the local AS and that of the destination AS will change. For the local AS, the TM will now be different because the egress point is different. If the IGP metric optimization process did not account for hot-potato routing

---

[1] This is why AS-path prepending can serve to detract traffic from particular links. A network provider can advertise a BGP prefix using an inflated AS-path, which will lead to it being a lower preference route by BGP, when compared against other shorter AS-paths

(as the one presented in Chapter 9), it will not expect the shift in egress. As a result, it may end up with more traffic load on the Miami to Seattle links than expected. Secondly, the destination AS will experience a different traffic load. Traffic that was ingressing its network in San Francisco is now ingressing in Seattle. Its TM has changed, and now the neighboring AS may also have to retune its IGP metrics.

It is important to measure the extent of this interaction using operational network scenarios to determine if further work is needed to solve global network optimization and coordination of local optimization among ISPs. To this end, we want to quantify two effects. (1) We want to identify by how much the maximum link utilization in the local AS changes as a result of hot-potato routing. IGP engineering optimizes metrics to reduce the maximum link utilization across the network because it makes the network more resilient to link failures and sudden traffic surges. If these new metrics cause hot-potato shifts that are not accounted for in the optimization process, the final link utilizations can be very different. The increase in maximum utilization is a measure of how important this problem is to the operation of an AS. (2) We want to find out how neighboring ASs are impacted. A large volume of traffic shift on the peering links to neighboring ASs can alter the internal link utilizations of those ASs.

10.2   Collection of all required inputs

Our analysis methodology and data have several components. We use the same Tabu-ISIS heuristic presented in Section 9.3 to lead the IGP link weight optimization process. In this context we do not consider the presence of logical failures in our framework (obtained by setting $W = 0$ in the objective function used by TabuISIS, see Equation (9.18)). In order to carry out a proper analysis, we collect inputs such as (i) a more detailed IP router-to-router topology, (ii) traffic data to build an accurate TM, (iii) BGP routing data and (iv) SLA constraints. We now describe each of these in detail.

The IP router-to-router topology is a list of every link inside the AS that the IS–IS or OSPF protocol encompasses. Each link is annotated with the source router name and the sink router name. Each link is characterized by its delay and capacity. We obtain the topology and the link capacities from the router configurations of all the routers in the Sprint IP network. The link capacity is specified by the router line card type or a lower configured limit. We obtain the delay by performing extensive measurements on the Sprint IP backbone by exchanging ICMP messages between routers. The current IGP metric settings are also in the router configurations. The SLA constraints are also needed and set to 50 ms for the North America Sprint IP backbone and 20 ms for the European Sprint IP backbone. This is a list of the average delay guarantees contracted with customers between different cities in the Sprint IP network.[2]

---

[2] An overview of the SLA for the Sprint IP network may be found at http://www.sprintworldwide.com/english/solutions/sla/

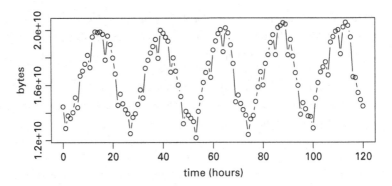

Fig. 10.2.   Total traffic entering Europe (one-hour bin).

We remind the reader that the TabuISIS heuristic optimizes for lowest link utilization. The utilization is calculated by considering the shortest paths in the network[3] based on the current IGP link weights and the input TM. In this context, the TM has been extended to be a point-to-multipoint TM. Each row represents a flow, defined by the ingress router, traffic volume and all the possible egress routers.

To build this TM, we use traffic data and BGP routing data. The methodology followed to assemble the TM itself, as well as more details on the data itself, have been provided in Sections 5.1.2 and 5.1.4. We aggregate the measurements into one-hour intervals. Due to hardware limitations of the operational routers, Cisco NetFlow could only be enabled on a fraction of ingress routers. We have NetFlow traffic information from all the ingress routers in the European portion of the Sprint network. Thus we can only build a partial TM. To work around this limitation we constrain our study of IGP optimization by allowing TabuISIS only to change the IGP metrics of the European links. We believe it would unfairly skew our results to change link metrics in the USA without the ingress traffic. However, the heuristic still considers the full network topology and exit points to capture all the hot-potato dynamics. This is still a sizeable problem – there are over 1300 links in the entire topology, 300 of which are in Europe. This approach of focusing on one continent of the network allows us to examine a scenario where we have complete network information and does not affect the validity of our findings for that portion of the network.

For the European NetFlow data, we have three months of ingress traffic data at our disposal. We need to pick a time period out of these three months to build a TM. In Figure 10.2, we plot the total volume of ingress traffic in bytes over a typical five-day period around February 12, 2004. As expected, there is a strong diurnal pattern. When optimizing IGP, network operation engineers typically consider peak-time utilization of the network instead of off-peak time. Their objective is to bound the worst-case

---

[3] Note that, in the Sprint IP network, the routers are configured to use ECMP. Between a single source router and single exit router, if multiple IGP paths exist with equal IGP cost, then traffic is split equally along both paths. This equal traffic split is performed at each next-hop instead of in an per path fashion. For ease of implementation, we use a per-path load balancing. Since we wish to model what occurrs in the network accurately, we account for ECMP in the heuristic.

performance of the network. Thus, we pick a one-hour window during one of the daily peaks. We have considered multiple one-hour peaks and the results are similar. Thus we present results from one representative period for conciseness.

This traffic data now comprise a point-to-point TM, where each flow entry gives the ingress Sprint router, the destination prefix and the number of bytes. We need to convert this into a point-to-multipoint TM, which lists the ingress Sprint router, all the candidate egress routers and the number of bytes. The candidate egress routers are all those that can reach the destination prefix.

For the same one-hour time period, we have BGP routing data from the Sprint IP network. We operate a software BGP router that connects to over 150 routers in the Sprint IP network. This collection router participates as a route reflector in the iBGP route reflector mesh [185] inside the Sprint IP network. Thus it sees all the candidate BGP routes that have reached the end of rule (7) in Box 10.1. We use a snapshot of the BGP routing table during the one-hour time period, which lists all the candidate egress routers for every destination prefix. We correlate this with the traffic to obtain a point-to-multipoint TM.

In our data set, there are over 30 700 entries in this TM. To reduce the processing time for our analysis, we consider a subset of this data. We consider the largest flows that collectively contribute at least 80% of the total traffic entering Europe. As has been noted in prior work [37, 157], the elephant and mice phenomenon exists in our data set – a few destination prefixes sink the majority of traffic. By considering the largest subset of traffic, we reduce the processing time significantly. This reduces our point-to-multipoint TM to 530 entries. For the remainder of this chapter, we treat this as the full input data set. We revisit this issue of running time in the conclusions.

10.3   Analysis

In this work, we want to study two issues. First we want to know how badly IGP optimization suffers when considering a point-to-point TM versus considering hot-potato shifts using a point-to-multipoint TM. Secondly, we want to know how traffic to neighboring ASs changes as a result of local hot-potato shifts during IGP engineering.

For the first issue, we operate TabuISIS in two modes: Traffic Engineering (TE) and BGPTE. TabuISIS-TE ignores hot-potato routing; it assumes the egress points do not change when IGP metrics change. It uses the egress points that the input state of the network would use; in essence, it is using a point-to-point TM. When running the heuristic in this mode, TabuISIS-TE performs exactly the same search executed by the TabuISIS presented in Chapter 9. We use the final output metrics from this mode to evaluate what the performance of the network would be without hot-potato shifts. We also evaluate how the network would actually perform with these new metrics after re-evaluating the final egress points. TabuISIS-BGPTE accounts for inter-domain traffic shifts by recalculating the egress point choice for every flow. It does this BGP route calculation in every iteration of the heuristic when evaluating a candidate set of IGP metrics (i.e. step named "Evaluation of each solution" in Section 9.3.1). We evaluate the performance of these metrics compared with the previous mode of operation. To study the impact on other ASs, we consider how traffic on links to large neighbors changes between the original IGP metrics and the final optimized IGP metrics.

10.4   Results

Based on the heuristic, the full Sprint IP network topology of over 1300 links, actual network link delays, actual SLA constraints and the point-to-multipoint TM that we generate from actual network data, we present results on how IGP engineering interacts with inter-domain routes and traffic.

10.4.1   Scope of problem

We begin by quantifying the extent of the hot-potato interaction problem, in terms of the number of prefixes and the amount of traffic affected. Such an effect is bound to be more visible when prefixes can be accessed through a large number of paths. In particular for our study, the destination prefixes that are reachable from multiple PoPs or egress cities in the Sprint topology are those that should be more vulnerable to hot-potato shifts.

In Figure 10.3, we plot the number of exit PoPs that prefixes in a typical routing table have. A typical iBGP routing table on a router inside the Sprint network will have about 150 000 prefixes. Each prefix may have multiple routes, but after applying BGP route selection up to rule (8) in Box 10.1, we are left with the candidates for hot-potato selection. Figure 10.3 considers only these remaining candidates for each prefix. We see that for about 40% of prefixes, there is only one egress point. That is, no matter how much IGP costs change inside the Sprint AS, those prefixes will always exit through the same PoP. However, for the remaining 60% of prefixes, there are two or more candidate routes where the IGP cost can play the determining factor in route selection. We see that for a significant percentage of routes, over six different exit cities exist! After examining these data more carefully, we have found that the prefixes with over six PoPs are behind large ISPs or "peer" ISPs. Typically, peer ISPs filter out MED values and BGP communities that modify BGP local preferences when receiving routes from each other.

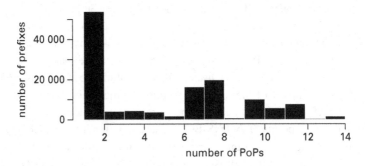

Fig. 10.3.   Distribution of destination prefixes by exit PoPs, February 12, 2004.

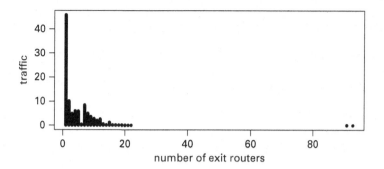

Fig. 10.4.   Distribution of European traffic by exit routers; February 12, 2004.

This is in accordance with their business agreements, which are different from the more common customer–provider relationship [83]. Thus, destinations behind large ISPs are more susceptible to hot-potato routing because of this policy to filter out certain BGP attributes. This leads us to consider the impact of IGP engineering on neighboring ASs, which we present later in this section.

In Figure 10.4, we plot the analog of Figure 10.3 in terms of traffic. We show the distribution of traffic entering the European portion of the network against the number of possible exit routers. As in Figure 10.4, we only consider the BGP routes where the next selection rule is IGP cost. We see that for about 45% of traffic, there is only one network egress point. However, the remaining 55% of traffic is susceptible to hot-potato changes. Examining these data further, we find that the traffic with around 90 exit routers is actually network management traffic for the Sprint network itself.

IGP changes can occur for a variety of reasons, including link failures, the addition of network links and IGP engineering. IGP changes in one part of the network can potentially cause BGP route selection to change across the network and thus cause large amounts of traffic to shift. Given the large number of prefixes and volume of traffic that are susceptible, it is now important to determine the extent of this interaction using operational network scenarios.

10.4.2   Impact of hot-potato shifts on IGP engineering

As described in Section 10.1, IGP engineering that uses a point-to-point TM and ignores inter-domain traffic shifts results in a simpler optimization algorithm with faster running time and less data collection constraints. While the resulting new IGP metrics may be optimal for the point-to-point TM, in reality they may cause BGP to recalculate egresses, causing them to be suboptimal. We want to measure how high the final traffic link loads can be.

We ran TabuISIS-TE, which ignores hot-potato shifts. In Figure 10.5, the points marked by crosses show the expected utilization of each link in the European topology with these new metrics. We sorted the links by utilization, which is the ratio of traffic flowing on a link to the capacity of the link. For ease of presentation, we show the 50 most utilized links out of the 300 links in Europe. For good network performance,

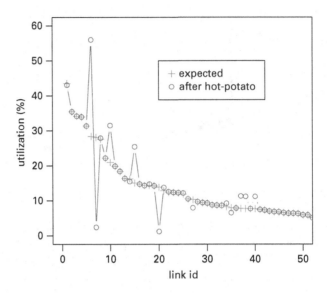

Fig. 10.5. Link utilization from TabuISIS-TE.

we want the highest link load on any link to be as low as possible, i.e. to minimize the effect of bottleneck links. TabuISIS-TE has optimized the network and expects the final worst-case link utilization to be 43.5%, which is on a high-capacity link from router $R10$ to $R11$. This corresponds to the first left point on the graph (link id 1). However, if we were to install these new metrics on the network, hot-potato shifts could result in different utilizations.

We re-evaluate these new IGP metrics while allowing BGP to change the exit points as needed, and show the resulting link utilizations as circles in Figure 10.5. We now see that the solution is actually significantly worse in reality – the highest utilization is 56.0%, on a low-capacity edge link from router $R4$ to $EX2$. This link was expected to be only 28.0% utilized. Also, 3.7% of the total volume of European traffic changed the exit point, which on individual links resulted in high utilization. Thus, in this data set, ignoring hot-potato shifts in IGP engineering resulted in a set of IGP metrics that have 12.5% higher network congestion.

Now we want to examine how much better IGP optimization can do if it uses a point-to-multipoint TM and recalculates BGP routes for every IGP metric change it considers. We ran TabuISIS-BGPTE using the same data set but allowing it to account for hot-potato shifts. The output set of IGP metrics causes the link utilizations shown in Figure 10.6. We see that now the maximum link load across the European links is 36.0%. The low-capacity edge link that suffered in the previous case now has a load of only 29.2%. Thus, in this data set, modeling hot-potato shifts in IGP optimization resulted in a set of final IGP metrics that have 20% lower maximum link utilization than when ignoring BGP route recalculation. Although the results presented are clearly related to the specific topology being used, similar behavior holds for other network topologies. Some networks will experience slightly better performance, while others would even suffer more.

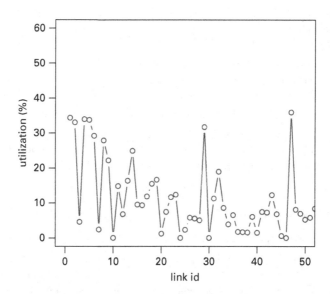

Fig. 10.6. Link utilization from TabuISIS-BGPTE.

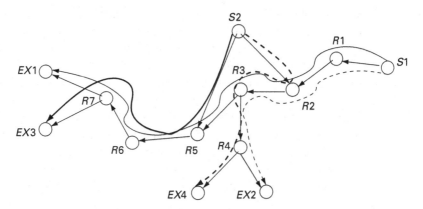

Fig. 10.7. Example of flow shifts when using the European topology.

By allowing the optimization heuristic to re-evaluate BGP routes in every iteration, we are allowing it to take advantage of hot-potato shifts to reduce link utilization further. To explain this, we present an example of traffic shift in Figure 10.7. The circles depict routers in the Sprint network; $S1$ and $S2$ are ingress routers; $EX1$ to $EX4$ are egress routers; the remaining $R1$ to $R7$ are internal routers. Recall that, in Figure 10.5, TabuISIS-TE expected the link $R4$ to $EX4$ to have only 28.0% utilization, but, due to hot-potato shifts, it ended up with 56.0% utilization. This is because, in the input TM, there are 11 flows going from $S1$ to $EX1$ and from $S2$ to $EX3$, as shown by the two solid lines in Figure 10.7. The original IGP metrics from the Sprint router configurations resulted in a hot-potato choice of sinking this traffic at $EX1$ and $EX3$, respectively. However, TabuISIS-TE reduced the link metric on $R3$ to $R4$. It continued to expect traffic to be sunk at $EX1$ and $EX3$. However, the destinations are multi-homed – one destination is behind both $EX3$ and $EX4$, and another is behind both $EX1$ and $EX2$. BGP re-evaluates its routes and picks $S1$ to $EX2$ and $S2$ to $EX4$ as the new routes. As a result, the link utilization of $R4$ to $EX4$ increases unexpectedly to 56.0%. When the IGP optimization was allowed to recalculate BGP routes at every iteration in TabuISIS-BGPTE, it discovered a set of IGP metrics where BGP selected a path for $S2$ to $EX3$ and a different path for $S1$ to $EX2$. This drove down the utilization of $R4$ to $EX4$. In essence, the heuristic discovered a solution where hot-potato-induced shifts reduced the link utilization.

10.4.3   Impact of IGP engineering on neighbors

Unfortunately, the downside of using heuristics that increase local AS performance by inducing hot-potato traffic shifts is that they may affect traffic going to neighboring ASs. In the previous example, $EX1$ and $EX2$ are ingress points for a neighboring AS, Y. Traffic has shifted between these two points due to the IGP engineering of the Sprint AS. As a result, the TM for AS Y is different, and its link utilizations will change. A significant enough change may require it to redo its own IGP engineering.

We have calculated how the traffic shifts for various large neighbors of the Sprint network. For example, a large peer ISP, A, connects to the Sprint network in seven locations. In Figure 10.8, we show for each prefix how many locations it is announced at; ISP A announces most prefixes at all seven peering points. In Figure 10.9 we show how much of the measured European traffic exits at each of these seven locations to ISP A. The dark bars show the distribution of traffic in the original network setting (i.e. with the IGP metrics from the router configurations). All seven locations are in the USA side of the topology, and routers 4 and 5 are on the East Coast, closest to the European continent. Since these two exit points are cheapest in terms of IGP cost from Europe, they sink the majority of traffic. The lighter bars show how the traffic to these two points would change after we apply the new IGP metrics proposed by TabuISIS-TE, when allowing it to account for BGP exit point shifts. We see now that router 5 sinks more traffic than router 4. In fact, about 22% of the traffic from Europe to this ISP changes its exit point. Due to privacy concerns, we cannot reveal the identity of the neighbor, nor

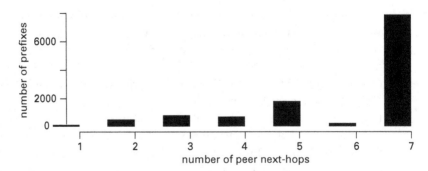

Fig. 10.8.   Prefix distribution to neighbor AS, A.

the volumes of traffic involved. Even though the BGP-aware IGP optimization reduced link utilization by 20%, it came at the cost of changing traffic going to a neighboring AS by 22%. As we described in Figure 10.3, traffic to "peer" ISPs is particularly vulnerable. This shift in egress points can potentially have a detrimental impact on the traffic distribution inside the neighboring network.

The extent of such an impact, however, depends significantly on the locations that an AS connects to Sprint and on the BGP policies that govern those connections. Consider another ISP, B, that has the egress traffic distribution in Figure 10.10. Of the eight peering locations, this ISP has two in Europe, five in the USA and one in Asia. In the extreme case, IGP costs to reach a destination will mimic relative geographic distances due to the SLA delay constraints. So, for our measured traffic entering Europe, any destinations in ISP B will naturally exit through one of the two European peering points. However, the Sprint router configurations have a BGP local preference for one of these peering points. As a result, no matter how much IGP metrics vary, hot-potato routing will not come into play because of this local BGP policy. Note that the small amount of traffic that appears to have shifted to this exit router comes from another AS.

As a third example, consider the traffic distribution for a peer ISP, C, shown in Figure 10.11. This large ISP has 17 peering locations, one of which is in Europe and

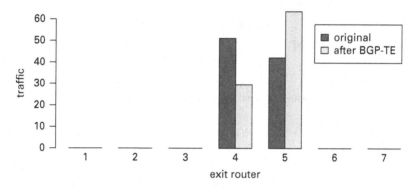

Fig. 10.9.   Traffic distribution to neighbor AS, A.

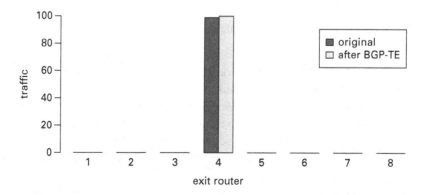

Fig. 10.10.   Traffic distribution to neighbor AS, B.

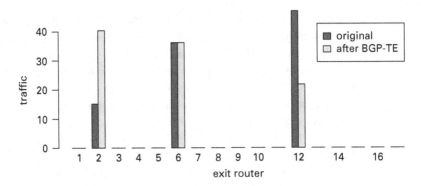

Fig. 10.11. Traffic distribution to neighbor AS, C.

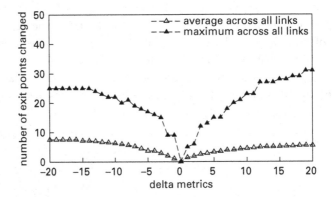

Fig. 10.12. Flows shifted with individual link metric perturbations.

the rest in the USA. Two of the USA locations are again on the East Coast. Here, the BGP route advertisements from ISP C do not match as well as they did for ISP A in Figure 10.8. As a result, router 6 in Europe sinks some traffic, while routers 2 and 12 on the East Coast of the USA sink the rest. IGP metric changes due to local optimization in Europe shift 25% of the traffic to ISP C between the two USA routers.

10.4.4 Impact of more frequent changes

Thus far, we have considered the case where a network operator chooses a set of IGP metrics for the entire network that optimizes utilization across all links. However, in practice, network-wide optimization is a rare occurrence. Typically an operator will apply a few, small link metric changes to react to sudden traffic surges or link additions or failures. They will incrementally adjust the metrics for a handful of links in a trial and error fashion, without considering the global optimum. From an operational standpoint, it is important to understand the impact of such changes in light of BGP hot-potato shifts.

In Figure 10.12, we show the number of exit points changed due to a perturbation in metric for a link. For example, consider the points at +20 units. For each link in the

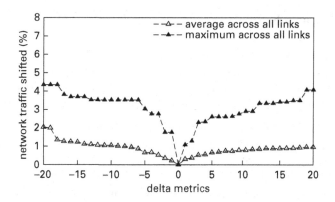

Fig. 10.13. Network traffic shifted with individual link metric perturbations.

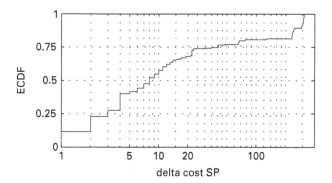

Fig. 10.14. Cumulative distribution of cost change required to cause hot-potato shifts.

network, we calculate which BGP paths would change the egress if this link's metric were 20 units higher, and then count the number of flows that go to these shifted egress points. We plot the number of flows shifted, averaged over all links, as well as the maximum over all links. Thus, if a random link is picked and its metric is increased by 20, it is likely that about eight flows will change their egress point; in the worst case, it can be as many as 32. In Figure 10.13, we show how much traffic this corresponds to. The vertical axis shows the percentage of the total volume of ingress traffic. This shows that as much as 4% of total network traffic can see hot-potato-induced shifts if any particular link's metric is changed by 0 to 20 units. These are relatively small metric changes – for the European network, the IGP link metrics are in the range of 1 to 60.

Hot-potato shifts can also be caused by link failures [134,192]. A link failure can only increase the cost of paths between two points that used that link, since IS–IS or OSPF will now have to pick a more expensive path around it. Using our final IGP metric solution, we calculate for each flow by how much the cost to reach the chosen egress has to change before inducing a hot-potato shift to a different egress point. We plot the cumulative distribution across all flows that have multiple egress points in Figure 10.14. We see that for 50% of flows, an increase of only 8 will cause them to shift their exit point.

10.5   Lessons learned

As the Internet matures, one can observe a higher degree of inter-connection between ASs. Such increased connectivity implies a greater number of paths to destinations throughout the Internet. This chapter makes the observation that such an effect may also be accompanied by a greater level of interaction between the interior and exterior routing protocols in use today, invalidating any assumptions on the isolation of the IGP in one AS and the IGP in a neighboring AS. Using a real network topology, link delays, link capacities, delay constraints, routing tables and TM, we have evaluated how inter-domain routing policies can interact with IGP engineering. We have found that, in our data set, ignoring the hot-potato routing policy can result in IGP metrics that are suboptimal by 20% in terms of maximal link utilization across the network, which is a significant amount for typical network operation.

However, to consider such hot-potato shifts, a point-to-multipoint TM has to be employed for IGP engineering. This is challenging because we cannot rely on SNMP-based TM estimation techniques to obtain it. Such a computation requires NetFlow information, which we could not even collect from the entire network due to different versions of deployed router operating systems. A second concern is the running time of IGP optimization. Hot-potato calculations in every iteration of the heuristic add significant complexity to the process. This has significantly increased the running time of the heuristic we run, even when considering only 80% of the European ingress traffic. In the point-to-point TM mode, the heuristic only takes 5 minutes to provide a solution on a dual 1.5-GHz Intel processor Linux PC with 2 GB of memory (running TabuISIS in TE mode). Using a point-to-multipoint TM for hot-potato shifts, it takes 55 minutes (running TabuISIS in BGPTE mode). Considering a smaller percentage of the traffic dramatically improves the running time since fewer flows are considered.

A bigger concern is the impact on neighboring ASs. Such shifts in traffic to neighboring ASs can impact their own network performance. In our study, we found cases where as much as 25% of traffic to a neighboring ISP shifts the egress point due to local IGP engineering. We found several other scenarios where local routing policy and peering locations played an important part in determining if a neighbor can be affected by local AS changes. This is an important issue because it further increases the inter-dependence of performance between neighboring ASs. One AS improving its performance can reduce a neighbor's performance. If the neighbor then recomputes its IGP metrics, it can lead to instability. However, coordinating IGP engineering between competing ISPs without revealing private information is challenging.

11 Capacity planning

Box 11.1. Network design process: Step 4

Goal Define a generic model for capturing the dynamics of Internet traffic and provide a robust methodology to predict *where* and *when* link upgrades (or addition) have to take place in the core of an IP network. In our context a link is meant to inter-connect logical adjacent PoPs.

Objective function Isolate the underlying overall trend in the traffic and the timescales that significantly contribute to its variability. Use such a trend to drive the capacity planning process.

Inputs (i) Logical topology, comprising nodes (i.e. PoPs) inter-connected to adjacent nodes via multiple parallel links, each of them with its associated capacity; (ii) SNMP data, collected over a long time period for all inter-PoP links constituting the logical topology; (iii) acceptable level of traffic per link (i.e. links characterized by an aggregated traffic exceeding those levels need an upgrade).

IP network capacity planning is a very important task that has received little attention in the research community. The capacity-planning theory for traditional telecommunication networks is a well explored area [120], which has limited applicability in a packet-based network such as the Internet. It normally depends on the existence of a Traffic Matrix (TM), identifying the amount of traffic flowing between any source to any destination of the network under investigation. Moreover, it requires accurate modeling of the incoming traffic, as well as accurate predictions of its future behavior. The above information is then combined in a network simulation to identify the points where future upgrades will be needed.

This approach cannot be used in the environment of an IP backbone network because: (i) direct measurement of an IP TM is not possible today and statistical inference techniques [77, 137] have not yet reached levels of accuracy that meet a carrier's target error rates (solutions to reduce those error rates were presented in Chapter 5); (ii) we do not

Portions reprinted, with permission, from Papagiannaki, K., Taft, N., Zhang, Z. and Diot, C. (2005). "Long-term forecasting of internet backbone traffic." *IEEE Transactions on Neural Networks. Special Issue on Adaptive Learning Systems in Communication Networks*, **16**: 5, 1110–1124.

really know how to model the incoming traffic of a backbone network; (iii) simulating such a large-scale network is typically not feasible.

The current best practice in the area is based on the experience and the intuition of the network operators. Moreover, it usually relies on marketing information regarding the projected number of customers at different locations within the network. Given provider-specific over-subscription ratios and traffic assumptions, the operators estimate the impact that the additional customers may have on the network-wide load. The points where link upgrades will take place are selected based on experience and/or current network state. For instance, links that currently carry larger volumes of traffic are likely to be upgraded first.

An initial attempt toward a more rigorous approach to long-term forecasting of IP network traffic is described in ref. [89]. The authors compute a single value for the aggregate number of bytes flowing over the NSFNET and model it using linear time series models. They show that the time series obtained can be accurately modeled with a low-order ARIMA[1] model, offering highly accurate forecasts (within 10% of the actual behavior) for up to two years into the future.

However, predicting a single value for the future network-wide load is insufficient for capacity-planning purposes. One needs to pinpoint the areas in the network where overload may occur in order to identify the locations where future provisioning will be required. Thus per-node or per-link forecasts are required. The authors of ref. [89] briefly address this issue, mentioning that initial attempts toward this direction did not prove fruitful.

Other work in the domain of Internet traffic forecasting typically addresses small timescales, such as seconds or minutes, that are relevant for dynamic resource allocation [32, 39, 52, 88, 172, 208].

To the best of our knowledge, the solution presented in this chapter is the first to model the evolution of IP backbone traffic at large timescales and to develop models for long-term forecasting that can be used for capacity-planning purposes.

In this chapter, we analyze three years of SNMP information collected throughout a major Tier-1 IP backbone (Box 11.1, Input:(ii)). Correlating those measurements with topological information (Box 11.1, Input:(i)), we calculate the aggregate traffic between any two adjacent PoPs and track its evolution over time. We explore the properties of these time series and present a methodology that can be applied to forecast network traffic volume months in the future. Traffic levels are predicted on a per-link basis. Whenever the predicted traffic level exceeds its acceptable value, as defined by the operator (Box 11.1, Input:(iii)), a link-capacity upgrade is needed.

The presented methodology relies on wavelet multi-resolution analysis and linear time series models. Initial observations on the traffic reveal strong periodicities, evident long-term trends and variability at multiple timescales. We use wavelets to smooth out

---

[1] The acronym ARIMA stands for Auto-Regressive Integrated Moving Average and, briefly, can be seen as the most general class of models for forecasting a time series which can be stationarized by transformations such as differencing and logging. The order(s) of differencing needed to stationarize the series and remove the gross features of seasonality is called the *order* of the ARIMA model.

the original signal until we identify the overall long-term trend. The fluctuations of the traffic around the obtained trend are further analyzed at multiple timescales. This analysis reveals that 98% of the energy in the signal is captured by two main components, namely the long-term trend and the fluctuations at the 12-hour timescale. Using the analysis of variance technique, we further show that a multiple linear regression model containing the two identified components can explain 90% of the signal's variance.

We model the weekly approximations of the two components using ARIMA models and develop a prediction scheme that is based on their forecasted behavior. We show that forecasting network backbone traffic based on the presented model can yield accurate estimates for at least six months in the future. Moreover, with a minimal computational overhead, and by modeling only the long-term trend and the fluctuations of the traffic at the 12-hour timescale, we produce estimates which are within 5 to 15% of the actual measured behavior.

The described methodology, combined with actual backbone traffic measurements, lead to different forecasting models for different parts of the network. Our results indicate that different PoP pairs exhibit different rates of growth and experience different types of fluctuations. This illustrates the importance of defining a methodology for deriving models, as opposed to developing a single model for inter-PoP aggregate traffic flows.

Lastly, we show that trends in Internet data may change over time for short or long periods. We acknowledge that forecasting a dynamic environment imposes challenges to forecasting techniques that rely on stationarity. Consequently, we complete this chapter by presenting a scheme for the detection of "extreme" (and perhaps erroneous) forecasts, which are due to short-lived (e.g. on the order of a few months) network changes. We provide the network operator with recommended values in place of these "extreme" forecasts and show that uncertainty in forecasts can be addressed through the analysis of historical trends, and impacts both the edge as well as the core of the network.

The remainder of the chapter is structured as follows. In Section 11.2 we present the data analyzed throughout the chapter and make some initial observations. Section 11.3 provides an overview of the wavelet multi-resolution analysis, along with results of its application on our measurements. Forecasts are derived using linear time series models, presented in Section 11.4. We evaluate our approach in Section 11.5. The trends modeled in the data may in fact change through time. In Section 11.6 we acknowledge the fact that the Internet is a dynamic environment, and we present a scheme to identify "extreme" forecasts due to changes in the network configuration. Section 11.7 concludes the chapter.

11.1 Objectives

The *capacity-planning* process consists of many tasks, such as addition or upgrade of specific nodes, addition of PoPs and expansion of already existing PoPs (Box 11.1, Goal). For the purposes of this work, we use the term *capacity planning* only to refer to the process of upgrading or adding links between two PoPs in the core of an IP network.

The core of an ISP backbone network is usually over-provisioned and consists of very high-speed links, i.e. OC-48, OC-192. These links are a rather large part of a network operator's investment and have a provisioning cycle between six and eighteen months. Therefore, the capability to forecast *when* and *where* future link additions or upgrades will have to take place would greatly facilitate network provisioning.

In order to address the issue of *where* upgrades or additions should take place, we measure and forecast aggregate traffic between adjacent PoPs. In that way carriers can determine which pair of PoPs may need additional inter-connecting capacity. There are a number of factors that influence *when* an upgrade is needed. These factors include SLAs with customers, network policies toward robustness to failures, the rate of failures, etc. We assume that carriers have a method for deciding how many links should inter-connect a given pair of PoPs and the acceptable levels of utilization on these links. Once carriers articulate a utilization threshold beyond which traffic levels between PoPs are considered prohibitive, one can schedule an upgrade before these levels are actually exceeded. Our task is to predict when in the future the traffic levels will exceed these acceptable thresholds.

In this chapter, we use historical information collected continuously since 1999 on the Sprint IP backbone network. There are many factors that contribute to trends and variations in the overall traffic. Our measurements come from a highly dynamic environment reflecting events that may have short- or long-lived effects on the observed behavior. Some of the events that may have a long-lived effect include changes in the network topology and the number of connected customers. These events influence the overall long-term trend and the bulk of the variability observed. Events that may have a short-lived effect include link failures, breaking news or flash crowd events, as well as denial-of-service attacks. These events normally have a direct impact on the measured traffic, but their effect wears out after some time. As a consequence, they are likely to contribute to the measured time series with values which lie beyond the overall trend. Given that such events are very hard to predict, and are already taken into account in the calculation of the threshold values that will trigger upgrades, as described earlier in this section, we will not attempt to model them in this chapter.

11.2   Measurements of inter-PoP aggregate demand

We now describe the measurements collected and analyzed throughout the chapter. We present some initial observations about Internet traffic at timescales longer than one hour. These observations motivate the approach used throughout the rest of the chapter.

11.2.1   Data collected and analysis

We collect values for two particular MIB (Management Information Base, see Section 3.2.1) objects, incoming and outgoing link utilization in bits per second, for all the links of all the routers in the Sprint IP backbone throughout a period from 1999 until the end of 2003. This operation yields traces from more than 2000 links, some of which may no longer be active. The values collected correspond to an exponentially weighted moving average computed on 10-second link utilization measurements. The exponential weighted average has an average age of 5 minutes and allows for more recent samples to be weighted more heavily than samples earlier in the measurement interval.[2]

Along with the SNMP data, we collect topological information. This information is collected several times per day by an agent downloading configuration information from every router in the network. It contains the names of the routers in each PoP, along with all their active links and their destinations. Therefore, it allows us to identify those links in the SNMP data set that inter-connect specific PoPs in the network.

We correlate the SNMP data with the topological information and derive aggregate demand, in bits per second, between any two adjacent PoPs. In this procedure we need to address two issues. First, the collection is not synchronized, i.e. not all links are polled at the same time to avoid overload at the collection station. Secondly, the collection is not reliable (SNMP messages use UDP as their transport protocol), i.e. we may not have one record for each 5-minute interval for every link in the network. As a consequence, the derivation of the aggregate demand is performed as follows.

- For each link in the SNMP data, we identify its source and destination PoP. We use the notation $l_{sd}(k)$ to denote the kth link connecting PoP s to PoP d.
- Time is discretized in 90-minute intervals. We denote time intervals with index t. The reasons why we selected intervals of 90 minutes are provided in Section 11.3.1.
- The aggregate demand for any PoP-pair (s, d) at time interval t is calculated as the sum of all the records obtained at time interval t from all links k in $\{l_{sd}(k)\}$, divided by the number of records. This metric gives the *average aggregate demand of a link* from PoP s to PoP d at time interval t.

This approach allows us to handle the case of missing values for particular links in the aggregate flow. Moreover, it does not suffer from possible inaccuracies in the SNMP measurements, since such events are smoothed out by the averaging operation. With the aforementioned procedure, we obtain 169 time series (one for each pair of adjacent PoPs in our network). For the remainder of the chapter we focus our discussion on eight of those. These are the longest traces at our disposal which also correspond to highly utilized paths throughout the network. In the following sections we look into their properties and devise techniques for forecasting their values in the medium (i.e. months ahead) and long-term future (i.e. 6 months).

11.2.2 Initial observations

In Figure 11.1 we present the aggregate demand for three PoP pairs in our network. The time series spanned the period from the end of 2000 until July 2002 and captured the activity of a multiplicity of links, the number of which may increase in time. Vertical bars identify the time when additional links became active in the aggregate. As can

---

[2] Because these objects belong to a proprietary MIB, we have no further information about how this average value is calculated.

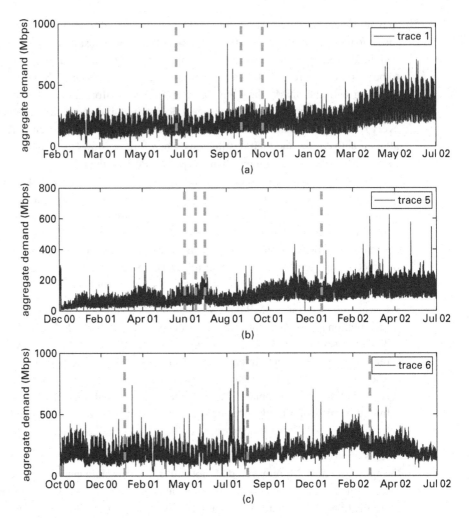

Fig. 11.1. Aggregate demand between two adjacent PoPs in Megabits per second for traces 1, 5 and 6. Plots (a) and (b) demonstrate an increase in the amount of traffic exchanged between the two PoPs. The vertical lines identify the points in time when additional links were provisioned for the inter-connection of the two PoPs, and are rarely preceded by evident increase in the carried traffic due to long provisioning cycles.

be seen, link additions are rarely preceded by a visible rise in the carried traffic. This behavior is due to the long provisioning cycles.

From the same figure we can see that different PoP pairs exhibit different behaviors as far as their aggregate demand is concerned. A *long-term trend* is clearly visible in the traces. For traces 1 and 5, this trend is increasing with time, while for trace 6 it looks more constant with a sudden shift in January 2002 that lasts for two months.

Shorter-term fluctuations around the overall long-term trend are also present across all traces, and manifest themselves in different ways. For instance, trace 1 shows an increasing *deviation* around its long-term trend. On the other hand, trace 6 exhibits smaller fluctuations that look consistent over time.

Regardless of the differences observed in the three traces, one common property is the presence of large spikes throughout them. These spikes correspond to average values across 90 minutes, which indicate a surge of traffic in that particular interval that is high or constant enough to have a significant effect on a 90-minute average. Such spikes may correspond to link failures, which reroute part of the affected traffic onto this particular path, routing changes, or even denial of service attacks. As mentioned in Section 11.2.1, we decide to treat those spikes as outliers. This does not mean we ignore the data, but simply that we do not attempt to model or predict these spikes.

In Figure 11.2 we present a detail of Figure 11.1, which corresponds to the month of May, 2002. This figure indicates the presence of strong daily and weekly cycles. The drop in traffic during the weekend (denoted by the vertical dashed lines) may be substantial, as in trace 1, smaller, as in trace 5, or even non-existent, as in parts of trace 6.

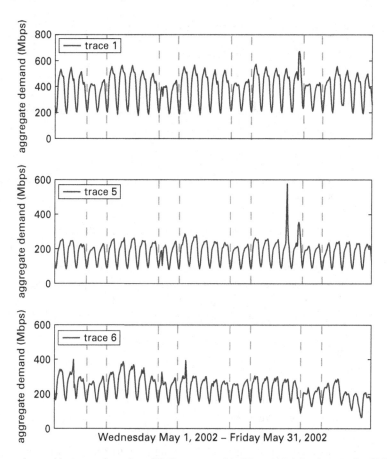

Fig. 11.2.   Aggregate demand between the pairs of PoPs presented in Figure 11.1 for the month of May, 2002 (traces 1, 5 and 6). Traffic exhibits strong diurnal patterns. Weekend traffic, for the two days between dashed lines, may differ from weekday traffic, but such a phenomenon is not evident across all three traces.

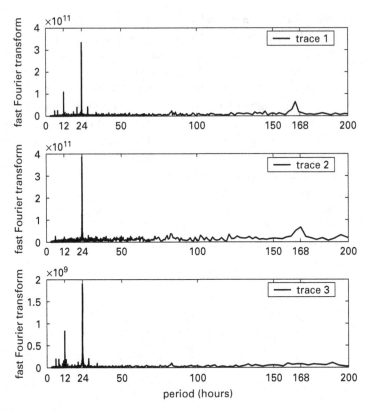

Fig. 11.3.   Fast Fourier transform of aggregate demand for traces 1, 2 and 3. The strongest cycle in the data is diurnal, e.g. 24 hours. Certain traces may also exhibit strong periodicities at the weekly or 12-hour cycle.

From the previous observations it is clear that there are strong periodicities in the data. In order to verify their existence, we calculate the Fourier transform for the eight traces at our disposal. Results indicate that the most dominant period across all traces is the 24-hour one. Other noticeable periods correspond to 12 hours (i.e. semi-daily) and 168 hours (i.e. weekly period). Figure 11.3 presents the fast Fourier transform for three of the traces, demonstrating that different traces may be characterized by different periodicities.[3]

In summary, initial observations from the collected time series lead to three main findings:

(1) there is a multi-timescale variability across all traces (traces vary in different ways at different timescales);
(2) there are strong periodicities in the data;
(3) the time series exhibit evident long-term trends, i.e. non-stationary behavior.

---

[3] Traces 5 and 6, presented in the previous figures, exhibit similar behavior to trace 1. However, for trace 6 the weekly period is stronger than the 12-hour one.

Such findings can be exploited in the forecasting process. For instance, periodicities at the weekly cycle imply that the time series behavior from one week to the next can be predicted. We address these three points in Section 11.3.

11.3 Multi-timescale analysis

In this section, we analyze the collected measurements at different timescales. We show that using the wavelet multi-resolution analysis we can isolate the underlying overall trend, and that those timescales that significantly contribute to its variability.

11.3.1 Wavelet MRA overview

The wavelet Multi-Resolution Analysis (MRA) describes the process of synthesizing a discrete signal by beginning with a very low-resolution signal (at the coarsest timescale) and successively adding on details to create higher-resolution versions of the same signal [65, 133, 200]. Such a process ends with a complete synthesis of the signal at the finest resolution (at the finest timescale). More formally, at each timescale 2^j, the signal is decomposed into an *approximate* signal (or, simply, *approximation*) and a *detailed* signal through a series of scaling functions $\phi_{j,k}(t)$ and wavelet functions $\psi_{j,k}(t)$, where $k \in Z$ is a time index at scale j. The scaling and wavelet functions are obtained by dilating and translating the mother scaling function $\phi(t)$, $\phi_{j,k}(t) = 2^{-j/2}\phi(2^{-j}t - k)$, and the mother wavelet function $\psi(t)$, $\psi_{j,k}(t) = 2^{-j/2}\psi(2^{-j}t - k)$. The approximation is represented by a series of (scaling) coefficients $a_{j,k}$, and the detail by a series of (wavelet) coefficients $d_{j,k}$.

Consider a signal (time series) $x(t)$ with N data points at the finest timescale. Using MRA, $x(t)$ can be written as follows:

$$x(t) = \sum_{k \in Z} a_{p,k}\phi_{p,k}(t) + \sum_{0 \le j \le p} \sum_{k \in Z} d_{j,k}\psi_{j,k}(t), \quad (11.1)$$

where $p \le \log N$. The sum with coefficients $a_{p,k}$ represents the approximation at the coarsest timescale 2^p, while the sums with coefficients $d_{j,k}$ represent the details on all the scales between 0 and p.

Using signal processing parlance, the roles of the mother scaling and wavelet functions $\phi(t)$ and $\psi(t)$ can be described and represented via a *low-pass* filter h and a *high-pass* filter g [200]. Consequently, the multi-resolution analysis and synthesis of a signal $x(t)$ can be implemented efficiently as a filter bank. The approximation at scale j, $\{a_{j,k}\}$, is passed through the low-pass filter h and the high-pass filter g to produce the approximation, $\{a_{j+1,k}\}$, and the detail, $\{d_{j+1,k}\}$, at scale $j + 1$. Note that, at each stage, the number of coefficients at scale j is decimated into half at scale $j + 1$, due to down-sampling. This decimation reduces the number of data points to be processed at coarser timescales, but also leaves some "artifacts" in coarser timescale approximations.

More recently, the so-called *à-trous wavelet transform* has been proposed, which produces "smoother" approximations by filling the "gap" caused by decimation, using redundant information from the original signal [149, 178]. Under the à-trous wavelet transform, we define the approximations of $x(t)$ at different scales:

$$c_0(t) = x(t), \tag{11.2}$$

$$c_j(t) = \sum_{l=-\infty}^{\infty} h(l)c_{j-1}(t + 2^{j-1}l), \tag{11.3}$$

where $1 \le j \le p$ and h is a low-pass filter with compact support. The detail of $x(t)$ at scale j is given by

$$d_j(t) = c_{j-1}(t) - c_j(t). \tag{11.4}$$

Let $d_j = \{d_j(t), 1 \le t < N\}$ denote the wavelet coefficient at scale j and let $c_p = \{c_p(t), 1 \le t < N\}$ denote the signal at the lowest resolution, often referred to as the residual. Then the set $\{d_1, d_2, \ldots, d_p, c_p\}$ represents the wavelet transform of the signal up to the resolution level p, and the signal $x(t)$ can be expressed as an expansion of its wavelet coefficients:

$$x(t) = c_p(t) + \sum_{j=1}^{p} d_j(t). \tag{11.5}$$

At this point we can justify our decision about averaging our measurements across 90-minute intervals. We know that using the wavelet MRA we can look into the properties of the signal at timescales 2^j times coarser than the finest timescale. Furthermore, the collected measurements exhibit strong periodicities at the cycle of 12 and 24 hours. Using 1.5 hours as the finest timescale allows us to look into the behavior of the time series at the periods of interest by observing its behavior at the third ($2^3 \times 1.5 = 12$) and fourth ($2^4 \times 1.5 = 24$) timescale.

11.3.2   MRA application on inter-PoP aggregate demands

For the smoothing of our data we chose as the low-pass filter h in equation (11.3) the B_3 spline filter, defined by (1/16, 1/4, 3/8, 1/4, 1/16). This is of compact support (necessary for a wavelet transform) and is point-symmetric. Symmetric wavelets have the advantage of avoiding any phase shifts; the wavelet coefficients do not "drift" relative to the original signal. The B_3 spline filter gives, at each resolution level, a signal which is much smoother than the one at the previous level, without distorting possible periodicities in the data, and preserving the original structure. The B_3 spline filter has been previously used in time series smoothing [28, 184, 214].

In order to understand how $c_j(t)$ is computed at each timescale j, we schematically present in Figure 11.4 how $c_1(5)$, $c_2(5)$ and $c_3(5)$ are calculated according to equation (11.3), and the B_3 spline filter. Element $c_1(5)$ is computed based on the values $c_0(t) = x(t)$ at times $(5-2)$, $(5-1)$, 5, $(5+1)$, and $(5+2)$. Then we can calculate $c_2(5)$, based on $c_1(1)$, $c_1(3)$, $c_1(5)$, $c_1(7)$ and $c_1(9)$. Note that moving toward coarser levels of resolution we need values from the previous resolution level which are farther apart from each other. For this reason, this wavelet transform is called the à-trous wavelet

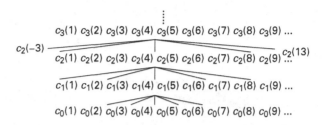

$$c_2(-3) \underbrace{\qquad c_3(1)\ c_3(2)\ c_3(3)\ c_3(4)\ c_3(5)\ c_3(6)\ c_3(7)\ c_3(8)\ c_3(9) \ldots \qquad}$$

Fig. 11.4.   Illustration of the à trous wavelet transform. Each level corresponds to a particular timescale; the first level corresponds to the timescale of the original signal and the approximation becomes coarser as we move upward in the figure. The coefficients of a particular level are computed based on the elements of the previous level. As the timescale becomes coarser, the elements used from the previous level are farther apart from each other.

transform, à-trous meaning "with holes." One important point we should make is that $c_p(t)$ is defined for each $t = 1, 2, \ldots, n$, where n corresponds to 1.5-hour intervals and is limited by the size, N, of the original signal. According to equation (11.3), computing $c_p(n)$ requires values of c_{p-1} until time $n+2^p$, which iteratively requires values of c_{p-2} until time $n+2^{p-1}$, etc. As a consequence, the calculation of $c_p(n)$ requires the original time series $x(t)$ to have $n + \sum_{j=1}^{j=p} 2^j$ values. Given that our original signal contains N values, our wavelet coefficients up to the sixth resolution level will contain n values, where $n + \sum_{j=1}^{j=6} 2^j = N$ or $n = N - 126$.

In Figures 11.5 and 11.6, we present the approximation and detail signals for trace 5 at each timescale, analyzed up to resolution level $2^6 = 96$ hours. We chose to use the sixth timescale as our coarsest timescale because it provides a sufficiently smooth approximation signal. In addition, given that it is the greatest power of 2 that leads to a number of hours smaller than the number of hours in a week (i.e. 168 hours), it captures the evolution of the time series from one week to the next without the effect of the fluctuations at the 12- and 24-hour timescale. Figure 11.5 clearly shows how the wavelet MRA smooths out the original signal. Visual inspection of the derived detail signals in Figure 11.6 further suggests a difference in the amount of variability that each one contributes.

Given the derived decomposition, we calculate the energy apportioned to the overall trend (c_6) and each one of the detail signals. The energy of a signal $y(t)$, $1 \leq t \leq N$, is defined as $E = \sum_{t=1}^{N} y^2(t)$. Table 11.1 shows that the overall trend c_6 accounts for 95 to 97% of the total energy. We then subtract the overall trend c_6 from the data, and study the amount of energy distributed among the detail signals. Figure 11.7 shows that, across all eight traces in our study, there is a substantial difference in the amount of energy in the detail signals. Moreover, the maximum amount of energy in the details is always located at the third timescale, which corresponds to the fluctuations across 12 hours. Approximating the original signal as the long-term trend, c_6, and the fluctuations at the 12-hour timescale, d_3, further accounts for 97 to 99% of the total energy (Table 11.1). In the next section, we look into the properties of the signals derived from the wavelet MRA with respect to the variance they account for in the overall signal.

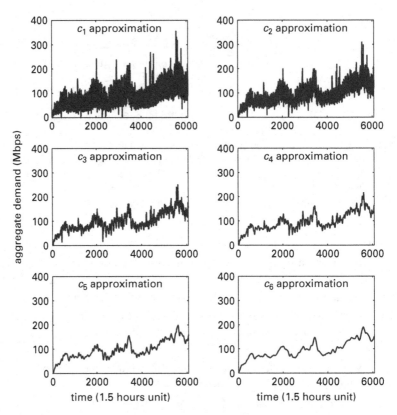

Fig. 11.5.   The six approximation signals for the inter-PoP aggregate demand of trace 5. The signals have been obtained through wavelet MRA based on the à-trous wavelet transform. The sixth timescale corresponds to 96 hours.

11.3.3   Analysis of variance

As explained in Section 11.3.1, the original signal can be completely reconstructed using the approximation signal at the sixth timescale and the six detail signals at lower timescales. The model defined in equation (11.5) can also be conceived as a multiple linear regression model, where the original signal $x(t)$ is expressed in terms of its coefficients.

The ANalysis Of VAriance (ANOVA) technique is a statistical method used to quantify the amount of variability accounted for by each term in a multiple linear regression model [101]. Moreover, it can be used in the reduction process of a multiple linear regression model, identifying those terms in the original model that explain the most significant amount of variance.

Using the ANOVA methodology, we calculate the amount of variance in the original signal explained by the sixth approximation signal and each one of the detail signals. The results indicate that the detail signals d_1, d_2, d_5 and d_6 contribute less than 5% each in the variance of the original signal.

Table 11.1. Percentage of total energy in c_6 and $c_6 + d_3$

Trace	1	2	3	4
c_6	96.07%	97.20%	95.57%	96.56%
$c_6 + d_3$	98.10%	98.76%	97.93%	97.91%

Trace	5	6	7	8
c_6	95.12%	95.99%	95.84%	97.30%
$c_6 + d_3$	97.54%	97.60%	97.68%	98.45%

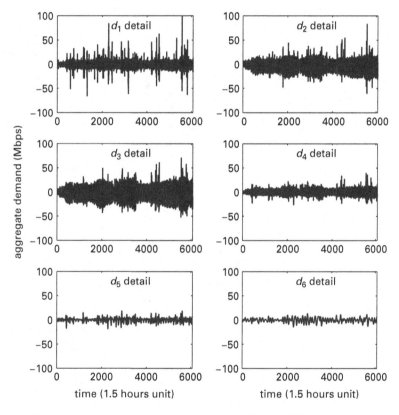

Fig. 11.6. The six detail signals for the inter-PoP aggregate demand of trace 5. The signals have been obtained through wavelet MRA based on the à-trous wavelet transform. The sixth timescale corresponds to 96 hours.

Ideally, we would like to reduce the model of equation (11.5), to a simple model of two parameters: one corresponding to the overall long-term trend and a second one accounting for the bulk of the variability. Possible candidates for inclusion in the model, except from the overall trend c_6, are the signals d_3 and d_4. We know that the detail signal d_3 carries the majority of the energy among all the detail signals. Thus, one possible reduced model is the following:

Table 11.2. ANOVA results for all eight traces

Trace	1	2	3	4
β	2.09	2.06	2.11	2.23
R^2	0.87	0.94	0.89	0.87
Trace	5	6	7	8
β	2.12	2.18	2.13	2.16
R^2	0.92	0.80	0.86	0.91

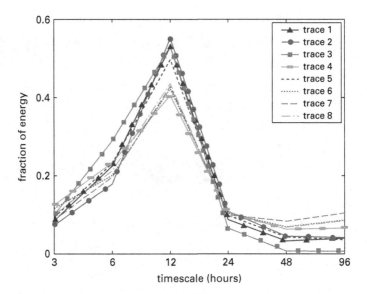

Fig. 11.7. Energy distribution for the detail signals. The majority of the energy accumulated in the detail signals is concentrated in the detail signal at the 12-hour timescale.

$$x(t) = c_6(t) + \beta \, d_3(t) + e(t). \tag{11.6}$$

Using least squares, we calculate the value of β for each one of the traces at our disposal. All traces lead to a β estimate between 2.1 and 2.3 (Table 11.2). Using ANOVA, we test how representative the model of equation (11.6) is with respect to the proportion of variance it explains [101].

If $x(t)$ is the observed response and $e(t)$ is the error incurred in equation (11.6), we define the sum of squared errors as $SSE = \sum_{t=1}^{n} e(t)^2$. The total sum of squares (SST) is defined as the uncertainty that would be present if one had to predict individual responses without any other information. The best one could do is predict each observation to be equal to the sample mean. Thus we set $SST = \sum_{t=1}^{n}(x(t) - \bar{x})^2$. The ANOVA methodology partitions this uncertainty into two parts. One portion is accounted for by the model. It corresponds to the reduction in uncertainty that occurrs when the regression model is used to predict the response. The remaining portion is the uncertainty that remains even after the model is used. We define SSR as the difference between SST and SSE. This difference represents the sum of the squares explained by the regression. The fraction of the variance that is explained by the regression, SSR/SST, determines the goodness of the regression and is called the "coefficient of determination," R^2. The model is considered to be statistically significant if it can account for a large fraction of the variability in the response, i.e. yields large values for R^2. In Table 11.2, we present the results obtained for the values of β and R^2 for all eight traces.

The reduced model is capable of explaining 80% to 94% of the variance in the signal. Moreover, if we decide to include the term d_4 in the model of equation (11.6), the results on R^2, presented in Table 11.2, are only marginally improved, and are increased by 0.01 to 0.04.

11.3.4   Summary of findings from MRA and ANOVA

From the wavelet MRA, we draw three main conclusions:

- there is a clear overall long-term trend present in all traces;
- the fluctuations around this long-term trend are mostly due to the significant changes in the traffic bandwidth at the timescale of 12 hours;
- the long-term trend and the detail signal at the third timescale account for approximately 98% of the total energy in the original signal.

From ANOVA, we further conclude that:

- the largest amount of variance in the original signal can be explained by its long-term trend c_6 and the detail signals d_3 and d_4 at timescales of 12 and 24 hours, respectively;
- the original signal can be sufficiently approximated by the long-term trend and its third detail signal. This model explains approximately 90% of the variance in the original signal.

Based on these findings, we derive a generic model for our time series, presented in equation (11.7). This model is based on equation (11.6), where we set $\beta = 3$, for a model valid across the entire backbone. We use a value for β that is slightly greater than the values listed in Table 11.2 since slight overestimation of the aggregate traffic may be beneficial in a capacity-planning setting:

$$x'(t) = c_6(t) + 3d_3(t). \tag{11.7}$$

11.3.5   Implications for modeling

For forecasting purposes at the timescale of weeks and months, one may not need to model accurately all the short term fluctuations in the traffic. More specifically, for capacity-planning purposes, one only needs to know the traffic baseline in the future along with possible fluctuations of the traffic around this particular baseline.

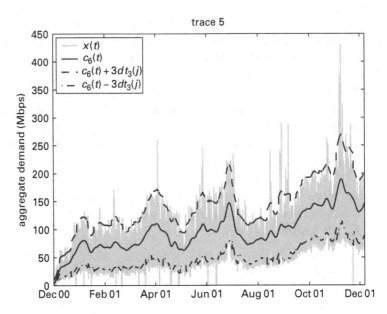

Fig. 11.8. Approximation of the original signal $x(t)$ using $c_6(t)$, which captures the long-term trend, and $dt_3(j)$, which captures the average daily standard deviation within a week. Signal $c_6(t) \pm 3dt_3(t)$ exposes the weekly variation in traffic without including outliers that mainly correspond to unpredictable events. The time series is defined for each 90-minute interval.

Component $d_3(t)$ in the model of equation (11.7) is defined for every 90 minute interval in the measurements capturing in time the short-term fluctuations at the timescale of 12 hours. Given that the specific behavior within a day may not be that important for capacity-planning purposes, we calculate the standard deviation of d_3 within each day. Furthermore, since our goal is not to forecast the exact amount of traffic on a particular day months in the future, we calculate the weekly standard deviation $dt_3(j)$ as the average of the seven values computed within each week. Such a metric represents the fluctuations of the traffic around the long-term trend from day to day within each particular week.

In Figure 11.8, we show the aggregate demand for trace 5, as calculated from the SNMP data. In the same figure we plot the long-term trend in the data, along with two curves showing the approximation of the signal as the sum of the long-term trend plus or minus three times the average daily standard deviation within a week. We see that approximating the original signal in such a way exposes the fluctuations of the time series around its baseline with very good accuracy.

Note that the new signal dt_3 features one value every week, exposing the average daily standard deviation within the week. Similarly, we can approximate the long-term trend with a more compact time series featuring one value for every week. Given that the sixth approximation signal is a very smooth approximation of the original signal, we calculate its average across each week and create a new time series $l(j)$ capturing the long-term trend from one week to the next. The forecasting process will have to predict the behavior of

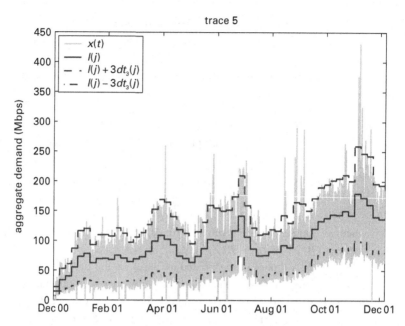

Fig. 11.9. Approximation of the signal using the average weekly long-term trend $l(j)$ and the average daily standard deviation within a week $dt_3(j)$. Signal $l(j) \pm 3dt_3(j)$ tracks the changes in the original signal $x(t)$ while operating at the timescale of one week. Compact representation of the collected time series allows for efficient processing.

$$\hat{x}(j) = l(j) + 3dt_3(j), \qquad (11.8)$$

where j denotes the index of each week in our trace.

The resulting signal is presented in Figure 11.9. We confirm that approximating the original signal, using weekly average values for the overall long-term trend, and the daily standard deviation results in a model that accurately captures the desired behavior.

In Section 11.4, we introduce the linear time series models and show how they can help derive forecasts for the two identified components. Once we have these forecasts, we compute the forecast for the original time series and compare it with collected measurements.

11.4   Time series analysis using the ARIMA model

11.4.1   Overview of linear time series models

Constructing a time series model implies expressing X_t in terms of previous observations X_{t-j} and noise terms Z_t, which typically correspond to external events. The noise processes Z_t are assumed to be uncorrelated with a zero mean and finite variance. Such processes are the simplest processes and are said to have "no memory," since their value at time t is uncorrelated with all the past values up to time $t - 1$.

Most forecasting models described in the literature are linear models. From those models, the most well known are the "AutoRegressive" (AR), "Moving Average" (MA) and "AutoRegressive Moving Average" (ARMA) models.

A time series X_t is an ARMA(p, q) process if X_t is stationary and if, for every t,

$$X_t - \phi_1 X_{t-1} - \cdots - \phi_p X_{t-p} = Z_t + \theta_1 Z_{t-1} + \cdots + \theta_q Z_{t-q},$$

where $Z_t \sim WN(0, \sigma^2)$ and the polynomials $(1 - \phi_1 z - \cdots - \phi_p z^p)$ and $(1 + \theta_1 z + \cdots + \theta_q z^q)$ have no common factors [40]. If $p = 0$, then the model reduces to a pure MA process, while, if $q = 0$, the process reduces to a pure AR process.

This equation can also be written in a more concise form as follows:

$$\phi(B)X_t = \theta(B)Z_t, \tag{11.9}$$

where $\phi(\cdot)$ and $\theta(\cdot)$ are the pth and qth degree polynomials and B is the backward shift operator ($B^j X_t = X_{t-j}$, $B^j Z_t = Z_{t-j}$, $j = 0, \pm 1, \ldots$).

The ARMA model fitting procedure assumes the data to be *stationary*. If the time series exhibits variations that violate the stationary assumption, then there are specific approaches that could be used to render the time series stationary. The most common one is often called the "differencing operation." We define the lag-1 difference operator ∇ by

$$\nabla X_t = X_t - X_{t-1} = (1 - B)X_t, \tag{11.10}$$

where B is the backward shift operator as already introduced. If the non-stationary part of a time series is a polynomial function of time, then differencing finitely many times can reduce the time series to an ARMA process.

An ARIMA(p, d, q) model is an ARMA(p, q) model that has been differenced d times. Thus it has the following form:

$$\phi(B)(1 - B)^d X_t = \theta(B)Z_t, \; Z_t \sim WN(0, \sigma^2). \tag{11.11}$$

If the time series has a non-zero average value through time, then equation (11.11) also features a constant term, μ, on its right-hand side.

11.4.2   Time series analysis of the long-term trend and deviation

In order to model the obtained components $l(j)$ and $dt_3(j)$ using linear time series models, we have to separate the collected measurements into two parts: (1) one part used for the estimation of the model parameters and (2) a second part used for the evaluation of the performance of the selected model. Since our intended application is capacity planning, where traffic demand has to be predicted several months ahead, we select the estimation and evaluation periods such that the latter contains six months of data.

For each one of the analyzed traces, we use the measurements collected up to January 15, 2002, for the modeling phase, and the measurements from January 16, 2002, until July 1, 2002, for the evaluation phase. Given that not all time series are of the same duration, the isolation of the last six months for evaluation purposes may lead to specific traces featuring a small number of measurements for the estimation phase. Indeed, after posing this requirement, three out of the eight traces in our analysis (traces 2, 3 and 7) consist of less than six months of information. Such limited amount of information in the estimation period does not allow for model convergence. As a consequence, we continue our analysis on the five traces remaining.

We use the Box–Jenkins methodology to fit linear time series models [40]. Such a procedure involves the following steps: (i) determine the number of differencing operations needed to render the time series stationary; (ii) determine the values of p and q in equation (11.9); (iii) estimate the polynomials ϕ, and θ; and (iv) evaluate how well the derived model fits the data. For the model fitting we can use both Splus [196], ITSM [40] and the Matlab System Identification Toolbox. All tools provided similar results in our case. The estimation of the model parameters is performed using *maximum likelihood estimation*. The best model is chosen as the one that provides the smallest AICC, BIC and FPE measures[4] [40], while offering the smallest mean square prediction error six months ahead. For more details about the metrics used in the quality evaluation of the derived model, we refer the reader to ref. [40]. One point we should emphasize is that metrics such as AICC and BIC not only evaluate the fit between the values predicted by the model and actual measurements, but also penalize models with a large number of parameters. Therefore, the comparison of the derived models against such metrics leads to the most parsimonious models fitting the data.

11.4.3   Models for $l(j)$ and $dt_3(j)$

The computed models for the long-term trend $l(j)$ indicate that the first difference of those time series (i.e. the time series of their changes) is consistent with a simple MA model with one or two terms (i.e. $d = 1, q = 1$ *or* $d = 1, q = 2$), plus a constant value μ (Table 11.3). The need for one differencing operation at lag 1, and the existence of term μ across all the models, indicate that the long-term trend across all the traces is a simple exponential smoothing with growth. The trajectory for the long-term forecasts will typically be a line whose slope is equal to μ. For instance, for trace 1 the long-term forecast will correspond to a weekly increase of 0.5633 Mbps. This forecast corresponds to the average aggregate demand of a link in the aggregate. The weekly increase in the total demand between two adjacent PoPs can thus be estimated through the multiplication of this value with the total number of active links in the aggregate. Given the estimates of μ across all models in Table 11.3, we conclude that all traces exhibit upward trends but grow at different rates.

Applying the Box–Jenkins methodology on the deviation measurements, we see that, for some traces, the deviation $dt_3(j)$ can be expressed with simple AR models (traces 4 and 6), while the remaining can be accurately modeled as MA processes after one differencing operation (Table 11.4). Therefore, the deviation for traces 1, 5 and 8 increases with time (at rates one order of magnitude smaller than the increase in their

---

[4] AICC = Akaike Information Criterion; BIC = Bayesian Information Criterion; FPE = Find Prediction Error.

long-term trends), while the deviation for traces 4 and 6 can be approximated with a weighted moving average, which indicates slower evolution. These results confirm earlier observations from Figure 11.1 in Section 11.2.2.

From Tables 11.3 and 11.4 we see that one cannot come up with a single network-wide forecasting model for the inter-PoP aggregate demand. Different parts of the network grow at different rates (long-term trend) and experience different types of variation (deviation from the long-term trend). The presented methodology extracts those trends from historical measurements and can identify these PoP pairs in the network that exhibit higher growth rates and thus may require additional capacity in the future.

At this point we should note that the Box–Jenkins methodology could also have been applied on the original time series $x(t)$. However, given the existence of three strong periods in the data (which would require a seasonal ARIMA model with three seasons [40]), the variability of the time series at multiple timescales, the existence of outliers and the size of the original time series, such an approach leads to highly inaccurate forecasts, while being extremely computationally intensive. Our technique is capable of isolating the overall long-term trend and identifying those components that significantly contribute to its variability. Predictions based on weekly approximations of those components provide accurate estimates with a minimal computational overhead. All forecasts were obtained in seconds.

In Section 11.5, we use the derived models for the weekly prediction of the aggregate traffic demands. Then forecasts are compared against actual measurements.

Table 11.3. ARIMA models for the long-term trend

Trace	Order	Model	μ	σ^2
1	(0,1,2)	$X(t) = X(t-1) + Z(t) - 0.1626Z(t-1)$ $-0.4737Z(t-2)$	0.5633e+06	0.2794e+15
4	(0,1,1)	$X(t) = X(t-1) + Z(t) + 0.4792Z(t-1)$	0.4155e+06	0.1339e+15
5	(0,1,1)	$X(t) = X(t-1) + Z(t) + 0.1776Z(t-1)$	0.2301e+07	0.1516e+15
6	(0,1,2)	$X(t) = X(t-1) + Z(t) - 0.3459Z(t-1)$ $-0.4578Z(t-2)$	0.7680e+06	0.6098e+15
8	(0,1,1)	$X(t) = X(t-1) + Z(t) + 0.2834Z(t-1)$	0.2021e+07	0.1404e+16

Table 11.4. ARIMA models for the weekly deviation

Trace	Order	Model	μ	σ^2
1	(0,1,1)	$X(t) = X(t-1) + Z(t) - 0.6535Z(t-1)$	0.3782e+05	0.2024e+14
4	(2,0,0)	$X(t) = 0.8041X(t-1) - 0.3055X(t-2)$ $+Z(t)$	0.1287e+08	0.7295e+13
5	(0,1,1)	$X(t) = X(t-1) + Z(t) - 0.1493Z(t-1)$	0.3094e+06	0.8919e+13
6	(3,0,0)	$X(t) = 0.3765X(t-1) - 0.1964X(t-2)$ $-0.2953X(t-3) + Z(t)$	0.2575e+08	0.3057e+14
8	(0,1,1)	$X(t) = X(t-1) + Z(t) - 0.5565Z(t-1)$	0.3924e+05	0.4423e+14

11.5   Evaluation of forecasts

Using our models we predict a baseline aggregate demand for a particular week in the future, along with possible deviations around it. The overall forecast for the inter-PoP aggregate demand is then calculated based on equation (11.8). For the remainder of this section we focus on the upper limit of the obtained forecasts, since this is the value that would be used for capacity-planning purposes.

In Figure 11.10, we present the time series collected for trace 5 until July 1, 2002. On the same figure we present the modeled behavior in the estimation period and the forecasts in the evaluation period.[5] From visual inspection of the presented plot, one can see that the presented methodology behaves very well for this particular trace.

In order to be able to quantify the quality of the predictions, we have to compare the forecasts against the behavior we model in the measured signal. We thus proceed as follows.

- We apply the MRA on the measurements in the evaluation period.
- We calculate the long-term trend $l(j)$ and weekly deviation $dt_3(j)$ for each week in the same period.

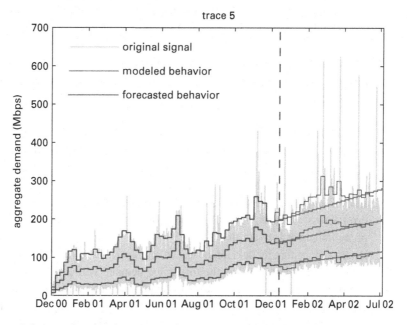

Fig. 11.10.   Six-month forecast for trace 5 as computed on January 15, 2002. Forecasts are provided for (i) the long-term trend and (ii) the modeled signal, which also takes into account 95% of the variability in the original signal. The time series model is trained on data from December, 2000, until December, 2001, capturing the behavior of the signal without the effect of outliers.

---

[5] In Figure 11.10 the vertical dashed line indicates the beginning of the forecasting period. The three lines in the evaluation period correspond to the forecast of the long-term trend $l(j)$ and the forecasts for $l(j) + 3dt_3(j)$, $l(j) - 3dt_3(j)$.

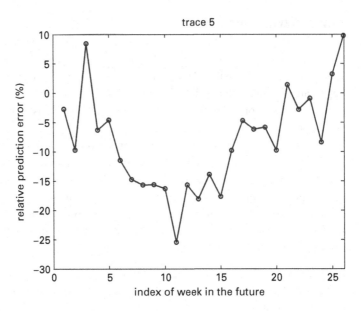

Fig. 11.11.   Weekly relative prediction error for trace 5 when forecasts are generated for six months in the future. The prediction error fluctuates around zero, indicating that there is no consistent under- or overestimation effect. The average absolute error is approximately 10%.

- We compute $\hat{x}(j)$ based on equation (11.8).
- Lastly, we calculate the error in the derived forecast as the forecasted value minus $\hat{x}(j)$, divided by $\hat{x}(j)$.

In Figure 11.11 we present the relative error between the derived forecast and $\hat{x}(j)$ for each week in the evaluation period. Negative error implies that the actual demand was higher than the one forecasted. As can be seen from the figure, the forecasting error fluctuates with time, but is centered around zero. This means that, on average, we neither underestimate nor overestimate the aggregate demand. The average prediction error across weeks is −3.6%. Lastly, across all five traces, the average absolute relative prediction error is lower than 15%.

Our forecasting models can be used to predict demand for more than six months in the future and to identify when the forecasted demand will exceed the operational thresholds that will trigger link upgrades (as explained in Section 11.1). In that case, though, forecasts should be used with caution. As is the case with any forecasting methodology, the farther ahead in the future one attempts to predict, the larger the error margin that should be allowed.

In Figure 11.12, we present the yearly forecast for trace 5 along with the measured aggregate traffic flowing between this particular pair of adjacent PoPs. Forecasts, as computed on January 15, 2001, are highly accurate until September 2002, when they divert for the remainder of the year. The reason behind this behavior is that traffic flowing between the analyzed adjacent PoPs experiences significant growth after September 2002.

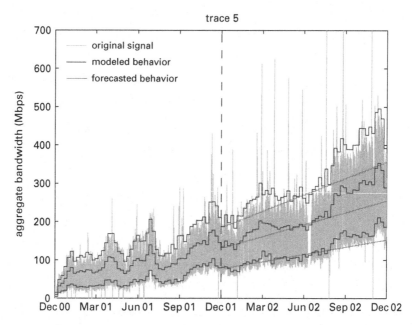

Fig. 11.12.   Yearly forecast for trace 5 as computed on January 15, 2002. Forecasts are provided for (i) the long-term trend and (ii) the modeled signal, which also takes into account 95% of the variability in the original signal. The time series model is trained on data from December, 2000, until December, 2001, capturing the behavior of the signal without the effect of outliers.

In terms of relative prediction errors, as shown in Figure 11.13, our forecasts are highly accurate for 36 weeks into the future. The average relative prediction error made during this period is approximately −5%. However, our methodology underestimates the amount of aggregate traffic between the two PoPs for the last 10 weeks in the year by approximately 20%, due to the significant increase in traffic. Performing this same type of analysis on all five traces resulted in an average absolute relative prediction error of 17% for the yearly forecasts. Thus, our yearly forecasts are slightly worse than the six-month forecasts. The reason for that is that the stationarity assumption is more likely to be invalidated across longer periods of time, for example, it is more likely to observe a change in the network environment in the long-term future.

11.6   Forecasting a dynamic environment

The accuracy of the computed traffic forecasts depends upon several factors. First, if the input itself is very noisy, then the long-term and deviation signals will be hard to model, i.e. the trends in this signal will be rather hard to capture in a consistent fashion. Secondly, sources of error can come from new behaviors, or changes, in the underlying traffic itself. Recall that any forecast is based on models developed from the already seen data. If new behaviors were to emerge, then these might not be captured by the model and could lead to errors in the forecast. There are typically two types of changes

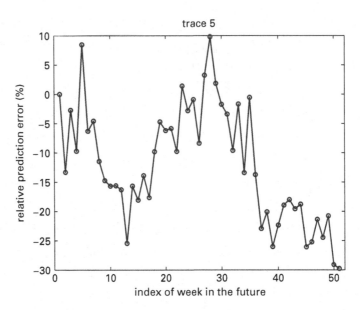

Fig. 11.13.   Weekly relative prediction error for trace 5 when forecasts are generated for twelve months in
the future. The prediction error fluctuates around zero, but appears to deviate significantly
toward the end of the year.

that can surface; those that are short-lived (also called *transient* changes) and those
that are long-lived (considered more *permanent* changes). In this context, short-lived
changes would be those that are on the order of a few weeks, while long-lived changes
refer to those that are on the order of several months or longer. An Internet backbone
is a highly dynamic environment because the customer base is continuously changing
and because any ISP's network is under continuous expansion, either via upgrades or
the addition of new equipment. If some of these factors occur with some regularity,
then they will be reflected in our forecasting model and should not generate prediction
errors. However, many of these factors would not necessarily be captured by the model
and could generate errors. Hence the accuracy of our forecasts very much depends on
whether the captured trends can still be observed in the future.

Before assessing the impact of such changes on our predictions, we give some examples of these sources of change in dynamic IP backbones. Short-lived changes could
be due to changes in routing, whereas long-lived changes would be generated from
topological changes inside the network. When failures such as fiber cuts happen in the
Internet, a common reaction among operators is to shift around the assigned link weights
that are used by the IGP protocol for shortest-path routing. Such moves offload traffic
from some links and, shift it elsewhere, thereby increasing the load on other links. Such
a routing change can last for a few hours or even a few days, depending upon how long
the fiber cut takes to fix. Large amounts of traffic can be shifted around when carriers implement changes to their load balancing policies. Shifting traffic around means
that the composition of flows contributing to the aggregate traffic is altered and so the
aggregate behavior could change as well. Finally, BGP announcements from the edge

of the network could cause new internal routes to be selected when a BGP next-hop is altered. Such a change could easily last a few weeks. More permanent changes come from topological changes that usually reflect network expansion plans. For example, the addition of new links, new nodes, or even new customers, may lead to changes that may ultimately affect the traffic load and growth on inter-PoP links. Similarly, the removal of links or nodes that are decommissioned may impact the model and the forecast.

To ensure that operators can use our model as a tool, we need to be able to help them identify, when possible, those predictions whose errors might be large. In this section we define bounds on our predictions and consider predictions outside these bounds to be "extreme" forecasts. By identifying extreme forecasts, carriers can choose to ignore them, or not, using their own understanding of what is currently happening in the network. We also explore the issue of whether some links are more error-prone than others.

11.6.1   Identification of "extreme" forecasts

We now define a simple method that is to be coupled with the main forecasting method to identify those extreme forecasts that should be treated with skepticism. We have a large data set at our disposal because the described forecasting method has been running continuously since September, 2002, as part of a capacity-planning tool on the Sprint IP backbone network; it computes forecasts for all pairs of adjacent PoPs once a week. Recall that different parts of the network are modeled using different ARIMA models; in addition, the particular ARIMA model that best fits the collected measurements for a specific pair of adjacent PoPs can also change over time. As already illustrated, our models capture a large amount of the regular fluctuations and variability. What we want to identify here are those predictions that may result in errors, not because of traffic variability, but because of underlying network changes (e.g. routing and/or topological changes).

Each week the forecasting tool generates a forecast for the next six and twelve months, thus defining the forecasted growth trend. With over one year's worth of data, we can see how similar an ensemble of predictions are. If the forecasted trend aligns itself (i.e. is within reasonable bounds of) with previously forecasted trends, then our confidence on the new forecasts is higher. Otherwise, the new forecasts may be an artefact of short- or long-lived changes.

To devise a scheme for the identification of "extreme" forecasts we proceed as follows. Although we have almost two years' worth of data, we only use the weeks between September, 2002, and July, 2003, to generate forecasts – so that we can compare one-year forecasts with actual data. For this period, we compute the predicted aggregate demand for each pair of adjacent PoPs inside the network for the next six and twelve months. More specifically, for each week i since November, 2002, we compute a forecast l_{i+26} and l_{i+52}, defining the line of forecasted "growth."

In Figure 11.14 we present the six and twelve-month forecasts for trace 5 that were generated with our methodology for the period of March, 2003, and July, 2004. Note that in Figure 11.14 we simply denote each forecast with a single point instead of a trend

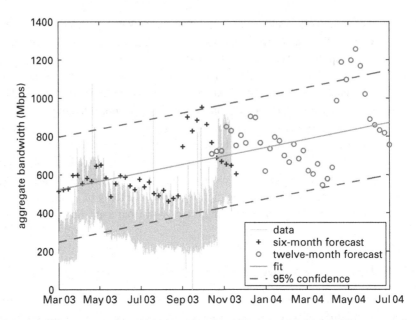

Fig. 11.14.   Visualization of the algorithm used for the detection of "extreme" forecasts. The measured traffic is presented with a solid line. Between March and November 2003 we generate two forecasts, one for the next six and one for the next twelve months. The linear trend in the forecasts along with its 95% confidence interval are able to pinpoint that the forecasts generated for the month of June in 2004 may significantly deviate from "typically observed behavior".

line as in previous figures. In essence, we have abstracted the forecasted trend with two points alone: one corresponding to the forecast after six months and one to the forecast after twelve months.[6] In the same figure we also present the actual measurements collected through SNMP with a continuous line. Figure 11.14 shows that forecasts until September, 2003, agree perfectly with the collected measurements. We note that forecasts generated through time may diverge from each other depending on the point in time when they were computed. For instance, the six-month forecasts obtained for October, 2003 (based on the data until April, 2003) significantly divert from the forecasts obtained for the end of August, 2003 (based on data until March, 2003). The forecasts generated for the month of October are the result of the spike in traffic observed six months earlier, in April, 2003, which lasted for two months.

"Extreme" forecasts are identified via divergence from typically observed behavior. To quantify what constitutes "typically observed behavior," we use a weighted least squares estimation to fit a polynomial through the historical forecasts. We have shown that our ARIMA models typically extract linear trends in the traffic exchanged between PoPs. Therefore, we set the degree of the polynomial function equal to one and compute the line $y = ax + b$ that best captures the trend among historical forecasts, along

---

[6] Note that the cyclic behavior observed in the figure is due to the fact that, for each point in time, we generate two forecasts corresponding to the next six and twelve months, thus obtaining a measure for the forecasted growth.

with its corresponding 95th confidence interval. We identify extreme forecasts as those that are outside the bounds defined by the 95th confidence interval. As can be seen in Figure 11.14, forecasts generated for this particular pair of adjacent PoPs typically follow an increasing trend. The 95th confidence bounds are wide around the fitted line, indicating that typical historical forecasts do fluctuate a good deal over time. After having identified an extreme forecast, a network operator can replace this estimate with a slightly more conservative one, i.e. the upper bound for the 95th confidence interval. In the example in Figure 11.14, the forecasts for the month of July, 2004, are considered "extreme" and could be replaced with the upper 95th confidence bound of 1.1 Gbps.

Note that our model adapts itself as it observes more data and thus can accommodate (over time) either short-lived or long-lived changes. If an underlying change persists, this will be reflected in the data, and ultimately in the model. Thus forecasts which are considered "extreme" at one moment in time, may in fact start aligning with later forecasts. With persistent change, the slope of the line fitted through the data will adjust its slope and so will the 95th confidence interval bounds. It can happen that forecasts considered "extreme" at some previous point in time become "typical" at some later point in time. This is an advantage of adaptive models.

11.6.2 Uncertainty in forecasts as a network artifact

It is now natural to ask whether there is some intrinsic property of a backbone network that makes specific areas harder to forecast. Alternatively, we would like to examine whether specific links inside the network are more error-prone than others, exhibiting trends that are harder to capture consistently over time.

To address these questions, we first explain why one might intuitively think that some links are more susceptible than others. Consider the examples we mentioned above regarding changes in traffic load induced by either routing or topological changes. In an IP network, traffic flowing between two adjacent PoPs is the result of the multiplexing of different Origin–Destination (OD) PoP-to-PoP flows, as dictated by routing. An OD PoP-to-PoP flow captures the total amount of traffic that originates at one particular PoP and departs the network at another specific PoP. Each such flow uses at least one path through the network for its traffic. These paths are determined using the intra-domain routing protocol in effect (in our case IS–IS). Consequently, the traffic we forecast in this work is the result of the superposition of multiple individual OD flows. At each point in time, and depending on the state of the routing regime, the aggregate demand between adjacent PoPs consists of different OD flows whose routes traverse those two particular adjacent PoPs. Therefore, transient changes can occur on an inter-PoP link if its traffic has changed in terms of its constituent OD flows. Routing changes propagated through BGP or changes in the customer population in specific network locations can actually lead to significant changes in the amount of traffic a particular PoP sources or sinks.

If a link is located at the edge of the network, then routing or topological changes across the network will only impact it if they are specific to its edge PoP. On the other hand, if a link is located in the core of the IP network, then its load can be affected by routing changes both inside and at the edge of the network. One might thus postulate that internal links are more error-prone, or that their forecasts are more uncertain. However, one might also conjecture that internal links have a greater amount of statistical multiplexing going on and that this smooths out uncertainty. To examine whether any of these effects impact one type of link more than another, we define loose metrics to describe the position of a link and its forecasting uncertainty, and we examine these two metrics in Sprint's IP backbone.

Position of a link in the network

For the PoP topology of the entire Sprint IP network we compute all routable paths across the network; i.e., for each PoP, we compute all paths to every other PoP. Then, for each link inside the network, we compute the position of this link inside every path. If the position of a link is consistently at the beginning or the end of a routable path, we call the link an "edge link;" otherwise, we call it an "internal link." In essence, an "internal link" is any link that is used to transit traffic between any two PoPs that are not its endpoints.

Uncertainty in forecasts

To assess whether one can generalize if one of these two broad classes of links is inherently more difficult to forecast, we assemble the 6- and 12-month forecasts for *all* 169 pairs of adjacent PoPs in our study. Note that in our data set (containing just a little less than two years' worth of data), numerous routing and topological changes did occur at various places. Thus by studying *all* inter-PoP links, these changes should be perceptible somewhere. Using the same approach as above, we again fit a line through the forecasts and compute the "growth slope" (i.e. the value of a). If the forecasts exhibit significant variations, the "growth slope" a will be accompanied by a large confidence interval. If the forecasts show good alignment, then the confidence interval for the "growth" is going to be more contained. If we denote the confidence interval of a as $[a - da, a + da]$, we define "uncertainty in forecast" to be the fraction $2da/a$. Note that this metric is less sensitive to the actual size of the forecast, in contrast to other possible uncertainty indices, such as the width of the 95th confidence interval band shown in Figure 11.14.

We first examine our "uncertainty" metric for two particular links inside the network; see Figures 11.15 and 11.16. The first link is an "internal" link and the second is an "edge" link. The forecasts obtained for the "internal" link appear to be less volatile than the forecasts obtained for the approximately equal-activity "edge" link. In this particular case, the uncertainty metric does differentiate the uncertainty of the two different links and indicates that the edge link (with an uncertainty of 0.89) is more volatile in terms of its forecasts than the internal link (with an uncertainty of 0.32).

Although our metric does differentiate these two particular links in terms of their uncertainty, we now show that one should be careful not to generalize too quickly, as this observation does not hold when examining *all* links. In Figure 11.17 we show two sets of results; one curve for the cumulative density function of the uncertainty of all "internal" inter-PoP links, and a second curve for the "edge" links. We note that, in

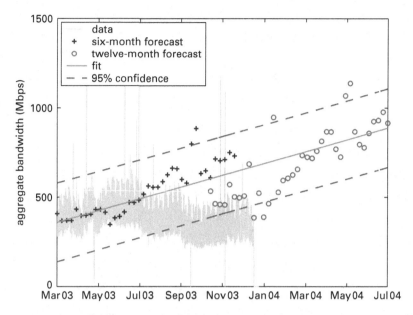

Fig. 11.15.   Uncertainty in forecast generated for an internal link (uncertainty $= 0.32$). The 95% confidence inteval around the trends in the forecasts is tight.

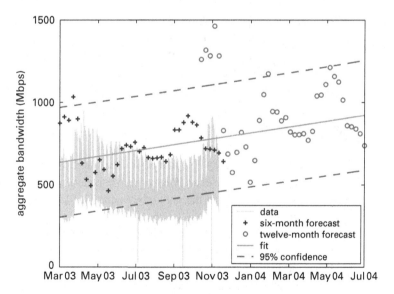

Fig. 11.16.   Uncertainty in forecast generated for an edge link (uncertainty $= 0.89$). Forecasts fluctuate with time and significantly exceed the 95% confidence interval in the forecasted trend in November, 2003. The confidence interval is wide around the fitted trend in the forecasts.

general, there are no significant differences in the forecasting uncertainty of edge and internal links. Hence the composite effect of the factors inducing change (change in routing, topology and customer base) and factors such as statistical multiplexing do not

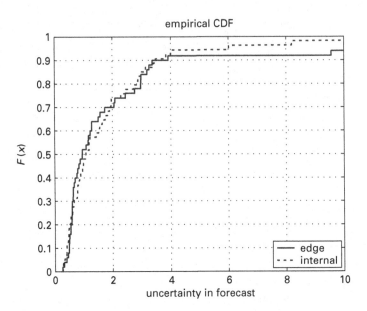

empirical CDF

Fig. 11.17.   Empirical cumulative density function for the forecasting uncertainty for "edge" and "internal" links. There appears to be no significant difference between the behavior of the two types of links.

seem to make one type of link more error-prone than another.[7] This also illustrates a strength of our method because it shows that our approach is not biased toward one type of link or another. The accuracy of the forecasts does not depend upon the position of the link inside the network.

11.7   Lessons learned

We presented a methodology for predicting *when* and *where* link upgrades/additions have to take place in the core of an IP network. We measured aggregate demand between any two neighboring PoPs in the core of a major Tier-1 IP network and analyzed its evolution at timescales larger than one hour.

We showed that the derived time series exhibit strong periodicities at the cycle of 12 and 24 hours, as well as one week. Moreover, they experience variability at multiple timescales and feature distinct overall long-term trends.

Using wavelet MRA, we isolated the *overall long-term trend* and analyzed variability at multiple timescales. We showed that the largest amount of variability in the signal comes from its *fluctuations at the 12-hour timescale*. Our analysis indicates that a parsimonious model consisting of those two identified components is capable of capturing

---

[7] We acknowledge that one could do a much more detailed study of the impact of such changes, and thus our exploration of this issue is preliminary in that we only examined one metric. However, our exploration of this metric is comprehensive since we examined *all* inter-PoP links in a large backbone.

98% of the total energy in the original signal, while explaining 90% of its variance. The resulting model is capable of revealing the behavior of the network traffic through time, filtering short-lived events that may cause traffic perturbations beyond the overall trend.

We showed that the weekly approximations of the two components in our model can be accurately modeled with low-order ARIMA processes. Our results indicate that different parts of the network grow at different rates and may also experience increasing deviations from their overall trend as time progresses. We further showed that calculating future demand based on the forecasted values for the two components in our traffic model yields highly accurate estimates. Our average absolute relative forecasting error is less than 15% for at least six months in the future, and 17% across a year.

Acknowledging the fact that the Internet is a dynamic environment, we then addressed the sensitivity of our forecasting scheme to changes in the network environment. We showed that the forecasting models obtained in an operational Tier-1 network may in fact vary in time. In addition, they may capture trends that may not persist. To address this issue, we presented a scheme for the identification of "extreme" and possibly erroneous forecasts and recommended alternative forecasts in their place. Forecasting a dynamic environment, like the one of an IP backbone network, imposes challenges which are mainly due to topological and routing changes. We showed that such changes appear to impact forecasting of backbone traffic, both at the edge as well as the core of the network.

As a concluding remark, we emphasize that, due to the properties of the collected time series, direct application of traditional time series analysis techniques proves cumbersome, computationally intensive and prone to error. The presented methodology is simple to implement and can be fully automated. Moreover, it provides accurate forecasts for at least six months in the future with a minimal computational overhead. In this chapter, we demonstrated its performance within the context of capacity planning. However, multi-resolution analysis of the original signal and modeling of selected approximation and detail signals using ARIMA models could possibly provide accurate forecasts for the behavior of the traffic at other timescales, such as from one day to the next or at a particular hour on a given day in the future. These forecasts could be useful for other network engineering tasks, such as scheduling of maintenance windows or large database network backups.

Part III

From bits to services

12 From bits to services: information is power

The convergence of traditional network services to a common IP infrastructure has resulted in a major paradigm shift for many service providers. Service providers are looking for profitable ways to deliver value-added, bundled, or personalized IP services to a greater number of broadband users. As cable operators and Digital Subscriber Line (DSL) providers capitalize on IP networks, they need to create higher-margin, higher-value premium services, such as interactive gaming, Video-on-Demand (VoD), Voice-over-IP (VoIP) and broadband TV (IPTV). The missing element of the current strategy is service differentiation, i.e. the ability to understand at a granular level how subscribers are using the network, identify what applications or services are being consumed, and then intelligently apply network resources to applications and attract subscribers that promise the highest return on investment. Operators need to manage and control subscriber traffic. This can be accomplished by introducing more intelligence into the network infrastructure, which enhances the transport network with application and subscriber awareness. Such unique visibility into the types of bits carried allows the network to identify, classify, guarantee performance and charge for services based on unique application and subscriber criteria. Instead of underwriting the expenses associated with random and unconstrained data capacity, deployment and consumption, this new wave of network intelligence allows operators to consider Quality-of-Service (QoS) constraints while enabling new possibilities for broadband service creation and new revenue-sharing opportunities. The same is true with third-party service providers, who may, in fact, be riding an operator's network undetected. Deep visibility into Internet traffic is bound to benefit further management tasks such as SLA monitoring, network security, traffic optimization and network management.

To make things a little more concrete, let us look into one particular example. Peer-to-Peer (P2P) and other broadband applications, i.e. applications characterized by high data rates and associated to large bandwidth consumption, represent today both a challenge and a new business opportunity for operators. P2P applications are increasing broadband demand as more subscribers use the Internet to download music, video, games and other compelling content. The P2P revolution is rapidly penetrating mainstream applications ranging from business-collaboration tools to distributed computing, gaming and voice services. P2P is by far the biggest consumer of network resources, with 70% and more of all broadband data throughput consumed by these applications. But traffic patterns for P2P applications vary dramatically from their client–server counterparts, causing a significant change in upstream data requirements, time-of-day activity and use of expensive international transit links. Service providers recognize that, left unmanaged, P2P can become a financial burden, since network resources are consumed, forcing constant investment in network capacity without any additional revenue. Failure to manage P2P traffic accurately also leads to an increase in number of customer support load and subscriber churn as network congestion degrades the performance of other revenue-generating applications. Similar considerations can be made for other applications such as IPTV, VoIP, on-line gaming, etc.

At the same time, operators have to deal with malicious threats hidden in those packet streams that are carried from one end point to another. Today, network security resides at the heart of the network infrastructure. We have moved from an Internet of implicit trust to an Internet of pervasive trust, where security policies are the arbiter and no packet, service or device can be trusted until it is inspected. The lack of security-conscious home users and the open nature of the Internet create a breeding ground for network security threats impacting both service providers and subscribers. Recent threats have created "security storms" resulting from popular viruses as Sasser, Slammer and Blaster. Additionally, as more "IP-enabled" handsets and Personal Digital Assistants (PDAs) become a target for hackers, service security turns out to be a paramount issue for operators on all fronts. Increased network traffic caused by increased numbers of infected hosts results in increased costs and technical support calls, while operators seek to track, disable and block the spread of a malicious activity. Infected machines generate network congestion as they attempt to propagate a viral infection, resulting in performance degradation for all users. Traffic visibility is essential to operators to react promptly to these threats and thus defend their infrastructures and services from being hijacked causing loss of revenue.

By looking at these few examples, one can see how "information" means "power" these days. As a consequence, we will focus this last part of the book on providing answers to how to get our arms around this important concept and deal specifically with (i) how to classify accurately traffic applications and protocols, referred to as the *traffic classification* problem, and (ii) how to detect efficiently malicious threats that use Internet applications as vehicles of delivery, referred to as the *network security* problem.

12.1   Building intelligence: extracting information from bits

In addition to the fundamental and traditional purposes of traffic analysis such as network planning, network problem detection and network usage reporting, traffic monitoring and analysis is required in many areas to improve network service quality, such as in abnormal traffic detection and usage-based accounting. The ability of a network operator to classify network traffic accurately into different applications (both known and unknown) directly determines the success of many of the above network management tasks. Hence it is imperative to develop traffic classification techniques that are fast, accurate, robust and scalable in order to meet today's and tomorrow's ISPs' needs. In order to deal with numerous Internet and network applications and protocols, a methodical and systematic identification process must be employed. Similar to a law enforcement process in which fingerprints are used to identify the involvement of individuals in specific incidents, signatures are used to identify applications and protocols. In their most broad sense, signatures are pattern recipes which are chosen to identify an associated application (or protocol) uniquely. When a new application or protocol is encountered, it is analyzed and an appropriate signature is developed and added to a database. Application signatures must be checked on a regular basis as they tend to vary as new application updates or protocol revisions occur. For example, BitTorrent, eMule and Skype tend to upgrade their client software on a regular basis and encourage (and in some cases even force) users to move on to the new release. The use of these new releases with non-up-to-date signatures will dramatically affect the classification accuracy of existing systems.

There are several possible methods used to identify and classify traffic. These range from the mapping of traffic to applications using the transport layer *port* numbers, identifying an application using *string matching* that uniquely maps to a single application, studying *numerical properties* of the traffic itself, which disambiguate its nature, and observing *pattern behavior* that may transcend the use of numerical properties, since it does not only rely on traffic statistics, but also is capable of extracting and focusing on the feature of interest.

- Identification through the use of transport layer *port numbers* is probably the easiest and most well known form of signature analysis. The reasoning is the simple fact that many applications use their default ports or some chosen ports in a specific and deterministic manner. A good example is POP3 used by email applications. Incoming POP3 connections typically use port 110 and port 995 when employing security mechanisms. The outgoing SMTP connection is always towards port 25. However, the ease of application identification through the use of transport layer port numbers has recently become a weakness to this approach, particularly because many current applications disguise themselves as other applications. The most notorious example is that of port 80, where many applications camouflage as pure HTTP traffic in order to go through firewalls. As noted above, some applications select random ports instead of using fixed default ports. Sometimes, there is a deterministic pattern in how the specific application chooses the ports to use, while in other cases the entire port selection process may be completely random. For all these reasons, it is often not feasible to use analysis by port as the only tool for identifying applications, but rather as a form of analysis to be used together with other tools.
- Identification through *string matching* involves the search for a sequence of textual characters or numeric values within the contents of the packet, also known as packet payload. Furthermore, string matches may consist of several strings distributed within a packet or several packets within the same flow. For example, many applications still declare their names within the protocol itself, as in Kazaa, where the string "Kazaa" can be found in a typical HTTP GET request. If analysis is performed by transport layer ports alone, then port 80 may indicate HTTP traffic and the GET request will further corroborate this observation. Hence this analysis will result in an inaccurate classification, i.e. HTTP and not Kazaa. However, the proliferation of proprietary protocols and applications makes

this approach not scalable (too many payload signatures to develop and check on a daily basis). At the same time, the increased usage of encryption techniques that scramble payload information makes the usefulness of such an approach very questionable in the years to come. This simple example emphasizes once again that several analysis tools are required to ensure a proper classification.

- Identification through the analysis of *numerical properties* involves the investigation of arithmetic and numerical characteristics within a packet and of a packet or several packets. Some examples of properties analyzed include payload length, the number of packets sent in response to a specific transaction and the numerical offset of some fixed string (or byte) value within a packet. For example, consider the process for establishing a TCP connection using some UDP transactions in Skype (version prior to 2.0). The client sends an 18 byte message, expecting in return an 11 byte response. This is followed by the sending of a 23 byte message, expecting a response which is 18, 51 or 53 bytes. Unfortunately, several applications might be characterized by similar pattern transactions, making this approach unreliable if used as a stand-alone tool. Similar to the identification by port and string matching, the identification using numerical properties alone is insufficient, and can often lead to a high level of inaccuracy.
- Identification through the analysis of *pattern behavior* refers to the way a protocol acts and operates. Pattern behavioral analysis typically boils down to the extraction of statistical parameters of examined packet transactions. Often, pattern behavioral analysis and port-based analysis are combined to provide improved assessment capabilities. For example, pattern behavioral analysis can use packet-length distribution to distinguish pure HTTP traffic from P2P file sharing. Indeed, pure HTTP packets tend to concentrate around a few hundred bytes in length (usually 250–350 byte packets on average), while P2P control layer information tends to use shorter packet lengths (usually 80–110 byte packets). In this way, by examining some short-term statistics, it is possible to conclude whether a port 80 connection carries pure HTTP traffic or other P2P-related traffic.

In summary, there is no single technique today that will correctly classify traffic applications as a stand-alone methodology. Although pattern behavior and port-based techniques represent a powerful combination to deal with the complexity of today's environment, major efforts from both the research and vendor communities are still required to identify correctly which information to extract from IP packets and how to consume it in a scalable manner. The development of techniques robust to evasion further form the focal point of research, both in the academic and industrial, as well as the operational, worlds.

12.2   Bridging the gap between network infrastructure and applications: cleaning the bits

A fundamental challenge faced by today's service providers is maintaining service predictability in the presence of an outbreak of malicious traffic from one/multiple end points spread across one/multiple network boundaries. In today's terms, this type of behavior has been considered equivalent to threats such as DDoS attacks, turbo worms, email and VoIP spam, phishing and viruses. The amount of traffic generated by infections and subsequent outbreaks can disrupt the normal operation and may compromise the performance of network elements that are responsible for basic routing and switching of packets. Although security attacks have always happened in the Internet history, today we are witnessing a key transformation. While attacks were once primarily the work of hackers who wanted to temporarily take well known sites off-line to get media attention, they are increasingly being used as the foundation of elaborate extortion schemes or are motivated by political or economic objectives, costing business and service providers millions of dollars each year. This transformation and its financial ramifications now require that service providers pay more attention to protect their networks and services. At the same time, due to the diversity of methods used malevolently to access or attack organization's computers and the ever increasing sophistication of how they are implemented, an ISP's customers turn toward their service providers for help. Hence, network security has become today a great venue of generating revenue for ISPs that must equip themselves with the right knowledge, hardware and software platforms to clean the customers' bits before being delivered.

Many organizations think that if they have a firewall or Intrusion Detection System (IDS) in place, then they have covered their security bases. Experience bears out that this is not the case. While important, a firewall alone cannot provide 100% protection. In addition, the rate at which new security threats are unleashed into the public Internet is phenomenal. Yesterday's security architectures and general-purpose tools cannot keep up. For this reason, incumbent network security policies and systems require ongoing scrutiny in order to remain effective. As Internet access continues to expand and the quantity and speed of data transfer increases, new security measures must be adapted. In order to understand how best to secure the ISP infrastructure against newer generation security threats, it is first necessary to understand the nature of the problem.

12.2.1   Families of security threats

The threats to an organization's network are varied and numerous. In the following we first introduce some of the most deadly families of attacks targeting network infrastructure and its end systems, such as DoS/DDoS, network intrusion, worms/viruses, malicious routes and bogus prefixes. We remind the reader that this list is not exhaustive and other threats like email and VoIP spam (known as SPIT), phishing and click-fraud represent a constant burden for ISPs and their customers.

Denial of Service

DoS is one of the more prevalent methods currently used by hackers. The goal of a DoS attack is to flood the network and tie up mission-critical resources used to run Web sites, VoIP applications or other enterprise applications. In some cases, vulnerabilities in the Unix and Windows operating systems are exploited intentionally to crash the system, while in other cases large amounts of apparently valid traffic are directed at sites until they become overloaded and fail. Another form of DoS is the *Distributed DoS attack*, i.e. DDoS. This uses an array of systems connected to the Internet to stage a flood attack against a single site. Once hackers have gained access to vulnerable Internet systems, software is installed on the compromised machines that can be activated remotely to launch the attack. We remark to the reader that any layer of the OSI stack can be a target of DoS and DDoS attacks, not just the routing and transport layers.

Network intrusion

This usually takes place as the preparation stage of a DDoS attack. Hackers first gain access to the systems unwillingly drafted for these malicious purposes. Network intrusion methods come in many forms. For example, *network scanners* and *sniffers* are used by hackers to discover vulnerabilities that may exist in network hardware and software configurations; *password cracking* and *guessing tools* are used for dictionary cracking and brute-force cracking methods to discover network passwords; *IP spoofing* is used to gain unauthorized access to computers, whereby the intruder sends messages to the target computer after forging the IP address of a trusted host by modifying the packet headers so it appears that the packets are coming from a different host (IP spoofing is a key tool used by hackers to launch DoS attacks); *Trojan horses* can pose as benign applications, often attached to email messages, and may be launched remotely by a hacker for destructive purposes.

Worms and viruses

These represent the second most deadly method used by hackers. They are malicious programs or pieces of code usually propagated via email or HTTP packets. After being loaded onto an unsuspecting recipient's computer, a virus can often replicate itself and manipulate the computer's email system to disseminate the virus to other users. Some viruses are capable of transmitting themselves across networks and bypassing security systems. Although data corruption tends to be the end goal of most viruses, a simple virus that reproduces itself over and over can be dangerous because it will quickly use all available system memory and will crash the computer. A worm is a special type of virus that can replicate itself and use up memory, but cannot attach itself to other programs. The most disconcerting characteristic about this specific family of attacks is related to its ever increasing virulence and sophistication over time; each new epidemic has demonstrated increased speed over its predecessors. While the Code Red worm took over 14 hours to infect its vulnerable population in 2001, estimated to 360 000 hosts, the Slammer worm, released some 18 months later, did the same in under ten minutes. Worse, worms are showing an always growing trend of sophistication that we expect to lead in the near future to an extensive usage of polymorphic techniques and encryption to prevent discovery.

Malicious routes and bogus prefixes

These are widely reported and today constitute a serious problem for the entire Internet user population. The goal of the attack is to gain control of entire packet data-streams directed toward a legitimate destination or set of destinations by announcing bogus routes. In some parts of the Internet, the false route replaces the authentic route to a prefix. As a result, traffic that follows the false path will eventually be dropped or delivered to someone who is pretending to be the legitimate destination. This attack is easy to launch since the BGP routing infrastructure was designed based on an implicit trust between all participating ASs and border routers. Thus, no mechanism is employed for authenticating the routes injected into or propagated through the system. An attacker can simply gain control of a BGP router due to the fact that most ISPs leave the management console password unchanged for a long time. Alternatively, attackers may not even need to break into a router. They could quite simply register as an ISP, purchase upstream access from a carrier and inject prefixes at will.

We conclude this section by remarking that a single attack can use more than one of the above families of threats during its life-cycle. For example, an attacker can start by stealing a block of prefixes, use them to search for end hosts having specific ports open, i.e. ports associated to applications for which the attacker has identified a vulnerability to exploit, infect the machines with malicious code and then launch a DDoS, spam or worm attack. Thus, solutions specialized in protecting the network infrastructure from a specific family of threats only will miss the end-to-end dynamics of the attack. Such solutions will always play a reactive role in the security life-cycle and never be in a position to identify the attack proactively and thus stop the attack at the very early stage, i.e. during the preparation mode.

12.2.2   Limitations of existing security solutions

No matter which tools a hacker group may employ to launch a malicious attack on an organization, the means to achieve the desired result are the same. Given the mission-critical nature of today's business networks, most organizations have recognized the need for firewalls and IDSs to secure their networks against the underlying mechanisms common to most attacks. However, what many companies fail to recognize is that isolated solutions, specialized to watch for a *one-threat-only*, do not provide absolute security. They address only specific parts of the entire security problem and therefore require complementary technology to guard against attackers' intent on harming a business. When looking within ISP networks, it is very common to find firewalls and IDSs deployed at the network edge as the first and second lines of defense in protecting ISPs' customer networks. But something is still missing! In order to understand the "why" and "what," let's first understand what these current security solutions offer today.

Firewalls and IDSs: merits and weaknesses

Firewalls are designed primarily to prevent unauthorized access to private networks by analyzing packets entering the network and blocking those that do not meet predefined security criteria. For this reason, no business on the Internet today should be without firewall protection. One of the key roles of a firewall is specifically to prevent unauthorized Internet users from accessing private networks, especially intra-nets, via Internet connections. Firewalls are generally considered to be the first line of defense in protecting network infrastructures.

IDS solutions detect suspicious activity in real time by cross-analyzing network activity against a database of traffic profiles associated with different attack methods and patterns of "normal" activity. Intrusion detection systems are generally installed in series with firewalls, helping to measure the effectiveness of the security system and to identify necessary enhancements. IDSs are generally considered to be the second line of defense in protecting network infrastructures.

Unfortunately, these solutions are limited in terms of *functionality*, *capacity* and *effectiveness*. First, firewalls and IDSs provide only part of the overall security system needed to secure today's business networks. These systems are focused on detecting and blocking worms and viruses or DoS/DDoS and are unable to detect a wider spectrum of threats. As a consequence, new appliances specialized in phishing, spam filtering, URL blocking, prefix hijacking, etc., will be required to be deployed as third, fourth and fifth levels of defense making the end-to-end management of the security infrastructure a complete nightmare for ISPs. Moreover, firewalls and IDS solutions are, respectively, host- and network-segment-aware, but not network-aware, i.e. they lack the perspective of the overall network state. As a consequence, they are lacking the end-to-end visibility into the attack pattern, i.e. how the attack has been set up, who is the mastermind, how the attack has propagated into the network, and are only focused on the end-systems they are offering protection to. We believe that this constitutes a major weakness in this approach to security.

Secondly, most firewalls use packet-filtering techniques to accept or deny incoming packets based on information contained in the packets' TCP and IP headers, such as source address, destination address, application, protocol, source port number or destination port number. Today's complex structure of Internet traffic, coupled with the growing frequency at which new threats and vulnerabilities are discovered, make this approach alone inadequate to guarantee a high level of security. Most advanced firewalls and IDS solutions take this to the next level, featuring increased intelligence and looking more deeply into packets, thus targeting more granular application traffic analysis. Although they are more careful in selecting which packets to filter out based on either complex packet filtering criteria or advanced traffic-behavioral anomaly detection engines, they face a key challenge due to the fact that they often introduce network latency problems. Indeed, in order to determine if a flow is legitimate traffic, an intelligent firewall or IDS needs to analyze a series of incoming packets in sequence before allowing or blocking a packet's entry into the network. This can take its toll on the security server processor. If the server has to wait for five or six packets to line up before making the appropriate determination for each packet, this creates a 500% to 600% increase in network latency. This traffic flow slowdown counteracts the effort to provide increased bandwidth across today's high-speed environments. Even when deployed at the ISP edge, these security systems do not have the throughput to keep up with the pace of today's Internet traffic.

Thirdly, existing firewalls and IDS solutions are not able to entrap the network intruder data and redirect it to a secure location in order to perform forensic analysis. They are specialized at blocking the malicious traffic, but are not able to learn the intruder identity and redirect the malicious streams somewhere else. As a result, they face extreme operational challenges when a severe large volume DoS/DDoS attack takes place, just because the attackers can count on many more CPUs than the defense system.

To guard against the types of problems faced by firewalls and IDSs, any complementary security solution needs to detect (i) a wider number of threats while being able to (ii) scale with network traffic and network segments and (iii) being robust to large volume attacks. Only very recently a new family of solutions has appeared in the market that tries to bring together the three aspects listed above. The market has coined a new term for such solutions: *Network Behavior Anomaly Detection Systems* (NBA-DS).

An NBA-DS solution collects packet and flow information from several data points in the network and correlates them to gain network traffic awareness. All information is processed in real time with advanced network-behavior anomaly detection algorithms. These first two steps are done in real time and in a completely passive fashion. Thus the system is not intrusive and does not impact the regular flow of packet streams in the network. Scalability is achieved through the efficient detection of threats, which relies on the application of the appropriate statistical tools on a small number of traffic features extracted from flows. As soon as a threat is detected, the system further explores the details of the pattern communication being noted as behaving abnormally, and packet payloads are carefully analyzed. Actions are then taken in a plurality of different forms, ranging from traffic blocking or rate-limiting or redirecting the malicious streams to safety zones for further forensic testing.

In summary, cyber-security is changing the way people think and act. Although many solutions have been proposed in the market, the jury is still out on qualifying the effectiveness of those solutions. Only recently, NBA-DS solutions appeared in the market that promise to fulfil the technology gaps of firewalls and IDSs with their unique network traffic awareness, broadly accepted today as a key requirement for security products. Although the architecture of NBA-DS solutions look promising, how to consume this huge amount of information and detect stealthy threats still remain open problems.

12.3 Summary

In this chapter, we have described the importance for ISPs to gain a deeper and global visibility into the applications that flow on top of their infrastructure. Several traditional network management tasks will benefit from such intelligence, which will definitely play a key role in the entire life-cycle of new IP services, from the design, monitoring and final launch onto the market. Unfortunately, the proliferation of proprietary protocols, the exponential growth of peer-to-peer and multi-media applications and the extensive usage of encryption makes this problem much tougher than ever. While describing the most accepted and used families of techniques, we highlight how pattern behavior seems to be the most promising approach to deal with the complexity of today's environment. As a consequence, in Chapters 13 and 14 we will introduce new pattern behavior techniques and show how the proper combination of the right traffic features and powerful statistical algorithms can lead to results of unprecedented accuracy.

Similarly to traffic classification, network security represents a key operational task for ISPs. The ever increasing list of exploited vulnerabilities and, the increase of virulence and sophistication of the new threats make network security a nightmare for operators. NBA-DSs have been elected by the market as the next-generation solutions for security that will raise the bar. How to baseline traffic correctly and differentiate good bits from bad bits still remain problems. As a consequence, Chapters 15 and 16 will introduce a pool of new behavior anomaly detection algorithms to be used within an NBA-DS for detecting a variety of different threats. Chapters 13–16 continue utilizing an extensive amount of operational measurements to improve the state-of-the-art performance of large-scale IP backbone networks.

13 Traffic classification in the dark

Classifying traffic flows according to the application that generates them is an important task for (a) effective network planning and design and (b) monitoring the trends of the applications in operational networks. However, an accurate method that can reliably identify the generating application of a flow is still to be developed. In this chapter and the next, we look into the problem of traffic classification; the ultimate goal is to provide network operators with algorithms that will provide a meaningful classification per application, and, if this is infeasible, with useful insight into the traffic behavior. The latter may facilitate the detection of abnormalities in the traffic, malicious behavior or the identification of novel applications.

13.1   State of the art and context

Currently, application classification practices rely to a large extent on the use of transport-layer port numbers. While this practice may have been effective in the early days of the Internet, port numbers currently provide limited information. Often, applications and users are not cooperative and, intentionally or not, use inconsistent ports. Thus, "reliable" traffic classification requires packet-payload examination, which is scarcely an option due to: (a) hardware and complexity limitations, (b) privacy and legal issues, (c) payload encryption by the applications.

Taking into account empirical application trends [104] and the increasing use of encryption, we conjecture that traffic classifiers of the future will need to classify traffic "in the dark." In other words, the traffic classification problem needs to incorporate the following constraints: (i) no access to user payload is possible; (ii) well known port numbers cannot be assumed to indicate the application reliably; (iii) we can only use the information that current flow collectors provide. Clearly, there may be cases where these constraints may not apply, which would make the classification easier. However, we would like to develop an approach that would be applicable and deployable in most practical settings.

Portions of this chapter are reprinted or adapted from Karagiannis, T., Papagiannaki, K. and Faloutsos, M. (2005). BLINC: multilevel traffic classification in the dark. In *ACM SIGCOMM*, Philadelphia, PA, August, 2005, pp 229–240, (http://doi.acm.org/10.1145/1080091.1080119).

Some novel approaches have been taken to consider application classification as a statistical problem. These approaches develop discriminating criteria based on statistical observations and distributions of various flow properties in the packet traces. Typically, such discriminating criteria refer to the packet-size distribution per flow, the inter-arrival times between packets, etc. However, for the most part, these methods do not exploit network-related properties and characteristics, which we believe contain a lot of valuable information. In addition, the validation of a classification method is a challenge. The effectiveness of most of the current approaches has not been validated at a large scale, since a reference point or a benchmark trace with known application consistency does not exist.

In this chapter, we will describe a novel approach for the flow classification problem as defined above, which we call BLINd Classification or *BLINC* for short. The novelty of this approach is twofold. First, we shift the focus from classifying individual flows to associating Internet hosts with applications, then classifying their flows accordingly. We argue that observing the activity of a host provides more information and can reveal the nature of the applications of the host. Second, *BLINC* follows a different philosophy from previous methods attempting to capture the inherent behavior of a host at three different levels: (a) the social level, (b) the network level and (c) the application level.

Combining these two key novelties, we classify the behavior of *hosts* at three different levels. While each level of classification provides increasing knowledge of host behavior, identifying specific applications depends on the unveiled "cross-level" characteristics (i.e. a single level *cannot* reveal the generating application by itself).

- At the *social level*, we capture the behavior of a host as indicated by its interactions with other hosts. First, we examine the popularity of a host. Second, we identify communities of nodes, which may correspond to clients with similar interests or members of a collaborative application.
- At the *functional level*, we capture the behavior of the host in terms of its functional role in the network, namely whether it acts as a provider or consumer of a service, or both in the case of a collaborative application. For example, hosts that use a single port for the majority of their interactions with other hosts are likely to be providers of the service offered on that port.
- At the *application level*, we capture the transport-layer interactions between particular hosts on specific ports with the intent to identify the application of origin. First, we provide a classification using only 4-tuples (source address, destination address, source port and destination port). Then, we refine the classification further by exploiting other flow characteristics such as the transport protocol or the average packet size.

Tunability

A key feature of the presented methodology is that it provides results at various levels of detail and accuracy. First, *BLINC* analyzes traffic at the aforementioned three levels. Second, the classification criteria are controlled by thresholds that, when relaxed or tightened, achieve the desired balance between an aggressive and a conservative classification. The level of accuracy and detail may be chosen according to: (a) the goal of the study and (b) the amount of exogenous information (e.g. application specifications).

The highlights of this chapter can be summarized as follows.

- *Developing a classification benchmark.* We provide a comparison benchmark for flow classification. We collect *full*-payload packet traces, and we develop a payload classification methodology. While this methodology could be of independent interest, we use it here to evaluate *BLINC*.
- *Identifying patterns of behavior.* We identify "signature" communication patterns, which can help us identify the applications that a host is engaged in. Using these patterns, we develop a systematic methodology to implement our multi-level approach.
- *Highly accurate classification.* We successfully apply *BLINC* to several real traces. While training was based on one of the data sets, *BLINC* manages to classify successfully 80–90% of the total traffic with more than 95% accuracy in *all* studied traces.
- *Detecting the "unknown."* We show how *BLINC* can help us detect: (a) unknown applications, such as a new p2p protocol, and (b) malicious flows, which emerge as deviations from the expected behavior. Note that these cases cannot be identified by payload-or port-based analysis.

This chapter presents *BLINC* as an example traffic classification methodology with a unique shift from characterizing flows by application to associating hosts with applications. As such, it also constitutes a first attempt at exploring the benefits and limitations of such an approach. Given the quality of the presented results, we find such an approach to offer great promise and to open interesting new directions for future research.

The remainder of the chapter is structured as follows. In Section 13.2, we motivate the problem and describe related work. In Section 13.3, we present the payload-based classification technique. *BLINC* is presented in Section 13.4, and its performance results are shown in Section 13.5. Section 13.6 discusses implementation details, limitations and future extensions to *BLINC*.

13.2   Background

Analysis of the application traffic mix has always been one of the major interests for network operators. Collection of traffic statistics is currently performed either by flow monitors, such as Cisco NetFlow, or by sophisticated network monitoring equipment, which captures one record for each (potentially sampled) packet seen on a link. The former produces a list of flow records capturing the number of bytes and packets seen, while the latter produces a list of packet records that can also be aggregated into 5-tuple flows (e.g. with the same source, destination IP address, source, destination port and protocol). For more details, please see Chapter 3. The subsequent mapping of flows to application classes is not as straightforward, and has recently attracted attention in the research community.

While port numbers were always an approximate, yet sufficient, methodology to classify traffic, port-based estimates are currently significantly misleading due to the increase of applications tunneled through HTTP (e.g. chat, streaming), the constant emergence of new protocols and the dominance of peer-to-peer (p2p) networking. Indeed, studies have confirmed the failure of port-based classification [143].

To address the inefficiency of port-based classification, recent studies have employed statistical classification techniques to assign flows probabilistically to classes, e.g. machine learning [128] or statistical clustering [124, 146]. In such approaches, flows are grouped in a pre-determined number of clusters according to a set of discriminants, which usually includes the average packet size of a flow, the average flow duration and the inter-arrival times between packets (or the variability thereof). The size of the first four packets in an application has been further shown to allow highly accurate classification, even when the traffic is encrypted through SSL [36]. Studies have also examined how the exact timing and sequence of packet sizes can describe specific applications in the slightly different context of generating realistic application workloads [93].

One of the most challenging application types is p2p traffic. Quantifying p2p traffic is problematic, not only due to the large number of proprietary p2p protocols, but also because of the intentional use of random port numbers for communication. Payload-based classification approaches tailored to p2p traffic have been presented in refs. [104 and 179], while identification of p2p traffic through transport-layer characteristics is proposed in ref. [104]. In the same spirit, Dewes *et al.* [66] look into the problem of identifying and characterizing *chat* traffic. The approach presented in this chapter goes beyond previous efforts aimed at classifying most of the applications that generate the majority of today's Internet traffic by describing their underlying behavior.

13.3   Payload-based classification

This section describes the payload classifier used for the derivation of the ground truth in application consistency of the analyzed traces and establishes a comparison reference point. The data used feature the unique property of allowing for accurate classification, since the monitors capture the *full* payload of each packet instead of just the header as is commonly the case. Such data allow us to move beyond simple port-based application classification and establish a comparison benchmark. To achieve efficient payload classification, we develop a signature-matching classifier able to classify the majority of current Internet traffic.

13.3.1   Payload packet traces

We use packet traces collected using a high-speed monitoring box [141] installed on the Internet link of two access networks. We capture every packet seen on each direction of the link along with its full payload.[1]

---

[1] Note that the data set used in this chapter is different to the traces used so far, since, in order to obtain the ground truth in application classification, one needs access to full payload, which was not possible for the Sprint traces.

Table 13.1. General workload dimensions of the studied traces

Set	Date	Day	Start	Duration	Source IP
GN	2003-08-19	Tue	17:20	43.9 h	1455 K
UN1	2004-01-20	Tue	16:50	24.6 h	2709 K
UN2	2004-04-23	Fri	15:40	33.6 h	4502 K

Set	Destination IP	Packets	Bytes	Av. utilization	Av. no. of flows
GN	14 869 K	1000 M	495 G	25 Mbps	105 K
UN1	2626 K	2308 M	1223 G	110.5 Mbps	596 K
UN2	5742 K	3402 M	1652 G	109.4 Mbps	570 K

Table 13.1 lists the general workload dimensions of the data sets: the counts of distinct source and destination IP addresses, the number of packets and bytes observed, the average utilization and the average number of flows per five-minute interval. Flows are defined according to their 5-tuple, e.g. source and destination IP address, source and destination port and protocol. In accordance with previous work [56], a flow is expired if it is idle for 64 seconds. We process traces with CAIDA's Coral Reef suite [142]. The two Internet locations we use are the following:

- *Genome campus.* This trace (GN in Table 13.1) reflects traffic of several biology-related facilities. There are three institutions on-site that employ about 1000 researchers, administrators and technical staff.
- *Residential university.* We monitor numerous academic, research and residential complexes on-site (UN1 and UN2 traces in Table 13.1). Collectively, we estimate a user population of approximately 20 000. The residential nature of the university reflects traffic covering a wider cross-section of applications.

The two sites and time-of-capture of the analyzed traces were selected so that the presented methodology could be tested against a variety of different conditions and a diverse set of applications. Indeed, the selected links reflect significantly different network types; this difference will become evident in the following section, in which we examine the application mix of these links. In addition, the two university traces were collected both during weekdays (UN1) and also at the beginning of the weekend (UN2) to capture possible weekday-to-weekend variation in application usage and network-traffic patterns. Finally, the traces were captured several months apart to minimize potential similarities in the offered services and client interactions. Such dissimilar traces were intentionally selected to stress-test the belief that the proposed approach models generic networking characteristics instead of link or network idiosyncrasies, ergo being applicable without requiring previous training in any type of network.

13.3.2   Payload classification

Even with access to full packet payload, classification of traffic is far from trivial. The main complication lies in the fact that payload classification of traffic requires a priori knowledge of application protocol signatures, protocol interactions and packet formats. While some of the analyzed applications are well known and documented in detail, others operate on top of non-standard, usually custom-designed, proprietary protocols. To classify such diverse types of traffic, we develop a signature-based classifier in order to avoid manual intervention, automate the analysis and speed up the procedure.

Our classifier is based on identifying characteristic bit strings in the packet payload that potentially represent the initial protocol handshake in most applications (e.g. HTTP requests). Protocol signatures were identified either from "Requests for Comments" (RFCs) and public documents in case of well documented protocols, or by reverse-engineering and empirically deriving a set of distinctive bit strings by monitoring both TCP and UDP traffic using tcpdump. Table 13.2 lists a small subset of such signature (bit) strings for TCP and UDP. The complete list of bit strings we used is included in Appendix B.

Once the signatures have been identified, we classify traffic using a modified version of the crl_flow utility of the Coral Reef suite [142]. Our technique operates on two different timescales and traffic granularities. The short timescale operates on a per-packet basis upon each packet arrival. The coarse timescale essentially summarizes the results of the classification process during the preceding time interval (we use intervals of five minutes) and assists in the identification of flows that potentially have remained unclassified during payload analysis.

Both operations make use of an {IP, port} pair table that contains records of the IP-port pairs that have already been classified based on past flows. These {IP, port} pairs associate a particular IP address and a specific port with a code reflecting its causal application. The {IP, port} table is updated upon every successful classification and consulted at the end of each time interval for evidence that could lead to the classification of unknown flows or the correction of flows misclassified under the packet-level operation. Since the service reflected at a specific port number for a specific IP does not change at the timescales of interest, we use this heuristic to reduce processing overhead.

Table 13.2. Application-specific bit strings at the beginning of the payload. "0x" implies Hex characters

Application	String	Transport protocol
eDonkey2000	0xe319010000	TCP/UDP
MSN messenger	"PNG"0x0d0a	TCP
IRC	"USERHOST"	TCP
nntp	"ARTICLE"	TCP
ssh	"SSH"	TCP

Table 13.3. Categories and, applications analyzed and their average traffic percentages in flows (bytes)

Category	Application/protocol	GN	UN1	UN2
Web	www	32% (14.0%)	31.8% (37.5%)	24.7% (33.5%)
p2p	FastTrack, eDonkey2000, BitTorrent, Gnutella, WinMX, OpenNap, Soulseek, Ares, MP2P, Dirrect Connect, GoBoogy, Soribada, PeerEnbler	0.3% (1.2%)	25.5% (31.9%)	18.6% (31.3%)
data (FTP)	ftp, databases (MySQL)	1.1% (67.4%)	0.3% (7.6%)	0.2% (5.4%)
Network Management (NM)	DNS, Netbios, SMB, SNMP, NTP, SpamAssasin, GoToMyPc	12.5% (0.1%)	9% (0.5%)	9.4% (0.2%)
Mail	mail (SMTP, POP, IMAP, Identd)	3.1% (3.4%)	1.8% (1.4%)	2.5% (0.9%)
News	news (NNTP)	0.1% (4.0%)	0% (0.3%)	0% (0.2%)
Chat/irc (chtirc)	IRC, MSN messenger, Yahoo Messenger, AIM	3.7% (0.0%)	1.8% (0.2%)	5.8% (0.7%)
Streaming (strm)	MMS (WMP), Real, Quicktime, Shoutcast, Vbrick streaming, Logitech Video IM	0.1% (0.8%)	0.2% (6%)	0.2% (6.8%)
Gaming (gam)	HalfLife, Age of Empires, etc.	–	0.3% (0.1%)	0.3% (0.3%)
Non-payload	–	45.3% (2.2%)	24.9% (0.5%)	30.9% (1.0%)
Unknown	–	1.3% (6.6%)	4.3% (11.9%)	7.3% (16.9%)

To avoid increasing memory requirements by storing an immense number of {IP, port} pairs, we only keep {IP, port} pairs that reflect known services such as those described in Table 13.3. Finally, to identify data transfer flows further, such as passive ftp, we parse the control stream to acquire the context regarding the upcoming data transfer, i.e. the host and port number where the follow-up data connection is going to take place.

Procedure *Classifier* (Box 13.1) captures the per-packet operation. The procedure simply examines the contents of each packet against the array of strings (in Appendix B), and classifies the corresponding flow with an application-specific tag in case of a match. Successful classification of a flow in one direction leads to the subsequent classification of the respective flow in the reverse direction, if it exists. Previously classified flows are not examined unless they have been classified as HTTP. This further examination allows identification of non-Web traffic relayed over HTTP (e.g. streaming, p2p, Web-chat, etc.). Finally if a flow is classified, we store the {IP, port} pair if the port number reflects a well known service.

Box 13.1.

```
(1)  procedure CLASSIFIER
(2)      Get pkt and find flow from 5-tuple;
(3)      if Is flow classified then
(4)          if Is flow classified as HTTP then
(5)              check payload
(6)              go to 11:
(7)          else
(8)              get next packet
(9)      else
(10)         check payload
(11)         if Is there a match then
(12)             clasify flow
(13)             classify reverse direction          ▷ where applicable
(14)             store {IP, port}pair
(15)             get next packet
```

At the end of each time interval, we simply compare all flows against the list of known {IP, port} pairs to classify possible unknown flows or correct misclassifications (e.g. a p2p flow that was classified under Web, because the only packet so far was an HTTP request or response).

13.3.3   Application breakdown

We classify flows in 11 distinct application-type categories. Table 13.3 lists these categories, the specific applications and their share of traffic as a percentage of the total number of flows and bytes (in parentheses) in the link. The *non-payload* category includes flows that transfer only headers and no user data throughout their lifetime, while the *unknown* category lists the amount of traffic that could not be classified.

As expected, the two network types (GN vs UN) are significantly different. The UN network is mostly dominated by *web* and p2p traffic, whereas GN contains a large portion of *ftp* traffic reflecting large data transfers of Genome sequences. Despite the difference in the day of capture and the large interval between the two UN traces, their traffic mix is roughly similar. Other interesting observations from these traces are the following.

- *Non-payload* flows account almost for one-third of all flows in both links. Examination of these flows suggests that the vast majority correspond to failed TCP connections on ports of well known exploits or worms (e.g. 135, 445). A large percentage of address-space scans is also implied by the large number of destination IPs, especially in the GN trace.
- *Unknown flows*. The existence of user payload data does not guarantee that all flows in the traces will be classified. Our analysis of the most popular applications cannot

possibly guarantee identification of all applications contributing traffic to the Internet. For example, 4–7% of all flows (10% in bytes) of the UN traffic cannot be identified. Note that a fraction of this unknown traffic is due to experimental traffic from *PlanetLab* [2], a global research network that supports the development of new network services (three *PlanetLab* nodes exist behind the monitoring point).

13.4 Transport-layer classification

This section describes a multi-level methodology, *BLINC*, for the classification of flows into applications without the use of the payload or "well known" port numbers. *BLINC* realizes a rather different philosophy compared with other approaches proposed in the area of traffic classification. The main differences are the following.

- We do not treat each flow as a distinct entity; instead, we focus on the source and destination hosts of the flows. We advocate that when the focus of the classification approach is shifted from the flow to the host, one can then accumulate sufficient information to disambiguate the roles of each host across different flows and thus identify specific applications.
- The proposed approach operates on flow records and requires no information about the timing or the size of individual packets. Consequently, the input to the methodology may potentially be flow record statistics collected by currently deployed equipment.
- The proposed approach is insensitive to network dynamics, such as congestion or path changes, which can potentially affect statistical methodologies that rely heavily on inter-arrival times between the packets in a flow.

13.4.1 Overview of BLINC

BLINC operates on flow records. Initially, we parse all flows and gather host-related information reflecting transport-layer behavior. We then associate the host behavior with one or more application types and thus indirectly classify the flows. The host behavior is studied across three different levels as follows.

- At the *social level*, we capture the behavior of a host in terms of the number of other hosts it communicates with, which we refer to as *popularity*. Intuitively, this level focuses on the diversity of the interactions of a host in terms of its destination IPs and the existence of user communities. As a result, we only need access to the source and destination IP addresses at this level.
- At the *functional level*, we capture the behavior of the host in terms of its functional role in the network; that is, whether it is a provider or consumer of a service, or whether it participates in collaborative communications. For example, hosts that use a single source port for the majority of their interactions are likely to be providers of a service offered on that port. At this level, we analyze characteristics of the source and destination IP address and the source port.

- At the *application level*, we capture the transport-layer interactions between hosts with the intent to identify the application of origin. We first provide a classification using only the 4-tuple (IP addresses and ports), and then we refine the final classification by developing heuristics that exploit additional flow information, such as the number of packets or bytes transferred, as well as the transport protocol. For each application, we capture host behavior using empirically derived patterns. We represent these patterns using graphs, which we call *graphlets*. Having a library of these *graphlets*, we then seek for a match in the behavior of a host under examination.

We want to stress that, throughout the described approach, we treat the port numbers as indexes without any application-related information. For example, we count the number of distinct ports a host uses, but we do not assume in any way that the use of port 80 signifies Web traffic.

While the preceding levels are presented in order of increasing detail, they are equally significant. Not only does the analysis at each level benefit from the knowledge acquired in the previous level, but also the classification depends on the unveiled "cross-level" characteristics. However, the final classification of flows into applications cannot be achieved without the examination of the application-level characteristics.

A key advantage of the proposed approach is its tunability. The strictness of the classification criteria can be tailored to the goal of the measurement study. These criteria can be relaxed or tightened to provide results at different points in the tradeoff between the completeness of the classification and its accuracy.

BLINC provides two types of output. First, it reports aggregate per-class statistics, such as the total number of packets, flows and bytes. Second, it produces a list of all flows (5-tuple) tagged with the corresponding application for every time interval. Furthermore, *BLINC* can detect unknown or non-conformant hosts and flows (see Section 13.5).

Throughout the rest of this section, we will present characteristic samples of behaviors at each level as seen in the traces. For the sake of presentation, we will only use examples generated for a single time interval of one of the traces (5 or 15 minutes); however, these observations were typical for all studied traces.

13.4.2   Classification at the social level

We identify the social role of each host in two ways. First, we focus on its *popularity*, namely the number of distinct hosts it communicates with. Second, we detect *communities* of hosts by identifying and grouping hosts that interact with the same set of hosts. A community may signify a set of hosts that participate in a collaborative application or offer a service to the same set of hosts.

Examining the social behavior of single hosts

The social behavior of a host refers to the number of hosts this particular host communicates with. To examine variations in host social behavior, Figure 13.1 presents

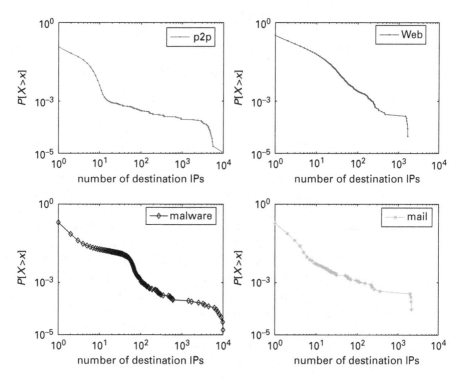

Fig. 13.1. Complementary cumulative distribution function of destination IP addresses per source IP for 15 minutes of the UN1 trace for four different applications. ©ACM, 2005.

the complementary cumulative distribution function (CCDF[2]) of the host popularity. Based on payload classification from Section 13.3, we display four different CCDFs corresponding to different types of traffic, namely *web*, *p2p*, *malware* (e.g. failed non-payload connections on known malware ports) and *mail*. In all cases, the majority of sources appear to communicate with a small set of destination IPs.

In general, the distribution of the host *popularity* cannot reveal specific rules in order to discriminate specific applications, since it is highly dependent upon the type of network, link or even the specific IPs. However, this distribution allows us to distinguish significant differences among applications. For example, hosts interacting with a large number of other hosts in a short time period appear to participate either in a p2p network or to constitute malicious behavior. In fact, the *malware* curve appears flat below 100 destination IPs, denoting the presence of a large number of possible address-space scans, where a number of sources has the same number of destination IPs during a specific time interval.

Detecting communities of hosts

Social behavior of hosts is also expressed through the formation of communities or clusters between sets of IP addresses. Communities will appear as *bipartite cliques* in the traces, like the one shown in Figure 13.2. The bipartite graph is a consequence of

---

[2] A CCDF plots the probability of the distribution exceeding the value on the x-axis.

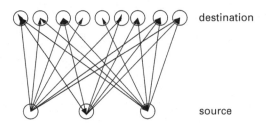

Fig. 13.2.   An example of a community in the traces: the graph appears as an approximate bipartite clique. ©ACM, 2005.

the single observation point. Interactions between hosts from the same side of the link are not visible, since they do not cross the monitored link. Communities in the bipartite graph can be either exact cliques, where a set of source IPs communicates with the exact same set of destination IPs, or approximate cliques, where a number of the links that would appear in a perfect clique is missing. Communities of interest have also been studied in ref. [20], in which a community is defined by either the popularity of a host or the frequency of communication between hosts. Our definition of community is targeted to groups of hosts per application type.

Identifying the communities is far from trivial, since identifying maximal cliques in bipartite graphs is an NP-Complete problem. However, there exist polynomial algorithms for identifying the *cross-associations* in the data mining context [50]. Cross-association is a joint decomposition of a binary matrix into disjoint row and column groups such that the rectangular intersections of groups are homogeneous or formally approximate a bipartite clique. In our case, this binary matrix corresponds to the interaction matrix between the source and destination IP addresses in the traces.

To showcase how communities can provide interesting features of host behavior, we apply the cross-association algorithm to gaming traffic for a small time period in one of the traces (a 5-minute interval of the UN1 trace) and we successfully identify communities of gamers. Specifically, Figure 13.3 presents the interaction matrix after the execution of the cross-association algorithm. The axes present source (x-axis) and destination (y-axis) IPs (350 total IPs), while the matrix is essentially the original interaction matrix shifted in such a way that strongly connected components appear clustered in the same area. The horizontal and vertical lines display the boundaries of the different clusters. Specifically, we observe two major clusters. First, three source IPs communicating with a large number of destination IP addresses although not an exact clique (at the bottom of Figure 13.3, x-axis: 0–280; y-axis: 347–350). Second, an exact clique of five hosts communicating with the exact same 17 destination IPs (x-axis: 280–285; y-axis: 300–317).

In general, we study three different types of communities, according to their deviation from a perfect clique as follows.

"Perfect" cliques: a hint for malicious flows
While the previous example displays a perfect clique in gaming traffic, we find that perfect cliques are mostly signs of malicious behavior. In our analysis, we identify

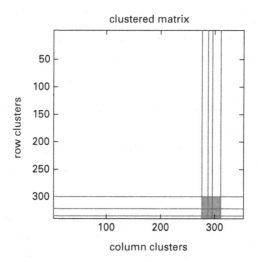

Fig. 13.3. Communities of on-line game players appear as highly connected clusters in the interaction matrix after applying the cross-associations algorithm (UN1 trace, 5 minutes). ©ACM, 2005.

a number of hosts communicating with the exact same list of IP addresses (approximately 250 destination IPs in 15 minutes). Further analysis revealed that these cases represented malicious traffic, such as flows belonging to the Windows RPC exploit and Sasser worm.

Partial overlap: collaborative communities or common-interest groups

In numerous cases, only a moderate number of common IP addresses appear in the destination lists for different source IPs. These cases correspond to peer-to-peer sources, gaming and also clients that appear to connect to the same services at the same time (e.g. browsing the same Web pages or streaming).

Partial overlap within the same domain: service "farms"

Closer investigation of partial overlap revealed hosts interacting with a number of IP addresses within the same domain, e.g. IP addresses that differ only in the least significant bits. Payload analysis of these IPs revealed that this behavior is consistent with service "farms": multi-machine servers that load-balance requests of a host to servers within the same domain. We find that service farms were used to offer *web*, *mail*, *streaming*, or even *DNS* services.

Conclusion and rules

Based on the above analysis, we can infer the following regarding the social behavior of network hosts. First, "neighboring" IPs may offer the same service (e.g. server farms). Thus, identifying a server might be sufficient to classify such "neighboring" IPs under the same service (if they use the same service port). Second, exact communities may indicate attacks. Third, partial communities may signify p2p or gaming applications. Finally, most IPs act as clients contacting a minimum number of destination IPs. Thus, focusing on the identification of the small number of servers can retrospectively pinpoint the clients and lead to the classification of a large portion of the traffic, while limiting the amount of associated overhead. Identification of server hosts is accomplished through the analysis of the functional role of the various hosts.

13.4.3   Classification at the functional level

At this level, we identify the functional role of a host: hosts may primarily offer services, use services, or both. Most applications operate within the server–client paradigm. However, several applications interact in a collaborative way, with p2p networks being the prominent example.

We attempt to capture the functional role by using the number of *source* ports a particular host uses for communication. For example, let us assume that host A provides a specific service (e.g. Web server), and we examine the flows where A appears as a source. Then, A is likely to use a single source port in the vast majority of its flows. In contrast, if A were a client to many servers, its source port would vary across different flows. Clearly, a host that participates in only one or a few flows would be difficult to classify.

To quantify how the number of used source ports may distinguish client from server behavior, we examine the distribution of the source ports a host uses in the traces. In Figure 13.4, we plot the number of flows (x-axis) versus the number of source ports

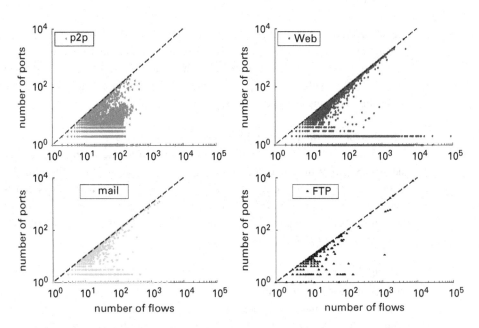

Fig. 13.4.   Number of source ports versus number of flows per source IP address in the UN1 trace for a 15-minute interval for four different applications. In client–server applications (Web, FTP, mail), most points fall on the diagonal or on horizontal lines for small values in the y-axis (number of used ports). In p2p, points are clustered in between the diagonal and the x-axis. ©ACM, 2005.

(y-axis) each source IP uses for 15 minutes in the UN1 trace.[3] Each subplot of Figure 13.4 presents traffic from a different application as identified by payload analysis. We identify three distinct behaviors.

Typical client–server behavior

Client–server behavior is most evident for Web traffic (Figure 13.4, top right), where most points fall either on the diagonal or on horizontal lines parallel to the x-axis for values of $y \leq 2$. The first case (where the number of ports is equal to the number of distinct flows) represents clients that connect to Web servers using as many ephemeral source ports as the connections they establish. The latter case reflects the actual servers that use one ($y = 1$, port 80, HTTP) or two ($y = 2$, port 80, HTTP and 443, HTTPS) source ports for all of their flows.

Typical collaborative behavior

In this case, points are clustered between the x-axis and the diagonal, as shown in the case in Figure 13.4 (top left), where discrimination between client and server hosts is not possible.

Obscure client–server behavior

In Figure 13.4, we plot the behavior for the case of *mail* and *FTP*. While *mail* and *ftp* fall under the client–server paradigm, the behavior is not as clear as in the Web case for two reasons.

(1) The existence of multiple application protocols supporting a particular application, such as *mail*. *Mail* is supported by a number of application protocols, i.e. SMTP, POP, IMAP, IMAP over SSL, etc., each of which uses a different service port number. Furthermore, mail servers often connect to *Razor* [12] databases through *SpamAssassin* to report spam. This practice generates a vast number of small flows destined to *Razor* servers, where the source port is ephemeral and the destination port reflects the SpamAssassin service. As a result, mail servers may use a large number of different source ports.

(2) Applications supporting control and data streams, such as *ftp*. Discriminating client–server behavior is further complicated in cases of separate control and data streams. For example, passive *ftp*, where the server uses as source ports a large number of ephemeral ports different than the service ports (20, 21), will conceal the FTP server.

Conclusion and rules

If a host uses a small number of source ports, typically two or fewer, for every flow, then this host is likely providing a service. Our measurements suggest that if a host uses only *one* source port number, then this host reflects a Web, chat or SpamAssassin server for TCP, or falls under the network management category for UDP.

---

[3] The source {IP, port} pair is used without loss of generality. Observations are the same in the destination {IP, port} case.

13.4.4   Classification at the application level

At this level, we combine knowledge from the two previous levels coupled with transport-layer interactions between hosts in order to identify the application of origin. The basic insight exploited by the presented methodology is that interactions between network hosts display diverse patterns across the various application types. We first provide a classification using only the 4-tuple (IP addresses and ports), and then we refine it using further information regarding a specific flow, such as the protocol or the average packet size.

We model each application by capturing its interactions through empirically derived signatures. We visually capture these signatures using *graphlet*s that reflect the "most common" behavior for a particular application. A sample of application-specific *graphlet*s is presented in Figure 13.5. Each *graphlet* captures the relationship between the use of source and destination ports, the relative cardinality of the sets of unique destination ports and IPs as well as the magnitude of these sets.

Having a library of these *graphlet*s allows us to classify a host by identifying the closest matching behavior. Since unknown behavior may match several *graphlet*s, and the success of the classification will then have to rely on operator-defined thresholds to control the strictness of the match.

In more detail, each *graphlet* has four columns corresponding to the 4-tuple source IP, the destination IP, the source port and the destination port. We also show some *graphlet*s with five columns, where the second column corresponds to the transport protocol (TCP or UDP) of the flow. Each node[4] presents a *distinct* entry to the set represented by the corresponding column, e.g. 135 in *graphlet* 13.5(a) is an entry in the set of destination ports. The lines connecting nodes imply that there exists at least one flow whose packets contain the specific nodes (field values). Dashed lines indicate links that may or may not exist and are not crucial to the identification of the specific application. Note that, while some of the *graphlet*s display port numbers, the classification and the formation of *graphlet*s do not associate in any way a specific port number with an application.

The order of the columns in the visual representation of each *graphlet* mirrors the steps of the multi-level approach. Our starting field, the source IP address, focuses on the behavior of a particular host. Its *social* behavior is captured in the fanout of the second column, which corresponds to all destination IPs this particular source IP communicates with. The functional role is portrayed by the set of source port numbers. For example, if there is a "knot" at this level, the source IP is likely to be a server, as mentioned before. Finally, application types are distinguished using the relationship of all four different fields. Capturing application-specific interactions in this manner can distinguish diverse behaviors in a rather straightforward and intuitive manner, as shown in Figure 13.5.

---

[4] The term "node" indicates the components of a *graphlet*, while the term "host" indicates an end point in a flow.

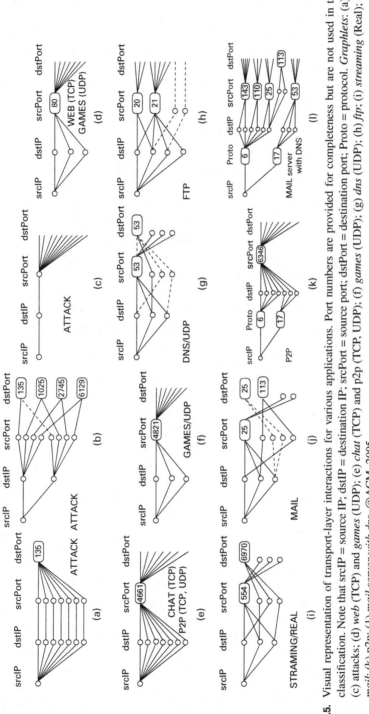

Fig. 13.5. Visual representation of transport-layer interactions for various applications. Port numbers are provided for completeness but are not used in the classification. Note that srcIP = source IP; dstIP = destination IP; srcPort = source port; dstPort = destination port; Proto = protocol. *Graphlets*: (a) – (c) *attacks*; (d) *web* (TCP) and *games* (UDP); (e) *chat* (TCP) and p2p (TCP, UDP); (f) *games* (UDP); (g) *dns* (UDP); (h) *ftp*; (i) *streaming* (Real); (j) *mail*; (k) p2p; (l) *mail server with dns*. ©ACM, 2005.

Let us highlight some interesting cases of *graphlet*s. The top row of Figure 13.5 displays three types of attacks (*graphlet*s (a)–(c)). Figure 13.5(a) displays a typical attack in which a host scans the address space to identify vulnerability at a particular destination port. In such cases, the source host may or may not use different source ports, but such attacks can be identified by the large number of flows destined to a given destination port. A similar but slightly more complicated type of attack common in the traces involves hosts attempting to connect to several vulnerable ports at the same destination host (Figure 13.5(b)). Similarly, we show the *graphlet* of a typical port scan of a certain destination IP in Figure 13.5(c).

The power of the presented method lies in the fact that we do not need to know the particular port number ahead of time. The surprising number of flows at the specific port will raise the suspicion of the network operator. Such behaviors are also identifiable by tools such as *AutoFocus* [75], which, however, do not target traffic classification.

In some cases, hosts offering services on certain ports exhibit similar behavior. For instance (the server side), *web* and *games* all result in the same type of *graphlet*: a single source IP communicates with multiple destinations using the same source port (the service port) on several different destination ports. In such cases, we need further analysis to distinguish between applications. First, we can use quantitative criteria such as the relative cardinality of the sets of destination ports versus destination IPs. As we will describe later in this section, the use of the transport protocol, TCP versus UDP, can further help to discriminate between applications with similar or complicated *graphlet*s depicted in Figure 13.5 (e)–(l).

Applications such as *FTP*, *streaming* or *mail* present more complicated *graphlet*s, exhibiting "cris-cross" flow interactions (Figure 13.5(h)–(j)). These *graphlet*s, have more than one service ports, or have both source and destination service ports. In the case of *ftp*, the source host provides the service at two main ports (control and data channel), whereas other source ports represent the case of *passive ftp*. *Streaming*, on the other hand, uses specific port numbers both at the source and the destination side. Streaming users (destination IPs in our case) connect at the service port (TCP) of the streaming server (control channel), while the actual streaming service is initiated by the server using an ephemeral random source port to connect to a pre-determined UDP user port. Similarly, *mail* uses specific port numbers at the source and destination side, yet all mail flows are TCP. *Mail* servers may further use port 25 both as source or destination port across different flows while connecting to other mail servers to forward mail. As previously noted, the specific port numbers are only listed to help with the description of these *graphlet*s and they are in no way considered in the algorithm.

Lastly, *graphlet*s become even more complex when services are offered through multiple application and/or transport protocols. As an example, Figure 13.5(l) presents a mail server supporting IMAP, POP, SMTP and *Ident*, while also acting as a DNS server. Knowledge of the role of the host may assist as corroborative evidence on other services offered by the same host. For instance identifying a host as an SMTP server suggests that the same host may be offering POP, IMAP, DNS (over UDP) or even communicate with SpamAssassin servers.

13.4.5 Heuristics

Here, we present a set of final heuristics that we use to refine the classification and discriminate complex or similar cases of *graphlets*. This set of heuristics has been derived empirically through inspection of interactions present in various applications in the traces.

Heuristic 1: using the transport-layer protocol

One criterion for such a distinction is the transport-layer protocol used by the flow. The protocol information can distinguish similar *graphlets* into three groups using: (a) *TCP*, which includes p2p, *web*, *chat*, *ftp* and *mail*, (b) *UDP*, which includes *Network Management traffic* and *games* and (c) both protocols, which includes p2p and *streaming*. For example, while *graphlets* for *mail* and *streaming* appear similar, mail interactions occur only on top of TCP. Another interesting case is shown in Figure 13.5(k), where p2p protocols may use both TCP and UDP with a single source port for both transport protocols (e.g. *Gnutella*, *Kazaa*, *eDonkey*, etc.). With the exception of *dns*, the traces suggest that this behavior is unique to p2p protocols.

Heuristic 2: using the cardinality of sets

As discussed earlier, the relative cardinality of destination sets (ports versus IPs) is able to discriminate between different behaviors. Such behaviors may be *web* versus p2p and *chat*, or *Network Management* versus *gaming*. Figure 13.6 presents the number of distinct destination IPs versus the number of distinct destination ports for each source

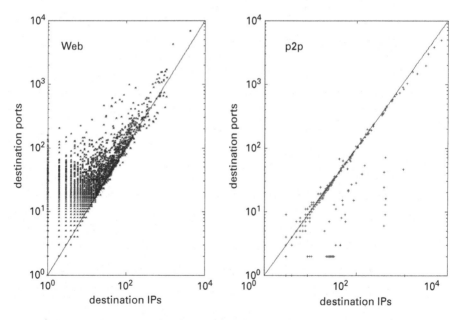

Fig. 13.6. Relationship between the number of destination IP addresses and ports for specific applications per source IP. The cardinality of the set of destination ports is larger than the one of destination IPs reflected in points above the diagonal for the Web. On the contrary, points in the p2p case fall either on top or below the diagonal. ©ACM, 2005.

IP in 15 minutes of the UN2 trace, for *web* and p2p. In the *web* case, most points concentrate above the diagonal, representing parallel connections of mainly simultaneous downloads of Web objects (many destination ports correspond to one destination IP). On the contrary, most points in the p2p case are clustered either close to the diagonal (the number of destination ports is equal to the number of destination IPs) or below (which is common for UDP communications, where the destination port number is constant for some networks).

Heuristic 3: using the per-flow average packet size

A number of applications display unique behavior regarding packet sizes. For instance, the majority of *gaming*, *malware* or *SpamAssassin* flows are characterized by a series of packets of constant size. Thus, constant packet size can discriminate between certain applications. Note that it is not the actual size that is the distinctive feature, but rather the fact that packets have the same size across all flows; in other words, we simply need to examine whether the average packet size per flow (e.g. the fraction of total bytes over the number of packets) remains constant across flows.

Heuristic 4: community heuristic

As discussed in the social behavior of network hosts, communities offer significant knowledge regarding interacting hosts. Thus, examining IP addresses within a domain may facilitate classification for certain applications. We apply the community heuristic to identify "farms" of services by examining whether "neighboring" IPs exhibit server behavior at the source port in question.

Heuristic 5: recursive detection

Hosts offering specific types of services can be recursively identified by the interactions among them (variation of the community heuristic). For example *mail* or *dns* servers communicate with other such servers and use the same service port both as source or destination port across different flows. Also, *SpamAssassin* servers should only communicate with mail servers.

Heuristic 6: *non-payload* flows

Non-payload or failed flows usually point to attacks or even p2p networks (clients often try to connect to IPs that have disconnected from the p2p network). The magnitude of failed flows can hint toward types of applications. As previously mentioned, the strictness of classification depends on operator-defined thresholds. These thresholds, which implicitly originate from the structure of the *graphlet*s and the heuristics, include:11

- the minimum number of distinct destination IPs observed for a particular host (T_d) required for *graphlet* matching (e.g. at least T_d IPs are needed for a host to match the p2p *graphlet*);
- the relative cardinality of the sets of destination IPs and ports (T_c) (e.g. for the p2p *graphlet*, it will define the maximum difference between the cardinalities of the two sets so that the *graphlet* is allowed to match and $T_c = 0$ indicates that the cardinalities must be equal);

- the number of distinct packet sizes observed (T_s) (defines the maximum number of distinct average packet sizes per flow below which Heuristic 3 is considered);
- the number of payload versus non-payload flows (T_p) (defines the maximum ratio for Heuristic 6 to be considered).

Note that these thresholds may be specific to the *graphlet*.

13.5 Classification results

We demonstrate the performance of the approach when applied to the traces described in Section 13.3. Overall, we find that *BLINC* is very successful at classifying accurately the majority of the flows in *all* studied traces.

We use two metrics to evaluate the success of the classification method. *Completeness* measures the percentage of the traffic classified by the described approach. In more detail, completeness is defined as the ratio of the number of classified flows (bytes) by *BLINC* over the total number of flows (bytes) indicated by payload analysis. *Accuracy* measures the percentage of the classified traffic by *BLINC* that is correctly labeled. In other words, accuracy captures the probability that a classified flow belongs to the class (according to payload) that *BLINC* indicates. Note that both these metrics are defined for a given time interval, which could be either over timescales of minutes or the whole trace, and can be applied to each application class separately or to the entire traffic.

The challenge for any method is to maximize both metrics, which, however, exhibit a tradeoff relationship. The number of misclassifications will increase depending on how aggressive the classification criteria are. These criteria refer to the thresholds discussed in the previous section and can be tuned accordingly depending on the purpose of the measurement study. In this work, the thresholds have been tuned in the *UNI* trace and applied as such in the rest of the traces. We examine the sensitivity of the presented approach relative to the classification thresholds in Section 13.5.2.

We use the payload classification as a reference point (Section 13.3) to evaluate *BLINC*'s performance. Given that the payload classifier has no information to classify *non-payload* flows, such flows need to be excluded from the comparison to level the field. Further, we have no way of characterizing "unknown" flows according to payload analysis. Consequently, the total amount of traffic used to evaluate *BLINC* for each trace does not include *non-payload* and *unknown* (according to payload) flows, which are discussed separately at the end of this section. However, our approach is even able to characterize flows where payload analysis fails.

13.5.1 Overall completeness and accuracy

BLINC classifies the majority of the traffic with high accuracy

In Figure 13.7, we plot the completeness and accuracy for the entire duration of each trace. In the *UN* traces, *BLINC* classifies more than 90% of the flows with approximately 95% accuracy. For the *GN* trace, *BLINC* classifies approximately 80% of the flows with 99% accuracy.

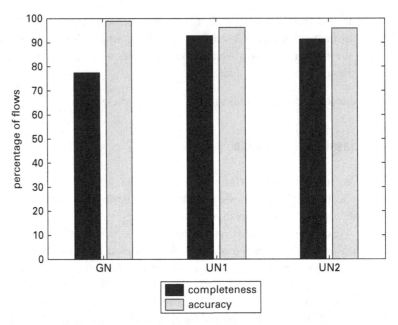

Fig. 13.7.   Accuracy and completeness of all classified flows. For UN traces more than 90% of the flows
are classified with approximately 95% accuracy. In the GN trace, we classify approximately
80% of the flows with 99% accuracy. ©ACM, 2005.

BLINC closely follows traffic variation and patterns in time

To stress-test our approach, we examine the classification performance across smaller
time intervals. In Figure 13.8, we plot flows (a) and bytes (b) classified with *BLINC* versus the payload classifier, computed over 5-minute intervals for the *UN1* data set. The
top line represents all classified flows according to the payload classifier, the middle
line represents flows classified by *BLINC*, and the bottom line represents flows classified correctly. The performance seems consistently robust over time. In terms of bytes,
completeness ranges from 70–85% for the *UN* traces and 95% for the *GN* trace with
more than 90% accuracy. It is interesting to note that the difference between *BLINC*
and payload in terms of bytes is due to a small number of large volume flows. In these
flows, both source and destination hosts do not present a sufficient number of flows in
the whole trace and thus cannot be classified with *BLINC* without compromising the
accuracy.

High per-application accuracy

Figure 13.9 presents the accuracy and completeness for each of the four dominant applications of each trace, collectively representing more than 90% of all the flows. In all
cases, accuracy is approximately 80% or more, and completeness in most cases exceeds
80%. Note that per-class accuracy and completeness depend on the total amount of traffic in each class. For example, *web*-related metrics always exceed 90% in *UN* traces
since *web* is approximately one-third of all traffic. In *GN*, where *web* is approximately
15% of the total bytes, completeness is approximately 70% (99% accuracy).

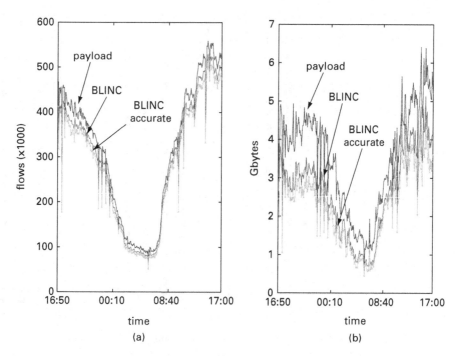

Fig. 13.8. Accuracy and completeness of *BLINC* in *UN1* trace in time (5-minute intervals). (a) Flows; (b) bytes. The top line represents flows (bytes) classified using the payload, the middle line represents flows (bytes) classified by *BLINC*; the bottom line represents flows (bytes) classified correctly by *BLINC*. The three lines coincide visually, indicating high completeness and accuracy. ©ACM, 2005.

13.5.2   Fine-tuning BLINC

The tradeoff between accuracy and completeness directly relates to the "strictness" of the classification criteria, as discussed in Section 13.4. Here, we study the effect of one of the thresholds we use in our approach. In classifying a host as a p2p candidate, we require that the host participates in flows with at least T_d distinct destination IPs. Setting T_d to a low value will increase completeness since *BLINC* will classify more hosts and their flows as p2p. However, the accuracy of the classification may decrease. In Figure 13.10, we plot the accuracy and completeness for p2p flows (left columns) and the total number of classified flows (right columns) for two different values of T_d: $T_d = 1$ and $T_d = 4$. We observe that by reducing the threshold, the fraction of classified flows increases, whereas the fraction of correctly identified flows drops from 99% to 82%. Note that the total accuracy is also affected (as previously "unknown" flows are now (mis)classified), but the decrease for total accuracy is much smaller than in the p2p case, dropping from approximately 98% to 93%. In all previous examples, we have used a value of $T_d = 4$, opting for accuracy.

This flexibility is a key advantage of the presented approach. We claim that, for a network operator, it may be more beneficial if *BLINC* opts for accuracy. Misclassified flows are harder to detect within a class of thousands of flows, whereas unknown flows can

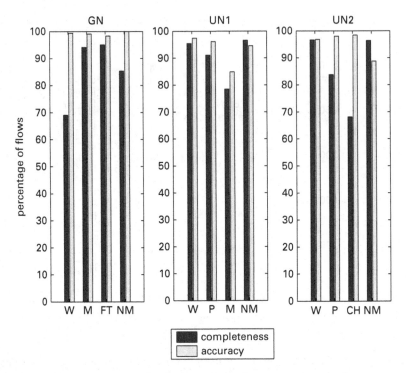

Fig. 13.9. Completeness and accuracy per application type. For each trace, we show the four most dominant applications, which contribute more than 90% of the flows. W: *web*, P: p2p, FT: *ftp*, M: *mail*, CH: *chat*, NM: *Network Management*. ©ACM, 2005.

potentially be examined separately by the operator by using additional external information such as *BLINC*'s social and functional role reports, or application specifications and discussions with other operators.

13.5.3   Characterizing "unknown" flows

In some cases, *BLINC* goes beyond the capabilities of the payload classification. Although we unavoidably use payload analysis as a benchmark, payload classification fails in two cases: (a) it cannot characterize non-payload flows (zero payload packets) and (b) it cannot profile traffic originating from applications that are not analyzed a priori. In contrast, *BLINC* has the ability to uncover transport-layer behavior that may allow for the classification of flows originating from previously unknown applications that fall under the *graphlet* modeled types (e.g. a new p2p protocol).

Non-payload flows

The multilevel analysis of *BLINC* highlighted that the vast majority of non-payload flows were due to IP address scans and port scans. Figure 13.11 presents the histogram of destination ports in the flows that were classified as address space scans for two different traces using the attack *graphlet*s (Figure 13.5 (a) – (c)). Inspecting the peaks of this histogram shows that *BLINC* successfully identified destination ports of well

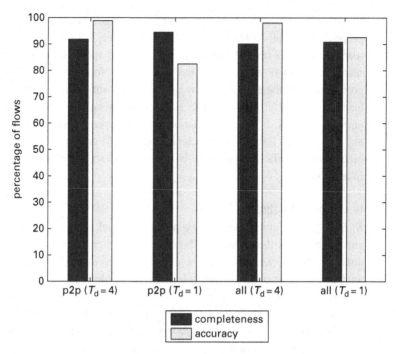

Fig. 13.10.   Tradeoff of accuracy versus completeness for p2p and the total number of flows. Decreasing the number of samples required to detect p2p behavior increases completeness but decreases accuracy. ©ACM, 2005.

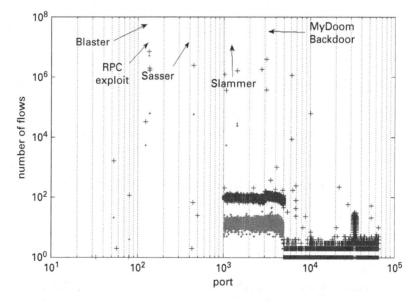

Fig. 13.11.   Histogram of destination ports for flows classified under address-space scans for *GN* and *UN2* traces. *BLINC* successfully discriminates major address-space scans at ports of "well known" worms or exploits. ©ACM, 2005

known worms or exploits, some of which are highlighted in the plot for visualization purposes. In total, *BLINC* classified approximately 26 million flows as address-space scans in the *UN2* trace. In addition, we classified approximately 100 000 flows as port scanning on 90 IP addresses in the same trace. Note that we did not need to use the port number of the exploits or any other external information. On the contrary, *BLINC* helped us identify the vulnerable ports by showing ports with unusually high traffic targeted at many different destination IPs. However, the presented approach cannot in any way replace IDS systems such as *SNORT* [15] or *Bro* [6]. *BLINC* can only provide hints toward malicious behavior by detecting destination ports with high activity of failed flows.

Unknown applications

BLINC has the ability to identify previously "unknown" protocols and applications, since it captures the underlying behavior of application protocols. Indeed, during our analysis, *BLINC* identified a new p2p protocol (classified as unknown with payload analysis) running on the *PlanetLab* network [2] (three *PlanetLab* nodes are behind the monitoring point). This p2p application corresponded to the *Pastry* project [161], which we identified after inspecting the payload, while we were examining the false positives. *BLINC* also identified a large number of gaming flows, which were classified as unknown by payload analysis.

13.6   Discussion

Implementing *BLINC* is not as straightforward as the presentation may have led us to believe. We present the implementation challenges and issues and discuss *BLINC*'s properties and limitations.

13.6.1   Implementation issues

We would like to highlight two major functions of the implementation: (a) the generation of graphlets and (b) the matching process of an unclassified host against the graphlets. The first function can be executed once in the beginning or periodically in an off-line fashion. Ideally, the second function should be sufficiently fast in order to enable the real-time monitoring of a network. This way, the processing of the data for a given time interval should complete before the data for the next interval become available. As we will see, the implementation is sufficiently fast for this purpose.

Creating the graphlets

In developing the graphlets, we used all possible means available: empirical observations, trial and error and hunches. An automated way of defining new graphlets is an interesting and challenging problem that is left for future work. In our experience, we typically followed the steps below for creating the majority of the graphlets: (i) detection of the existence of a new application (which could be triggered from unusual amounts of unknown traffic); (ii) manual identification of the hosts involved in the unknown activity; (iii) derivation of the graphlet according to the interactions observed; (iv) verification using human supervision and, partially, *BLINC*.

Matching process among different graphlets

The general idea here is to examine the unknown host against all graphlets and determine the best match. The approach we follow uses a carefully selected order of the graphlets from more specific to more general. This way, once a match is found the host is classified and the matching process is stopped. This ordered approach increases the speed of the matching process compared with examining all possible graphlets every time.

Extensibility: adding new *graphlets*

As mentioned in previous sections, *BLINC* is extensible by design. In fact, this was an exercise we had to go through ourselves in our attempt to develop graphlets to capture the majority of the applications in the traces. The addition of a graphlet requires careful consideration. Before the insertion of the new graphlet in the library, one needs find to the right place in the order and eliminate race conditions between the new graphlet and other graphlets with which they may compete for the same hosts. If the new graphlet is unique, attention needs to be paid regarding its position in the matching order. If the new graphlet presents significant similarities with other graphlets, then the order must be carefully examined and potentially additional distinguishing features need to be derived. Currently, the implementation of *BLINC* utilizes three special purpose data structures that capture the diverse application behavior across the graphlets in the library. The mapping algorithm then goes through each flow and maps it to the application corresponding to the graphlet that best matches the profile of a flow's source or destination host. To avoid breaking up the flow of this chapter, we present a description of the three structures in Appendix C along with the pseudo-code that performs the actual mapping of flows into applications.

Computational performance

The first version of *BLINC* shows great promise in terms of computational efficiency. Despite the fact that the current C++ implementation has hardly been optimized, *BLINC* classified the largest and longest (34-hour) UN2 trace in less than 8 hours (flow tables were computed over 5-minute intervals); processing took place on a DELL PE2850 with a Xeon 3.4 GHz processor and 2 GB of memory, of which maximum memory usage did not surpass 40%. Consequently, *BLINC* appears sufficiently efficient to allow for a real-time implementation alongside currently available flow collectors.

13.6.2 Limitations

Classifying traffic in the dark has several limitations. Note that many of these limitations are not specific to our approach, but are inherent to the problem and its constraints. *BLINC* cannot identify specific application subtypes: our technique is capable of identifying the type of an application, but may not be able to identify distinct applications.

For instance, we can identify p2p flows, but it is unlikely that we can identify the specific p2p protocol (e.g. *eMule* versus *Gnutella*) with packet header information alone. Naturally, this limitation could be easily addressed, if we have additional information such as the specifications of the different protocols or in the case of distinctive behavior at the transport layer. We believe that, for many types of studies and network management functions, this finer classification may not be needed. For example, the different instances of the same application type may impose the same requirements on the network infrastructure.

Encrypted transport-layer headers

Our entire approach is based on relationships between the fields of the packet header. Consequently, our technique has the ability to characterize encrypted traffic, as long as the encryption is limited to the transport-layer payload. Should layer-3 packet headers also be encrypted, our methodology cannot function. However, this is probably true for most classification methods.

Handling NATs

Note that BLINC may require some modification to classify flows that go through Network Address Translators (NATs). Some classification may be possible, since our method examines the behavior of IP, port pairs, and thus different flows sourcing behind the NAT will be discriminated through the port number. However, we have no results to argue one way or the other, since we have not encountered (or identified) any flows that pass through NATs in the traces.

13.7   Lessons learned

In this chapter, we presented *BLINC*, a traffic classification approach with significantly different philosophy compared with existing efforts. The novelty of *BLINC* lies in two key features. First, it classifies hosts by capturing the fundamental patterns of their behavior at the transport layer. Second, it defines and operates at three levels of host behavior: (i) the social level; (ii) the functional level; (iii) the application level. Additionally, *BLINC* is tunable, striking the desired point of balance in the tradeoff between the *percentage* of classified traffic and its *accuracy*.

Application of *BLINC* on three real traces resulted in very promising results. *BLINC* classified approximately 80–90% of the total number of flows in each trace, with 95% accuracy. In terms of individual application types, *BLINC* classified correctly more than 80% of the flows of each dominant application in the traces, with an accuracy of at least 80%. Finally, *BLINC* identified malicious behavior or previously "unknown" applications without having a-priori knowledge or port-specific information.

Practical impact and the grand vision

We envision this approach as a flexible tool that can provide useful information for research or operational purposes in an evolving network with dynamic application behavior. By focusing on the fundamental communication behavior, our approach provides the first step towards obtaining understanding of traffic traces that transcends the technical specifications of the applications. Lastly, application masquerading within this context is significantly harder than when using other application classification techniques. While applications may change their packet size or inter-packet timings, they will not be able to achieve their objective if they modify their fundamental communication pattern.

14 Classification of multimedia hybrid flows in real time

Since the 1950s, voice and video services such as telephony and television have established themselves as an integral part of everyone's life. Traditionally, voice and video service providers built their own networks to deliver these services to customers. However, tremendous technical advancements since the 1990s have revolutionized the mode of delivery of these services. Today, these services are delivered to the users over the Internet, and we believe that there are two main reasons for this: (i) delivering services over the Internet in IP packets is much more economical for voice and video service providers and (ii) the massive penetration of broadband (i.e. higher bandwidth) Internet service has ensured that the quality of voice and video services over the Internet is good enough for everyday use. The feasibility of a more economical alternative for voice and video services attracted many ISPs including Comcast, AT&T and Verizon, among several others, to offer these services to end users at a lower cost. However, non-ISPs, such as Skype, Google, Microsoft, etc. have also started offering these services to customers at extremely competitive prices (and, on many occasions, for free).

From an ISP's perspective, traffic classification has always been a critical activity for several important management tasks, such as traffic engineering, network planning and provisioning, security, billing and Quality of Service (QoS). Given the popularity of voice and video services over the Internet, it has now become all the more important for ISPs to identify voice and video traffic from other service providers for three reasons. (i) Voice and video services other than on ISP's own service will severely impact its revenues and hence ISPs may wish to shape/block flows from these services; (ii) from a traffic engineering perspective, ISPs may sometimes need to prioritize other more important traffic (e.g. VPN traffic) to ensure the promised QoS is met, allocating less resources to voice and video services from other service providers like Skype, Microsoft, and Google; (iii) from a security perspective, an ISP should always have the ability to identify accurately *all* flows and block all malicious ones.

Service providers like Skype, Microsoft (MSN) and Google (GTalk) have evolved from providing isolated services to bundled services. In other words, service providers initially offered only individual services (either voice, video, file transfer *or* chat) to end users. However, today they offer voice, video, file transfer *and* chat as one bundled service. This change in service paradigm has resulted in changes to the voice and video traffic/flow characteristics. In general, we categorize all multimedia (i.e. voice and video) flows into two types.

(i) *Homogeneous flows:* these are flows where every multimedia flow is either a voice or a video flow.
(ii) *Hybrid flows:* these are flows where voice and video streams are bundled with other services, such as file transfer and chat.

In other words, the same flow at layer 3/layer 4 can now carry multiple streams at the application layer.

Identifying voice and video traffic in both homogeneous and hybrid flows is a very challenging task. Initial approaches for voice detection relied on application payload signatures. For example, in ref. [127] the authors study the problem of identifying voice traffic that uses the standard H.323 protocol [3]. It identifies voice traffic by first recognizing the TCP setup phase of H.323 using payload signatures, and subsequently analyzing the UDP data to identify the associated RTP stream. Such techniques are simple, efficient and extremely accurate. However, service providers started using advanced encryption techniques for their services, thus making signature-based approaches obsolete. Hence, the research community started focusing on developing general classification techniques that rely solely on layer-3/layer-4 information, such as packet size and packet inter-arrival time. These techniques can be broadly classified into two: (i) techniques that classify multimedia (i.e. voice and video) from the rest of the Internet traffic [123, 148, 84, 105] and (ii) identifying traffic originating from specific applications; for example, ref. [218] tries to identify Skype voice flows, while ref. [138] tries to identify all Skype flows. These techniques have two main drawbacks when dealing with realistic traffic on the Internet. (i) None of these approaches can separate generic homogeneous voice and video flows from each other, irrespective of the application that generated them. (ii) When it comes to hybrid flows, there are no known approaches that can effectively identify the existence of hidden voice and/or video streams.

In this chapter, we address the above issues and propose a self-learning voice and video traffic classifier called *VOVClassifier*, that not only identifies voice and video traffic in both homogeneous and hybrid flows, but also labels these flows with the application that generated them. This classifier works in two phases: an *off-line training* phase and an *on-line detection* phase.

In the off-line training phase, a sample set of flows from applications of interest are passed as an input to the *VOVClassifier*. For example, if the application of interest is Skype, then in the training phase we feed the *VOVClassifier* with *homogeneous* Skype voice and video flows. Note that, irrespective of whether we are interested in classifying homogeneous or hybrid flows, the input to the *VOVClassifier* during the training phase is always *homogeneous* flows. Similar to other classifiers [124], the *VOVClassifier* also relies on two main characteristics of packets in voice and video flows: *packet size* and *packet inter-arrival time*. However, unlike other approaches that consider such metrics independently from each other, the presented methodology extracts the hidden temporal and spatial correlations of these features and studies its regularities in the frequency domain. *VOVClassifier* first models these features into a two-dimensional stochastic process, and then analyzes the properties of this process using the *Power Spectral Density* (PSD) analysis. This PSD analysis results in application fingerprints that can now be used for accurate classification of flows from the trained application. Note that this fingerprinting mechanism not only identifies voice and video traffic, but also clusters these voice and video flows into specific applications. Such a grouping will help differentiate between voice/video flows from different applications (for example, Skype voice traffic and MSN voice traffic), thus enabling ISPs to administer application priorities depending on the service level agreements.

In the on-line detection phase, *VOVClassifier* first computes the PSD fingerprint for the flow, and then compares it to the existing fingerprints to classify and subsequently label the flow as belonging to a particular application.

The main points of this chapter are as follows:

- We present a novel voice and video classifier called *VOVClassifier*, which is capable of identifying voice and video streams even if these streams are hidden inside bundled application sessions. This classifier works in two phases: an off-line training phase and an on-line detection phase. In the training phase, we extract fingerprints for voice and video flows, and in the detection phase we use these fingerprints to classify and label accurately flows in real-time.
- We present a novel methodology for extracting the fingerprint of voice and video flows. We first model the packet size and inter-arrival time of packets in a flow as a two-dimensional stochastic process, and subsequently use PSD analysis to extract the hidden regularities constituting the fingerprint of the flow. We show that these fingerprints are unique for voice and video flows (and also for each application that generates these flows) and can be easily clustered to create a voice and video subspace. These subspaces can be separated by a linear classifier.
- We use real packet traces containing voice, video and file transfer sessions in Skype, Google Talk and MSN to evaluate comprehensively the presented methodology. Our results show that such an approach is very effective and extremely robust to noise. In fact, we found that the detection rate for both voice and video flows in *VOVClassifier* was 99.5%, with a negligible false positive rate when considering only homogeneous flows. When considering hybrid flows, the voice and video detection rates were 99.5% and 95.5% for voice and video traffic, while still keeping the false positive rate close to zero.

The rest of the chapter is organized as follows. In Section 14.1 we provide some background on voice, video and file transfer traffic and describe the packet traces we have collected and used to validate the methodology presented. In Section 14.2 we show how simple metrics proposed in the literature, i.e. inter-arrival time and packet size distributions, can lead to a misclassification of voice and video flows if not combined in a proper way in both homogeneous and hybrid cases. We also present some key intuitions that constitute the essence of our methodology. In Section 14.3 we present the overall architecture of *VOVClassifier* and explain each main module in greater detail. Section 14.4 presents a variety of different results obtained using real packet traces, while Section 14.5 concludes the chapter.

14.1   Background and data description

Let us first present a brief introduction to voice, video and file transfer streams. Alongside we describe the respective data we collect for all discussed experiments.

14.1.1   Background: voice, video and file transfer streams

Voice stream

A key aspect of most voice streams on the Internet is the interactive behavior of users. For example, a Voice-over-IP (VoIP) phone call typically involves two parties communicating with each other in real time. Hence, one of the important parameters that characterizes voice traffic is the Inter-Packet Delay (IPD). Several standard codecs can be used for this voice communication, and each of these codecs specifies different IPD values. We list a few of these standard codecs in Table 14.1. VoIP service providers, such as Skype, Microsoft (or MSN) and Google (or GTalk), give users the ability to configure different codecs depending on the network conditions. However, by default most of these service providers use proprietary codecs whose specifications are not available to analyze. In our experiments, we noticed that Skype and MSN voice traffic use proprietary codecs that could either transmit packets every 20 ms or 30 ms. Although these codecs specify the IPD at the voice transmitter side, the packets that arrive at the receiver do not have this IPD. One of the primary reasons for this is the non-deterministic delay (due to router queues, packet paths, etc.), also referred to as jitter, experienced by the packets that traverse different links in the Internet. In order to minimize the impact of jitter on the quality of voice traffic, voice applications typically generate packets that are very small in size. Thus, despite the variations due to jitter, voice streams still exhibit strong regularities in the Inter-Arrival Time (IAT) distribution at the receiver side.

Video streams

Video applications (i.e. live streaming media) send images from a transmitter to a receiving device by transmitting frames at a constant rate. For example, the H.323 codec tries to dispense frames at constant rate of 30 frames per second. Typically there are two

Table 14.1. Commonly used speech codecs and their specifications

Standard	Codec method	Inter-packet delay (ms)
G.711	PCM	0.125
G.726	ADPCM	0.125
G.728	LD-CELP	0.625
G.729	CS-ACELP	10
G.729A	CS-ACELP	10
G.723.1	MP-MLQ	30
G.723.1	ACELP	30

types of frames, *Intra frames (I-frame)* and *Predicted frames (P-frame)*, that are transmitted/interleaved in a periodic fashion. Usually, the number of P-frames between two consecutive I-frames is constant. An I-frame is a frame coded without reference to any frame except itself and usually contains a wider range of packet sizes depending on the texture of the image to be transmitted. I-frames serve as starting points for a decoder to reconstruct the video stream. A P-frame may contain both image data and motion vector displacements and/or combinations of the two. Its decoding requires access to previously decoded I- and P-frames. Typically, each P-frame is contained in a small sized packet in the range 100–200 Bytes. Due to the above properties of video streaming, we expect to still see some regularities in the IAT distribution at the receiver side.

File transfer streams

File transfer applications deliver bulk data from a transmitter to a receiver as fast as possible, leveraging the network conditions. These applications typically partition a file into equal-sized segments and transfer one segment at a time to the receiver. Therefore, a file transfer flow is likely to be composed of equal-sized packets of large size, except for a few packets at the beginning and at the end of the data transfer. From our analysis, we observed that all the applications under investigation show that 90% of packets have size between 1400 and 1500 bytes (1397 bytes for Skype, 1448 bytes for MSN), while only 10% have size between 50 and 150 bytes.

14.1.2   Data sets

To collect data for our experiments, we set up multiple PCs (1.8 GHz Pentium 4 with Windows XP) in two different universities to run three applications, Skype, MSN and GTalk. These universities are located in different parts of the North American continent. We generated voice, video and file transfer streams between the end hosts over the timeframe of a week (May 15–22, 2007). We generated both homogeneous and hybrid flows. Since we manually generated all of the flows in the data set, we can easily label each of them as either voice, video or file transfer along with the application that generated them (Skype, MSN or GTalk). The average time and number of packets in each of the sessions were 8 minutes and 9500 packets, respectively. In total, we generated about 690 sessions, and a full breakdown of the sessions generated is given in Table 14.2.

As mentioned earlier in the chapter, our classifier works in two phases: training and detection. The input to the training phase comprises homogeneous flows with labels, while the input to the detection phase comprises both homogeneous and hybrid flows that need to be labeled. Hence we split the data set into two sets – the training set and the detection set. The training data set contained 90 homogeneous flows (45 of them were voice and 45 of them were video). We used this set to train the *VOVClassifier*. The detection set contained the other 600 flows, which include both homogeneous and hybrid flows. We used this set to evaluate the performance of the classifier.

Table 14.2. Number of flows, average number of packets per flow and the average duration of each flow (in seconds) collected for each application to evaluate the performance of *VOVClassifier*

	Homogeneous flows		Hybrid flows	
	voice	video	voice+file	video+file
Skype	54 / 20 000/ 10	42 / 125 000 / 25	88 / 220 000 / 60	84 / 478 000 / 60
MSN	82 / 58 000/ 20	58 / 68 000/ 20	80 / 428 000 / 60	93 / 420 000 / 60
GTalk	26 / 39 000 / 20	38 / 33 000 / 20	25 / 333 000 / 60	30 / 401 000 / 60

14.2 Challenges, intuition and discussion

14.2.1 Challenges

We start this section by considering the simpler case of homogeneous voice and video flows. Differentiating homogeneous voice flows from videos flows has been addressed as a research problem, with limited success [124]. Two features that have been repeatedly proposed for this are the *packet Inter-Arrival Time* (IAT) and the *Packet Size* (PS). Figure 14.1 shows the Cumulative Distribution Function (CDF) of the two metrics computed independently for a Skype voice and video homogeneous flow. The sharp knees in the CDF of the IAT shows that both Skype voice and video flows show regularity in the packet IAT (30 ms for voice and 5 ms for video). Similarly, the knees in the CDF of the PS shows that most of the packets in Skype voice flows are 120–160 bytes long, while most of the packets in Skype video flows are 480–520 bytes long. We find similar results for all of the Skype homogeneous flows in our data set. In order to differentiate between Skype voice and video homogeneous flows, one can use a simple filter that first checks regularity in the IAT, and subsequently separates voice and video flows based on the PS distribution of the flow. In other words, Figure 14.1 suggests that the IAT and the PS can be computed independently from each other, and we can use simple cut-off thresholds on the two distributions to distinguish voice from video flows.

Although such a simple technique works reasonably well for differentiating Skype traffic, it cannot be used in the general case for differentiating voice flows from video flows. For example, consider the case of MSN. Figure 14.2 shows the CDF of both the IAT and the PS for an MSN voice and video flow. We can see that voice packets tend to reach the destination at very regular time intervals of about 20 ms, with packets that are 105–120 bytes long. For the case of MSN video flows, the detection turns to be not as friendly as it was for Skype video flows. First, video packets do not exhibit a strong regularity in IAT, as shown by the complete absence of any knees in the CDF of the IAT. Video packets reach the destination in an almost complete random fashion, and thus can be easily interpreted as packets belonging to any other non-voice or non-video application. Secondly, even if we were able to find some regularity in the IAT for video packets, we cannot distinguish video flows from voice flows due to the significant overlap of the PS distribution. As a consequence, it is hard to draw a clear cut boundary

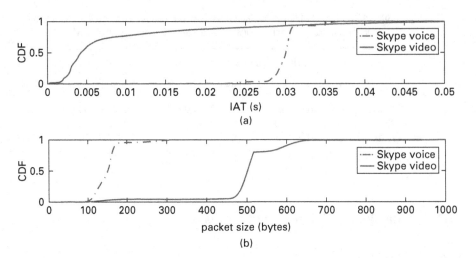

Fig. 14.1.   CDF of (a) the IAT and (b) the PS for a typical Skype homogeneous flow.

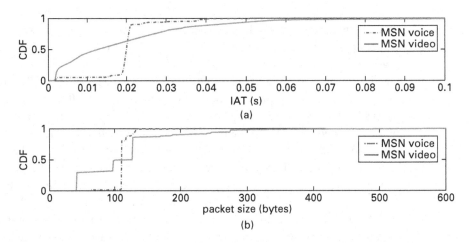

Fig. 14.2.   CDF of (a) the IAT and (b) the PS (on the bottom) for a typical MSN homogeneous flow.

between the two, thus making the simple approach proposed for distinguishing Skype voice and video flows not generalizable for other applications.

Next, we show how the strong regularity observed in the IAT distribution for both Skype and MSN homogeneous voice flows is severely affected when we consider hybrid flows.

Figure 14.3 shows the PS and IAT for all packets in a typical MSN homogeneous and hybrid voice flow. We plot this graph using the following technique. We traverse the flow from the first packet to the last packet. For each packet, P_i, encountered in the flow, we record its size, PS_i, and the associated IAT to the next packet, IAT_i. Each packet P_i is then represented by the pair $\langle PS_i, IAT_i \rangle$. In Figure 14.3 we plot all packets P_i belonging to an MSN homogeneous and hybrid voice flow. As we also noted previously in Figure 14.2, Figure 14.3 shows that MSN homogeneous voice flows have a

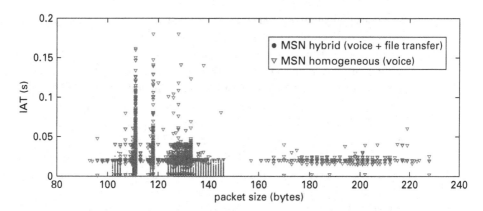

Fig. 14.3. IAT versus PS for a typical MSN hybrid video flow. We show the restricted range of packets between 80 and 240 bytes to highlight the impact on the IAT of voice when considering file transfer packets.

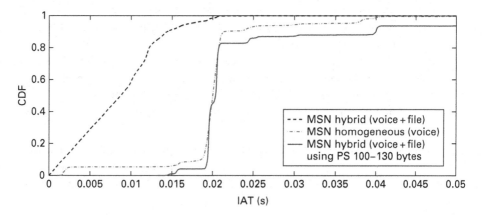

Fig. 14.4. CDF of IAT for a typical MSN homogeneous and hybrid voice flow.

strong regularity with an IAT of 20 ms, around which the majority of packets lie. However, in case of hybrid flows, we can clearly see that this regularity is lost. Packets of size 100–140 bytes (which represent the most common packet sizes used by MSN homogeneous voice flows) now span a large range of IAT values. As a consequence, such a flow does not exhibit any significant pattern that can reveal the presence of voice.

In order to quantify this, in Figure 14.4 we show the CDF of the IAT for an MSN homogeneous and hybrid voice flow. As we can see, the sharp knees that exist for homogeneous flows (around 20 ms range) completely disappear for hybrid flows, suggesting that these packets reach the destination in a purely random fashion.

14.2.2 Intuition and approach

As we saw in Figures 14.2 and 14.4, homogeneous and hybrid voice and video flows do not always exhibit regularities in IAT. Also, the PS distribution could vary considerably from one application to another. Hence, considering IAT and PS as independent metrics will not result in effective classification of voice and video.

Our hypothesis is, however, that *all* voice and video flows should exhibit some kind of regularity/pattern in IAT and PS. These patterns could be hidden in the packet stream. Next we show how analyzing the spatial and temporal correlation of these two features can reveal such hidden regularities. We carry out this analysis using MSN and show that such intuition has great potential in revealing the presence of voice and video streams in the context of both homogeneous and hybrid flows.

First, we consider the case of MSN homogeneous video flows for which no clear pattern can be observed when the two features (PS and IAT) were computed and analyzed independently (see Figure 14.2). For this case, we conduct two experiments using the original video flow:[1]

- We consider 128-byte packets in the flow along with their timestamps (i.e. the time at which we received the packet). Note that we discard all other packets in the flow, and the flow now has only 128-byte packets with their timestamps. We compute the CDF of the IAT of this new flow (Figure 14.5). We can now see that these packets exhibit regularities in their IAT.
- We consider sequences of packets; i.e., if two consecutive packets in the flow have sizes 128 and 42 bytes, respectively, then we extract those packets along with their timestamps and discard the rest. We now plot the CDF of the IAT of this series and find that these sequences of packets also exhibit regularity.

The main take-away point from the above experiments is that, although we do not find any significant patterns by considering the IAT and PS metrics individually, we find very strong patterns when we combine and analyze these metrics together. In other words,

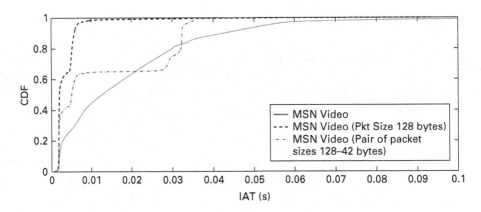

Fig. 14.5.   CDF of the IAT with certain PS grouping for MSN homogeneous video flows.

---

[1] Looking at the PS versus IAT graphs for MSN homogeneous video flows, we found that there was a large number of packets of size 42 and 128. Hence we pick these packets for these experiments.

the regularities of a video flow reside in specific combinations of packet sizes and inter-arrival times that are maintained for the entire duration of the session. We observe the same phenomenon for video traffic.

Next, we consider the case of an MSN hybrid voice flow. We apply the same concept as before and focus on packets of sizes 100–130 bytes (the range of most common packets observed in MSN homogeneous voice flows). In Figure 14.4, we report the CDF of the IAT for such a flow. As we can see, the shape of the CDF looks very similar to the original MSN homogeneous voice flow, and clearly is very different from the one computed by considering the two metrics independently.

This forms the basis of our approach. In fact, we combine these two metrics using a stochastic process that is further analyzed in the frequency domain using PSD analysis. Such a methodology searches for combinations of packet sizes and IAT pairs that occur more frequently than others and thus carry the majority of the energy of the stochastic process being created. Furthermore, we remark to the reader that, since different applications are likely to use different pair-wise combinations of packet sizes and IAT, it is unrealistic to explore exhaustively such a huge search space. As a consequence, it is imperative to provide a methodology that automates such a task.

14.2.3   Chat traffic versus file transfer traffic in hybrid flows

Hybrid multimedia flows can contain voice and/or video streams along with file transfer and/or chat streams. Both chat and file transfer streams do not show any hidden regularities in IAT distributions. Hence, from the perspective of our problem, both these streams represent pure noise to the voice and video stream classifier. From our data set, we see that the PS distribution of chat streams typically ranges from 50 to 600 bytes, whereas the same for file transfer streams ranges from 60 to 1500 bytes. While the PS distribution of chat streams can be randomly distributed in its range, the distribution of PS in file transfer streams is mainly concentrated in two ranges: 90% of packets are in 1400–1500 bytes and 10% in 60–110 bytes.

In this work, we only consider hybrid flows that contain *voice* or *video streams* along with *file transfer streams*. We do this for two main reasons.

(1) The data rate of chat streams is much lower than the data rates of file transfer, voice and video streams. During our experiments, we noticed that the number of chat packets observed in a 10-second observation window is negligible compared with the number of file transfer packets encountered. As a consequence, the presence of chat traffic can be interpreted as low-level random noise that minimally impacts the IAT regularities of voice and video streams. On the other hand, we can find a lot more file transfer packets in any observation window. File transfer packets are highly interleaved with voice and video packets, and thus have great potential to impact severely the IAT distribution of the overall flow (as we can see in Figure 14.4).

(2) Although the PS distribution of chat spans a wide range of packets, from 50 to 600 bytes, the average number of packets that fall in the same range of voice and video packets is minimal. On the other hand, the PS distribution of file transfer is

heavily centered around 1400–1500 bytes, and the remaining packets reside in the range 60–110 bytes, the typical range used by voice traffic. As a consequence, the PS distribution of voice traffic might be severely impacted by the presence of file transfer. Note that, although 10% is a small fraction of the overall file transfer PS distribution, it translates to a large number of packets (orders of magnitude more than the number of packets in chat streams), since the total number of packets in file transfer streams is very large.

14.3   *VOVClassifier* system

As we mentioned in Section 14.1, *VOVClassifier* is a self-learning voice and video classifier that works in two phases: an off-line training phase and an on-line detection phase. In the off-line training phase, the *VOVClassifier* requires several *labeled* voice and video homogeneous flows as the input. The system *automatically* learns the fingerprints of the flows and marks these signatures with the label that was provided during training. For example, if we provide Skype voice flows as the input to the *VOVClassifier*, then the classifier learns the signature for these flows and marks this signature as `<SkypeVoice>`. In the on-line detection phase, the *VOVClassifier* examines a new flow and computes its fingerprint. Based on the distance of the fingerprint from the existing clusters in the fingerprint database, the new flow is classified as either voice or video (or neither) and labeled with the application name. We wish to point out that the *VOVClassifier* requires *non-sampled* packet traces (just the layer-3 and layer-4 information) for both training and classification.

The off-line training phase has two modules: (i) *Feature Extractor* (FE) and (ii) *Voice/Video Subspace Generator* (VSG). Note that a flow in the training set is first processed by FE, which extracts the fingerprints, and subsequently by VSG, which clusters the fingerprints in an efficient way. The on-line detection phase encounters one more module, the *Voice/Video CLassifier* (VCL). The flow is first processed by FE, as in the training phase, which sanitizes the data, collects relevant metrics and extracts the fingerprints, and then is forwarded to the VCL. In the rest of this section, we will elaborate on each of these modules.

14.3.1   Training phase: FE

The FE module has two main goals: (i) to extract the relevant features from homogeneous voice and video flows in the training set and (ii) to generate the flow fingerprint based on the extracted features. As soon as the FE module encounters a flow, say $\mathcal{F}_S$, it extracts two features for every packet in the flow that are important for our fingerprint generation – packet sizes and inter-arrival times. In other words, the FE module internally describes every packet in the flow as: $\mathcal{F}_S = \{\langle \mathcal{P}_i, \mathcal{A}_i \rangle ; i = 1, \ldots, I\}$, where $\mathcal{P}_i$ and $\mathcal{A}_i$ represent the PS and the relative arrival time of the ith packet in the flow; I is the total number of packets in the flow.

To generate the fingerprint for a flow, the above information for every flow is modeled as a continuous stochastic process as follows:

$$P(t) = \sum_{\langle P, \mathcal{A} \rangle \in \mathcal{F}_S} P\delta(t - \mathcal{A}), \tag{14.1}$$

where $\delta(\cdot)$ denotes the delta function. Since digital computers are more suited to deal with a discrete-time sequence, we transform $P(t)$ to a discrete-time sequence by sampling at a frequency $F_S = 1/T_S$. The signal in equation (14.1) is a summation of delta functions and hence its spectrum spans the entire frequency domain. To reshape the spectrum of $P(t)$ in order to avoid aliasing when it is sampled at the interval T_S, we apply a low-pass filter, $h(t)$. Thus the discrete-time signal can be represented as follows:

$$P_d(i) = P_h(iT_S) = \sum_{\langle P, \mathcal{A} \rangle \in \mathcal{F}_S} Ph(iT_S - \mathcal{A}), \tag{14.2}$$

where $i = 1, \ldots, I_d$, $I_d = \mathcal{A}_{max}/T_S + 1$ and $\mathcal{A}_{max}$ is the arrival time of the last packet in the flow.

If we assume $\{P_d(i)\}_{i=1}^{I_d}$ to be a second-order stationary process, then its PSD can be computed as follows:

$$\psi(\varpi; P_d) = \sum_{k=-\infty}^{\infty} r(k; P_d) e^{-j\varpi k}, \quad \varpi \in [-\pi, \pi), \tag{14.3}$$

where $\{r(k; P_d); k = 1, \ldots, I_d\}$ represents the autocovariance sequence of the signal $\{P_d(i); i = 1, \ldots, I_d\}$. Recall that $\{P_d(i)\}_{i=1}^{I_d}$ is obtained by sampling a continuous time signal $P_h(t)$ at time interval T_S. Thus, the PSD in terms of real frequency f is given by

$$\psi_f(f; P_d) = \psi\left(\frac{2\pi f}{F_S}; P_d\right), \tag{14.4}$$

where $F_S = 1/T_S$. Equation (14.4) shows the relationship between the periodic components of a stochastic process in the continuous time domain and the shape of its PSD in the frequency domain. Note that $\psi_f(f; P_d)$ is a continuous function in the frequency domain. We assume that M frequencies, i.e. $0 \leq f_1 < f_2 < \cdots < f_M \leq F_S/2$, can be used to represent each feature vector. Finally, we define our PSD feature vector $\vec{\psi} \in \Re^M$ as follows:

$$\vec{\psi} = [\psi_f(f_1; P_d), \quad \psi_f(f_2; P_d), \quad \psi_f(f_M; P_d)]^T. \tag{14.5}$$

The above feature vectors uniquely identify voice and video stream. In Section 14.4, we show experimental results that clearly distinguish voice and video PSD fingerprints, thus validating our approach. In fact, these fingerprints can also be used to identify different applications that generate the voice and video flows.

14.3.2 Training phase: VSG

Once the feature vectors (i.e. the PSD fingerprints) are generated by the FE module, the first step in the VSG module is to perform *subspace decomposition*, i.e. cluster all the fingerprints such that flows with similar feature vectors are grouped with each other in the same cluster. This clustering mechanism is a key component of the *VOVClassifier* that helps to tailor the behavior of the classifier to the user's needs. For instance, if the user is interested in just distinguishing voice flows from video flows, then he/she specifies only two labels for all the flows in the training set. The clustering step in the VSG will result in three clusters: one for voice, one for video and the last one for all other flows. However, if the user is interested in differentiating different applications, say Skype voice, Skype video, MSN voice and MSN video flows, from each other, then they specify these four labels for the training set, and the VSG module thus yields five clusters, one for each of the applications and a last one for all others that do not belong to these applications. This feature of the VSG module makes the *VOVClassifier* extremely flexible and configurable.

The second step in the VSG module is to reduce the dimension of the feature vectors generated by the FE module. Note that the M frequencies used for the PSD fingerprint could be arbitrarily large, and hence in this step we reduce the number of frequencies required to capture the PSD fingerprint. Indeed, the fewer frequencies we have, the less memory and CPU cycles the system will consume during the on-line classification phase. We refer to this step as *bases identification*.

Subspace decomposition

Let N_1 and N_2 represent the number of voice and video flows, respectively, used for training the *VOVClassifier*. The FE module results in two sets of feature vectors:

$$Voice : \Psi^{(1)} \triangleq \left\{ \vec{\psi}_1(1), \vec{\psi}_1(2), \ldots, \vec{\psi}_1(N_1) \right\}, \qquad (14.6)$$

$$Video : \Psi^{(2)} \triangleq \left\{ \vec{\psi}_2(1), \vec{\psi}_2(2), \ldots, \vec{\psi}_2(N_2) \right\}. \qquad (14.7)$$

We can consider $\Psi^{(i)}$ as an $M \times N_i$ matrix, for $i = 1, 2$, where each column represents a specific feature vector. In other words, $\Psi^{(i)} \in \Re^{M \times N_i}$. We now apply the concept of minimum coding length proposed in ref. [96] to obtain non-overlapping partitions of the voice and video subspace:

$$\Psi^{(i)} = \Psi_1^{(i)} \cup \cdots \cup \Psi_{K_i}^{(i)}. \qquad (14.8)$$

Note that the feature vectors associated with the same application reside within the same subspace.

Subspace bases identification

As mentioned earlier, the feature vectors for voice and video, $\Psi^{(i)}$, can be considered as a matrix with dimensions $M \times N_i$, where M represents the number of frequencies in the PSD fingerprint and N_i represents the number of training vectors for voice and video. These dimensions can be very large, thus increasing the complexity of the on-line detection phase. Hence, it is imperative to reduce the high-dimensional structure in each subspace into a lower-dimensional subspace. We use PCA to identify *bases* that should be used for each subspace, $\Psi_k^{(i)}; k = 1, \ldots, K_i; i = 1, 2$.

Box 14.1. Function *IdentifyBases* identifies bases of each generic subspace

(1)   function $\left[\vec{\mu}, \hat{\mathcal{U}}, \hat{\Sigma}, \bar{\mathcal{U}}, \bar{\Sigma}\right] = IdentifyBases\left(\Psi \in \mathfrak{R}^{M \times N}, \delta\right)$

(2) $\vec{\mu} = \frac{1}{|\Psi|} \sum_{\vec{\psi} \in \Psi} \vec{\psi}$

(3) $\bar{\Psi} = \left[\vec{\psi}_1 - \vec{\mu}, \vec{\psi}_2 - \vec{\mu}, \ldots, \vec{\psi}_{|\Psi|} - \vec{\mu}\right]$

(4)   Do eigenvalue decomposition on $\bar{\Psi}\bar{\Psi}^T$ such that

$$\bar{\Psi}\bar{\Psi}^T = \mathcal{U}\Sigma\mathcal{U}^T, \qquad (14.9)$$

where $\mathcal{U} \overset{\triangle}{=} [\vec{u}_1, \cdots, \vec{u}_M]$, $\Sigma \overset{\triangle}{=} \mathrm{diag}\left([\sigma_1^2, \ldots, \sigma_M^2]\right)$ and $\sigma_1^2 \geq \sigma_2^2 \geq \cdots \geq \sigma_M^2$.

(5) $\mathcal{J} = \arg\min_{\hat{\mathcal{J}}} \sum_{m=1}^{\hat{\mathcal{J}}} \sigma^2(m) \geq \delta \sum_{m=1}^{M} \sigma^2(m)$

(6) $\hat{\mathcal{U}} = [\vec{u}_1, \vec{u}_2, \ldots, \vec{u}_{\mathcal{J}-1}]$

(7) $\bar{\mathcal{U}} = [\vec{u}_{\mathcal{J}}, \vec{u}_{\mathcal{J}+1}, \ldots, \vec{u}_M]$

(8) $\hat{\Sigma} = \mathrm{diag}\left([\sigma_1^2, \ldots, \sigma_{\mathcal{J}-1}^2]\right)$

(9) $\bar{\Sigma} = \mathrm{diag}\left([\sigma_{\mathcal{J}}^2, \ldots, \sigma_M^2]\right)$

(10)   end function

The basic idea is to identify uncorrelated bases and choose those bases with dominant energy. Box 14.1 presents the algorithm, where argument $\Psi \in \mathfrak{R}^{M \times N}$ represents the generic subspace $\Psi_k^{(i)}$ and δ is a user-defined parameter that specifies the percentage of energy to be retained by the new bases. Note that the larger the value of δ, the larger will be the number of bases selected by PCA since more details of the structure of the original subspace should be captured. Typically, δ is chosen between 90% and 95% so that the selected bases capture almost the entire energy in the subspace, while getting rid of the majority of bases that contribute to very little energy.

The PCA algorithm returns five variables that capture the properties of the low-dimensional subspace obtained. (i) $\vec{\mu}$ represents the sampled mean of all feature vectors residing in the subspace. It is the origin of the newly identified low-dimensional subspace. (ii) $\hat{\mathcal{U}}$ the columns of this matrix represent the bases with dominant energy (i.e. maximum variance). (iii) $\hat{\Sigma}$ represents the variance of the bases identified in the columns of $\hat{\mathcal{U}}$. (iv) $\bar{\mathcal{U}}$ the columns of this matrix are composed of the null subspace of the original higher-dimensional subspace. In other words, the columns refer to the dimensions that have been suppressed during the projection to the new lower-dimensional space as they capture the remaining $(1 - \delta)$ energy. (v) $\bar{\Sigma}$ represents the variance of the columns in $\bar{\mathcal{U}}$. These last two outputs are required to calculate the distance of a generic feature vector to this subspace, which will be described in Section 14.3.4.

14.3.3   Detection phase: FE

Once we train the *VOVClassifier* using the strategy discussed in Sections 14.3.1 and 14.3.2, the next step is to use this for on-line detection and classification of flows. As soon as a flow that requires classification comes into the *VOVClassifier*, it is processed by the FE module, which performs exactly the same tasks as the FE in the training phase described in Section 14.3.1. It extracts the PSD fingerprint of each flow and sends it to the VCL module.

14.3.4   Detection phase: VCL

This module receives the feature vector, $\vec{\psi}$, of the flow that needs to be classified from the FE module. The classification algorithm used in this module is shown in Box 14.2. The algorithm has three inputs, $\vec{\psi}$, θ_A and θ_V. The parameters θ_A and θ_V are user-defined distance thresholds for voice and video flow classification. In other words, they represent the maximum distance of a flow from the voice and video clusters to be classified as either a voice or video flow.

This algorithm first calculates the normalized distances between $\vec{\psi}$ and all low-dimension subspaces $\Psi_k^{(i)}$ that were calculated in the training phase (using subspace

---

Box 14.2. Function *voicevideoClassify* determines whether a flow with PSD feature vector $\vec{\psi}$ is of type voice or video or neither; θ_A are θ_V are two user-specified threshold arguments. Function *voicevideoClassify* uses function *NormalizedDistance* to calculate the normalized distance between a feature vector and a subspace.

(1)   function $[Voice, Video] = voicevideoClassify\left(\vec{\psi}, \theta_A, \theta_V\right)$

(2)   For $\forall i = 1, 2, \forall k = 1, \ldots, K_i$

$$d_k^{(i)} = NormalizedDistance\left(\vec{\psi}, \vec{\mu}_k^{(i)}, \vec{U}_k^{(i)}, \bar{\Sigma}_k^{(i)}\right)$$

(3)   For $\forall i = 1, 2,$

$$d_i = \min_k d_k^{(i)}$$

(4)   if $d_1 < \theta_A$ and $d_2 > \theta_V$

(5)   Label the flow as voice.

(6)   else if $d_1 > \theta_A$ and $d_2 < \theta_V$

(7)   Label the flow as video.

(8)   else

(9)   Label the flow as neither voice nor video.

(10)   end if

(11)   end function

(12)   function $d = NormalizedDistance\left(\vec{\psi}, \vec{\mu}, \vec{U}, \bar{\Sigma}\right)$

(13) $d = \left(\vec{\psi} - \vec{\mu}\right)^T \vec{U}\bar{\Sigma}^{-1}\vec{U}^T \left(\vec{\psi} - \vec{\mu}\right)$

(14)   end function

decomposition and PCA). Then, based on the distances from the different clusters, this algorithm labels the flow as either voice, video or neither. Note that, in this section, we have just considered voice and video as the two categories for classification; however, incorporating other categories such as Skype voice, MSN voice, Skype video, etc., is as straightforward as the above case.

14.4   Experimental results

In this section, we first validate the presented approach and then demonstrate the effectiveness and feasibility of *VOVClassifier*. In other words, we will first show how the FE module extracts fingerprints that are unique to voice and video flows. We then show how the VSG module generates voice and video subspaces that are distinct from each other, thus giving us an opportunity to use simple linear classifiers to separate voice and video flows. Finally, we will show the effectiveness of the entire system in clearly identifying voice and video streams in the context of hybrid flows and multiple applications.

14.4.1   Feature extractor module

As we explained in Section 14.3.1, the FE module takes homogeneous voice and video flows as the input to compute a stochastic process for a flow (equation (14.1)) using the IAT and packet sizes, and then extracts a fingerprint using the PSD distribution (equation (14.5)). Figure 14.6 shows the PSD of the stochastic process for voice and video flows. We randomly picked two Skype homogeneous voice and video flows from the data traces. The top and bottom graphs in Figure 14.6 show the PSD fingerprint generated by the FE module for the two voice flows and the two video flows, respectively. There are two main observations to be made on this figure. (i) The PSD fingerprints for

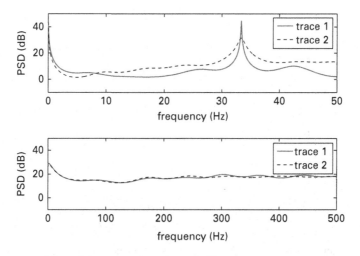

Fig. 14.6.   PSD fingerprints of Skype homogenenous voice and video streams.

voice and video flows are very distinct from each other. In other words, these finger-prints can be used to distinguish voice flows from video flows. (ii) The PSD fingerprint of the two voice flows (and the two video flows) are very close to each other. This implies that when these flows are clustered together by the VSG, *all* the voice flows can be clustered into a sphere in hyperspace with very small radius. This small and concen-trated cluster can be easily differentiated from other clusters using linear classifiers. We remark on the importance of such a result in the context of application classification. Being able to extract such similar fingerprints for voice and video flows, generated by the same application, makes the distinction between one application and another a very straightforward task. The same is true for video flows.

14.4.2   Voice and video subspace generator

The output from the FE module is used by the VSG module to generate voice and video subspaces (or clusters) that are far away from each other. The larger the distance between the two subspaces, the easier it is to classify any flows containing voice and video streams. Figure 14.7 shows the distance of a Skype homogeneous flow from the voice subspace on the x-axis and the distance from the video subspace on the y-axis. For this result, we first train the *VOVClassifier* using several homogeneous voice and video flows from our traces. After the training phase, we choose many hybrid flows (contain-ing voice, video and file transfer), for which we already know the correct classification, and feed it to the classifier to conduct on-line classification. We compute the distance of all of these flows in the two subspaces that we computed in the off-line training phase and plot them in Figure 14.7. We can clearly see that all the hybrid flows that contain voice are very close to the voice subspace compared with the hybrid flows that contain video streams (which are very far from the voice subspace). This implies that, as soon as the *VOVClassifier* computes the PSD fingerprint and calculates the distance from the existing subspaces, it can tell whether a flow contains voice or video streams using

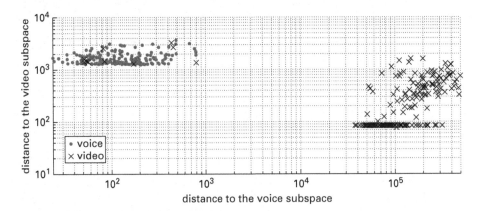

Fig. 14.7.   Output of the training phase when considering multiple Skype homogeneous voice and video flows.

a simple threshold-based scheme. This figure has two important take-away points. (i) The VSG module actually generates subspaces that are far away from each other, thus making it easy to classify traffic based on simple thresholds. (ii) The *VOVClassifier* can perform efficient and effective on-line classification.

Finally, we wish to point out a limitation in the presented system. In Figure 14.7, we can see that some of the video flows are clustered with voice flows. Our analysis shows that these video flows are hybrid flows that contain both voice and video. In other words, when a flow contains both voice and video streams, the *VOVClassifier* will classify this as only voice flows, but not video flows. The reason of this is the well known piggy-back methodology characteristic of voice and video applications that piggy-back voice and video packets together to reduce the overall overhead on the data rate.

14.4.3 Overall system

After the off-line training phase, the main goal of the *VOVClassifier* is to label all the incoming flows as containing voice, video or neither. Note that these labeling categories could be more specific than just voice and video. For example, we could train the *VOVClassifier* to label the flows as Skype voice, MSN voice, Skype video and MSN video. However, for ease of presentation in this section, we choose to label an incoming flow as one of the following: voice, video or neither.

Figure 14.8 and Figure 14.9 show the *Receiver Operating Characteristics* (ROC) curve for the *VOVClassifier*. In other words, the y-axis shows the probability of *correctly* labeling a flow, P_D, and the x-axis shows the probability of a false alarm or *incorrectly* labeling a flow, P_{FA}. We formally define P_D and P_{FA} as follows: when a flow with actual label X is sent through the *VOVClassifier* for classification, and the classifier labels the flow as $\hat{X}$, then

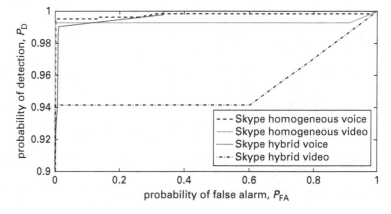

Fig. 14.8. ROC values for homogeneous and hybrid flows generated by Skype.

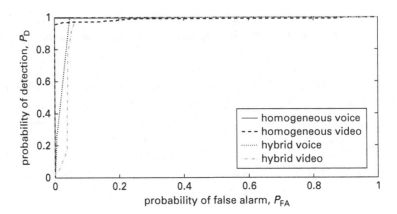

Fig. 14.9.   ROC values for homogeneous and hybrid flows generated by Skype, MSN and Google Talk.

$$P_{\mathrm{D}|X} \overset{\triangle}{=} P\left(\hat{X} = 1 \,\middle|\, X = 1\right), \tag{14.10}$$

$$P_{\mathrm{FA}|X} \overset{\triangle}{=} P\left(\hat{X} = 1 \,\middle|\, X = 0\right). \tag{14.11}$$

We use several hundred homogeneous and hybrid voice and video flows (as described in Section 14.1) as the input to our classifier. As the first step, 45 voice and 45 video flows are used to train the classifier. We then use the rest of the flows to test the system accuracy. Note that we already know the actual labels (i.e. the ground truth) for all of the input flows. We now let our system make a decision on these flows and label them.

First, we consider the case of flows generated by a single application only, i.e. Skype. In Figure 14.8, we can see that the detection rate of the Skype homogeneous voice and video flows is very high (over 99%) when the false alarm rate is very small (<1%). Comparable results were obtained for hybrid voice flows, while, for hybrid video flows, we can observe a slight drop in performance, with 94% detection rate for a false alarm rate of 1% or below. Similar results were obtained for MSN and Gtalk. This first set of results shows that the *VOVClassifier* can effectively and accurately classify voice and video flows generated by one single application, even when these flows are mixed with other streams like file transfer.

Next, we consider a more complex scenario in which homogeneous and hybrid flows are generated by a mixture of applications, i.e. Skype, MSN and Gtalk. In this case, the *VOVClassifier* is asked to (i) classify voice and video flows as before and (ii) label each flow with the associated application being used to generate such a flow. As a consequence, a homogeneous voice flow being generated by MSN but labeled as Skype homogeneous voice will be considered as a false positive. In Figure 14.9 we report the results of our experiment. As the reader can see, even in this extreme scenario, the detection rate of the homogeneous voice and video flows is very high (over 99%) when the false alarm rate is very small (<1%). However, when we input hybrid flows, the results are not as good. In other words, if we want to keep the false alarm rate to less than 1%, the detection rate will also be very low (between 20% and 30%). However, when the false positive rate is about 4%, the overall detection rate jumps beyond 99%, thus show-ing that the *VOVClassifier* performs very well, even when asked to reveal the application behind a voice or video flow.

14.5   Lessons learned

In this chapter, we presented a novel system, called *VOVClassifier*, which provides a robust and accurate detection of voice and video flows. We have shown that this system is able to detect the presence of voice and video data streams both in the con-text of homogeneous flows (flows containing either voice or video packets) and hybrid flows (voice and/or video traffic is bundled together with other types of traffic in the same flow).

To determine the existence of voice and video traffic, our system (i) creatively mod-els a network flow using a stochastic process that combines the *inter-arrival times* of packets within the flow and the associated *packet size*; (ii) extracts the hidden regu-larities of voice and video traffic and highlights their major differences by applying power spectral density analysis (their fundamental properties are captured into fea-ture vectors); (iii) groups feature vectors characterized by some degree of similarity (e.g. associated to the same type of traffic and application) into subspaces and reduces their high-dimension space to a more manageable one, applying principal component analysis; (iv) uses minimum coding length as the similarity metric to perform clas-sification. Results demonstrate the effectiveness and robustness of such an approach, able to achieve 99.5% detection rate (false positive close to zero) for both voice and video in the case of homogeneous flows, and 99.5% and 95.5% (false positive close to zero), respectively, for voice and video when dealing with the more complex scenario of hybrid flows.

This chapter concludes our work on traffic classification and serves as an illustration of how much visibility into IP services and applications can be obtained from a simple flow-level knowledge of IP traffic. Next, we consider the problem of how to use this visibility to discover malicious threats hidden behind normal traffic activity that aim at compromising machines attached to the Internet.

15 Detection of data plane malware: DoS and computer worms

Open, any-to-any connectivity is clearly one of the fundamentally great properties of the Internet. Unfortunately, the openness of the Internet also enables an expanding and ever-evolving array of malicious activity. During the early 1990s, when malicious attacks first emerged on the Internet, only a few systems at a time were typically compromised, and those systems were rarely used to continue or broaden the attack activity. At first, the attackers were seemingly motivated simply by the sport of it all. But then, as would seem to be the natural order of things, the miscreants were seized by the profit motive. Today, network infrastructure and end systems are constantly attacked with an increased level of sophistication and virulence.

In this chapter, we discuss and face two of the most dangerous threats known by the Internet community: *Denial of Service (DoS)* and *computer worms*. In the following we refer to them simply by DoS and Computer Worms. Those two families of threats have different goals, forms and effects than most of the attacks that are launched at networks and computers. Most attackers involved in cyber-crime seek to break into a system, extract its secrets, or fool it into providing a service without the appropriate authorization. Attackers commonly try to steal credit card numbers or proprietary information, gain control of machines to install their software or save their data, deface Web pages, or alter important content on victim machines. Frequently, compromised machines are valued by attackers as resources that can be turned to whatever purpose they currently deem important. DoS and worms are attacks launched against the network infrastructure and the end systems attached, respectively. In DoS attacks, breaking into a large number of computers and gaining malicious control of them is just the first step. The attacker then moves on to the DoS attack itself, which has a different goal, i.e. to prevent victim machines or networks from offering service to their legitimate users. No data are stolen, nothing is altered on the victim machines and no unauthorized access occurs. The victim simply stops offering service to normal clients because it is preoccupied with handling the attack traffic. Worms, a simple "slang" term for automated intrusion agents, are fast becoming a popular method to compromise networks and systems attached to them. Due to their speed and aggressiveness, worms represent a rising

Portions reprinted, with permission, from Ranjan, S., Shaleen, S., Nucci, A., Munafo', M. M., Cruz, R. and Muthukrishnan, S. M. (2007). "Dowitcher: effective worm detection and containment in the Internet core." In *IEEE Infocom*, Anchorage, AK, May, 2007.

concern for network operators. When used as stand-alone tools, worms can cause severe damage to the network infrastructure by eating up a large portion of bandwidth and thus causing degradation of quality of service to commercial applications and routing instability. Often worms have been used in the preliminary stage of a DoS when the attacker seeks to gain control over a large number of machines in a very short time frame (for example, Code Red [144]).

In the following we discuss how these threats work and introduce some commonly used terminology. Due to the importance of such a topic and the limited knowledge of techniques that can efficiently be used to fight those threats, we will present in this chapter a flexible framework for network behavior anomaly detection based on the novel concept of information entropy that can be of great assistance to IDSs and Firewalls in defending network infrastructures and customer network perimeters. We note that the material presented in this chapter is intended to be an introduction to the problem and should not be considered an exhaustive survey of the topic.

15.1 Understanding denial of service

A DoS attack is characterized by an explicit attempt to prevent legitimate use of service. There are two main approaches to denying a service: exploiting a vulnerability present on the target or sending a vast number of seemingly legitimate messages. The first kind of an attack is usually called a *vulnerability attack*, while the second is called a *flooding attack*.

Vulnerability attacks work by sending a few specifically crafted messages to the target application that posesses a vulnerability. This vulnerability is usually a software bug or a bug in a default configuration of a given service. Malicious messages by the attacker represent an unexpected input that the application programmer did not foresee. The messages cause the target application to: enter an infinite loop; severely slow down, crash, freeze, or reboot a machine; or consume a vast amount of memory and deny service to legitimate users. This process is called *exploiting a vulnerability*, and the malicious messages are called the *exploit*. In this chapter we will not focus on this type of DoS.

Flooding attacks work by sending a vast number of messages whose processing consumes some key resource at the target. For instance, complex messages may require lengthy processing, which takes CPU cycles, because large messages take up bandwidth and messages that initiate communication with new clients take up memory. Once the key resource is tied up by the attack, legitimate users cannot receive service. The crucial feature of flooding attacks is that their strength lies in the volume, i.e. not necessarily meant as bytes per second, rather than in the content. This has two major implications. First, the attackers can send a variety of packets. The attack traffic can be arbitrarily similar to the legitimate traffic, which greatly hinders defense, and attack the victim machines in many different forms. Second, the flow of traffic must be large enough to consume the target's resources. The attacker usually has to engage more than one machine to send out the attack traffic. This process is usually performed automatically through *scanning* of remote machines: looking for security holes that will enable subversion. Usually this phase is carried out through the use of *computer worms* and/or *Trojans*. Poorly secured machines that do not have recent patches and software updates, or are not protected by a firewall or other security devices, or their users have easy to guess passwords, are usually infected. We will discuss a few of those scanning techniques in Section 15.2, as they are commonly used by several instances of known worms. Flooding attacks are therefore commonly referred to as Distributed DoS, or *DDoS*. DDoS attacks manifest themselves in so many different forms as they can choose among a variety of victim types, i.e. application, host, resource, network and infrastructure.

Application attacks target a given application on the victim host, thus disabling legitimate use of that application and possibly tying up resources of the host machine. If the shared resources of the host machine are not completely consumed, other applications and services should still be accessible to the users. For example, a bogus signature attack on an authentication server ties up resources of the signature verification application, but the target machine will still reply to ICMP Echo requests, and other applications that do not require authenticated access should still work. Detection of application attacks is challenging because other applications on the attacked host continue their operations undisturbed and the attack volume is usually small enough not to appear anomalous. The attack packets are virtually indistinguishable from legitimate packets at the transport level, and the semantics of the targeted application must be heavily used for detection.

Host attacks disable access to the target machine completely by overloading or disabling its communication mechanism or making a host crash, freeze or reboot. An example of this attack is a TCP SYN attack. The victim host reserves some memory in a limited-size buffer for each new communication request, while the attacker can send out those requests without any memory cost. This asymmetry helps the attacker disable any new communication during the attack, while sending very few TCP SYN packets. This type of attack may manifest itself with an associated low, medium or high volume of traffic.

Resource attacks target a critical resource in the victim's network such as a specific WEB server, a router or a bottleneck link (i.e. local network interface that attaches the victim to the Internet). Depending on the target, the attacker can sometimes perpetrate an effective flooding attack with much smaller volumes. These types of DDoS attacks are also known as *stealthy DDoS*. An example of such an attack can be observed when the attacker wants to bring down a Web server. It can either flood the server with a huge number of HTTP requests, and thus consume the resources of the target, or it can monitor and thus discover the server activity over time identifying the few pages that request the largest consumption of the server resources to be processed. Web pages containing large video clips or many images are characteristic of such attacks. The attacker can then request access to these pages many times. In this case, although the total number of requests originated to the Web server under attack might still look very normal in terms of the total amount of incoming traffic, the server might experience a severe degradation in performance and eventually crash.

Network attacks consume the incoming bandwidth of a target network with attack packets whose destination address can be chosen from the target network's address space. These attacks can deploy various packets (since it is volume and not content that matters) and are easily detected due to their high volume. For example, if the attacker knows that the victim is attached to a 1 Gbps network segment, then they will send enough packets to the victim or other nodes on the segment to overwhelm it. Most networks become unusable as the traffic offered to them approaches their rated capacity; so little or no legitimate traffic will get through to the victim.

Infrastructure attacks target some distributed service that is crucial for global Internet operation. Examples include the attacks on Domain Name Service (DNS), large core routers, routing protocols, certificate servers, etc. The key feature of these attacks is not the mechanism they employ to disable the target, but the simultaneity of the attack on multiple instances of a critical service in the Internet infrastructure.

Denial of service is possible without using distributed techniques, but it poses a challenge for an attacker. For example, imagine that a DoS attack based on pure flooding originates at a single machine with a 10 Mbps link and that it is directed toward a victim machine that has a 100 Mbps link. In an attempt to overwhelm the victim's link, the attacker will flood his own network and deny service to himself. To disrupt the victim's communication successfully, the attacker must compromise an agent machine that has more network resources than the victim. It is very intuitive that this approach does not scale. In order to overcome the above issue, the attackers might perform the same actions, but in a pure distributed manner, say, using 100 machines. Each machine can now send 1 Mbps toward the victim. Assuming all 100 machines have a 10 Mbps link, none of them generates enough traffic to cause serious harm to its own local network. But the Internet delivers all attack traffic to the victim, overwhelming its link. Thus, the victim's service is denied, while the attackers are still fully operational. Those machines being used to launch a DDoS are called *zombies, daemons, agents,* or *slaves*. In this book we use the term *agents*. These agents, as mentioned above, are usually poorly secured machines with no recent patches and software updates, unprotected by firewalls or having easily guessed passwords. Their recruit is usually executed in an automated fashion, using either scanning techniques that search for specific vulnerabilities, i.e. computer *worms*, or through distributing attack software under the disguise of a useful application, i.e. *Trojan horses*.

The attacker can further hide their identity by deploying several layers of indirection between his machine and the agents. They may use one or several machines that deliver commands to the agents. These machines are called *handlers* or *masters*, and the attacker's machine is called the *controller*. In this book we use the term *handlers*. Another layer of indirection consists of the attacker's logging on to several machines in sequence, before accessing the handlers. These intermediary machines between the controller and the handlers are called *stepping stones*. Both handlers and stepping stones are used to hinder investigation attempts. If authorities located and examined an agent machine, all its communication would point to a stepping stone, and then to another stepping stone, and so on. If stepping stones are selected from different then countries and continents (and they usually are), it becomes very difficult to follow the trail back to the controller and unveil their identity. Another means of obscuring the attack is through the use of *IP spoofing*. Each packet in the Internet carries some control information preceding the data, i.e. an IP header. One field in the IP header specifies the address of the sender, i.e. the source IP field. This information is filled in by the machine that sends the packet (an action similar to putting a return address on a letter), and is used by the destination, or the routers on the path to the destination, to send replies back to the source. Attackers commonly forge this field to achieve impunity for the attacks and hinder the discovery of agent machines. Note that this approach can only be used if the attacker is not making use of bidirectional communication.

The issues described above make DDoS attacks a frightening possibility. Yet researchers in computer and network security are aware of many frightening possibilities that never come to pass. Are security researchers merely alarming the public with claims of the dangers of DDoS? Unfortunately, DDoS attacks are not speculation or fiction. A number of recent studies have demonstrated that DDoS attacks are extremely common in today's networks [9]. Given that they are usually quite effective and that perpetrators are rarely caught, there is reason to believe they will become even more popular in the future. Measuring the frequency of any form of attack in the Internet is difficult. Victims do not always realize that they are under attack. When a service does not respond correctly, people first blame their infrastructure, searching for network problems. Even if they do recognise an attack, they often fail to report it to any authority. A number of organizations use survey techniques to gain some insight into the prevalence of DDoS attacks. Interestingly enough, all of them seem to point to year 2001 as the time where DDoS attacks started to become popular in the Internet. Since then, DDoS attacks have evolved both in terms of size of the agent army (the average number of agents in 2001 was estimated to hundreds, while in 2007 the number increased to hundreds of thousands of agents), sophistication (attacks are increasing towards high-revenue services such as VoIP and IPTV) and obfuscation (it was in 2005 that the first case of DDoS using a mixture of stepping stones and IP spoofing techniques was observed in a commercial environment).

15.2   Understanding worms

A computer *worm* is a program that self-propagates across a network, exploiting security or policy flaws in widely used services. Worms are not a new phenomenon, having first gained widespread notice in 1988 [189]. We distinguish between worms and viruses in that the latter infect otherwise non-mobile files and therefore require some sort of user action to abet their propagation. As such, viruses tend to propagate more slowly. They also have more mature defenses due to the presence of a large anti-virus industry that actively seeks to identify and control their spread. In order to understand the worm threat, we divide its dynamics into several steps: *target discovery*, *carrier*, *activation* and *intent of the attack*.

15.2.1 Target discovery

For a worm to infect a machine, it must first discover that the machine exists, known as *target discovery*. There are a number of techniques by which a worm can discover new machines to exploit, including scanning, pre-generated target lists, external target lists and internal target lists. Worms can also use a combination of these strategies. If a defense blocks a given strategy, this can prevent an entire class of worms from propagating. *Scanning* entails probing a set of addresses to identify vulnerable hosts. Two simple forms of scanning are *sequential* (working through an address block using an ordered set of addresses) and *random* (trying addresses out of a block in a pseudo-random fashion). Due to its simplicity, it is a very common propagation strategy, and has been used quite extensively and successfully. Scanning worms spread comparatively slowly compared with a number of other spreading techniques, but, when coupled with automatic activation, they can still spread very quickly in absolute terms. There are several optimizations which apply to scanning worms. A highly effective optimization is a preference for local addresses. Although this may be slightly inferior for Internet-scale propagation, it enables the worm to exploit a single firewall breach to scan the entire local network. Using this technique, a single copy of the scanning program can compromise many vulnerable machines behind the firewall. *Permutation scanning* enables a worm to utilize distributed coordination to scan the net more effectively and to determine when the bulk of the network is infected. The efficiency of a scanning activity can be limited by either delay constraints or bandwidth constraints. Many worms, such as Code Red [145], use scanning routines which are limited by the latency of connection requests rather than the throughput by which requests can be sent. Indeed, Code Red required the transmission of a TCP-SYN packet, followed by a response or timeout. In principal, worms can compensate for this latency by invoking a sufficient large number of threats. However, in practice, context switching overhead is significant and there are insufficient resources to create enough threats to counteract the network delay, i.e. the worm quickly stalls and becomes latency limited. The alternative, a bandwidth-limited scanner, is substantially faster. Slammer/Sapphire [144] was inadvertently a bandwidth-limited scanner as a side-effect of its single-packet UDP design. It used a single 404 bytes packet (compared with the 4 Kbytes packet of Code Red and 60 Kbytes packet of Nimda) to UDP port 1434 (SQL Server vulnerability exploit). The worm was able to send these scans without requiring a response from the potential victim and thus was able to spread nearly two orders of magnitude faster than Code Red. In only ten minutes it was able to infect a similar size of hosts as Code Red, i.e. around 360 000, while its predecessor took over 12 hours [197].

In general, the speed of scanning worms is limited by a combination of factors, including the density of vulnerable machines, the design of the scanner and the ability of edge routers to handle a potentially significant increase in new, diverse communication. Pre-generated target lists can be used to accelerate a scanning worm. An attacker could obtain a target list in advance, creating a "hit-list" of probable victims. The biggest obstacle is the effort to create the hit-list itself. For a small target list, readily available public sources or open access points can be used to perform small-scale scans.

Comprehensive lists require more effort: either a distributed scan or the compromise of a complete database. Similar concepts hold in the case of an external target list. In this case, the attacker maintains a separate server to track the vulnerable population of hosts. The worm queries the external server in order to determine its new targets. These worms are known as *metaserver worms* and typically attack Web servers (for example using Google as a metaserver in order to find other Web servers to attack). Although we have not seen a metaserver worm in the wild, we want to emphasize the significant risks that are associated with this due to the great speed such a worm could achieve. A worm may use the information that many applications contain about other hosts in order to spread. Such target lists can be used to create *topological worms*, where the worm searches for local information to find new victims by trying to discover the local communication topology. The original Morris worm [189] used topological techniques including Network Yellow Pages, /etc/hosts, and other sources to find new victims. For applications that are fairly highly connected, such as chat, email or gaming, such worms can be incredibly fast.

15.2.2   Carrier

The means by which propagation occurrs can also affect the speed and stealth of a worm. A worm can either actively spread itself from machine to machine, or it can be carried along as part of normal communication. There are three types of carrier modes. A worm can actively transmit itself as part of the infection process, i.e. it delivers the worm body during the target discovery phase (see CRClean worm as an example [46]). Some worms, such as Blaster [29], require a secondary communication channel to complete the infection. Although the exploit uses the Remote Procedure Call (RPC), the victim machine connects back to the infecting machine using Trivial FTP (TFTP) to download the worm body, completing the infection process. Finally, a worm can send itself along as part of a normal communication channel, either appending to or replacing normal messages. As a result, the propagation does not appear as anomalous when viewed as a pattern of communication.

15.2.3   Activation

A key step of the worm contamination process is the activation of the worm body residing on the victim host. An efficient worm activation can drastically affect how rapidly a worm can spread because some worms can arrange to be activated nearly immediately whereas others may wait days or weeks to be activated. In general, the activation process can either be invoked (i) by the execution of a specific code by the end user, (ii) by some human activity not strictly related to the worm, (iii) according to a scheduled process or (iv) it can be completely automated with no kind of human involvement. The slowest activation approach, known as *human activation*, requires a worm to convince a local user to execute the local copy of the worm. Since most people do not want to have a worm executing on their system, these worms rely on a variety of social engineering techniques. Some worms, such as the Melissa email-worm [47], indicate urgency on the part of someone you know ("Attached is an important message for you"); others, such as the Iloveyou [48] attack, appeal to an individual's vanity ("Open this message to see who loves you"); and others, such as the Benjamin [190] worm appeal to greed ("Download this file to get copyrighted material for free"). Similarly, many worms are activated when the user performs some activity not normally related to the worm, such as resetting the machine or logging in, therefore executing scripts, etc. This activation approach is known as *human activity-based activation*. Another activation process being used is known as *scheduled process activation*. Such programs can propagate through mirror sites (e.g. OpenSSH Trojan, etc.), or directly to desktop machines. Many desktop operating systems and applications include auto-updater programs that periodically download, install and run software updates. Early versions of these systems did not employ authentication, so an attacker needed only to serve a file to the desktop system to infect the target (see ref. [78] for more details). Other systems periodically run backup and other network software that includes vulnerabilities. The fastest activated worms known so far are able to initiate their own execution by exploiting vulnerabilities in services that are always on and available (e.g. Code Red exploiting IIS Web servers) or in the libraries that the services use (e.g. XDR [49]). Such worms either attach themselves to running services or execute other commands using the permissions associated with the attacked service. Execution occurrs as soon as the worm can locate a copy of the vulnerable service and transmit the exploit code.

15.2.4   Intent of the attack

Last in the process is the *attack intent*, reflected in code carried in the packet payload of the worm. Different sorts of attackers will desire different payloads to further their ends. Code Red II opened a trivial-to-use privileged backdoor on victim machines, giving anyone with a Web browser the ability to execute arbitrary code. The Sobig worm [186] created an open-mail relay for use by spammers. Another common payload is a DoS attack. Code Red, Yaha [131] and others have all contained DoS tools, either targeted at specific sites or retargetable under the attacker's control.

In this chapter we introduce a general framework that can be used to detect a variety of different network anomalies, e.g. DDoS, worms, port scans, etc. The methodology is composed of two tiers. The first tier uses the powerful concept of information entropy to detect promptly network anomalies that alter the structure and randomness of Internet traffic. By monitoring the divergence of specific traffic features observed in real time, we are able to classify correctly the network anomalies in several families. The second tier gains a deeper understanding of the detected anomaly by collecting and processing key payload content information. This second module comes to be very helpful if the operator is interested in extracting more information from the ongoing attacks, such as the packet payload signatures in case of worms or the subject/body of an email or VoIP spam activity.

15.3   Related work: Worms and DoS

Various methods have been proposed in the literature to detect DoS and worms. An extensive survey of those techniques can be found in refs. [139] and [203]. We start by exploring the techniques proposed to detect worms. These can be divided into two major categories.

15.3.1   Worms

The first class is based on *content fingerprinting* using the layer-7 traffic information refs. [106, 112, 150, 180]. The primary intuition underlying this class is that an ongoing worm propagation should manifest itself in the presence of higher than expected *byte-level similarity* among network packets: the similarity arises because of the unchanging portions of the worm packet payload, something expected to be present even in poly-morphic or obfuscated worms (albeit spread out over the length of the packet or across several packets belonging to the flow). In particular, Earlybird [180] tries to collect efficiently fingerprints of fixed-size payload blocks from *all* the traffic crossing the net-work border and then checks the address dispersion for the content, reporting a worm when this dispersion is above a fixed threshold. Autograph [112] and Polygraph [150] use the opposite approach. A pool of suspicious flows is created, using the number of unanswered inbound SYN packets (hint of a port scanning activity) as a tentative and imperfect indicator[1] of suspect activity, and then fingerprinting is applied to short variable-length content blocks to identify content prevalence and report possible worms. All these approaches consider packet content as a *bag of substrings* (of either a fixed length [180] or a variable packet-content-based length [112, 150]). In ref. [106] the authors analyze the characteristics of the *inverse distribution, I(f)*, which tracks, for a given frequency f, the number of substrings that appear with that frequency and propose $I(f)$ as a new discriminator for the early detection of worms. Although the metric used is interesting and more robust than the previous ones, the approach must still inspect the payload of all packets passing through the link and hence is still not scalable.

In contrast to the aforementioned class of approaches based on packet-content anal-ysis, the second class refs ([117, 198, 204, 212]) consists of techniques which identify network anomalies by examining the layer-4 traffic distribution across a few features. The primary intuition underlying these approaches is that a worm manifestation breaks the statistical characteristics of Internet traffic; worm traffic is more uniform or struc-tured than normal traffic in some respects and more random in others. These approaches propose various techniques, based primarily on information-theoretic measures, such as information entropy [117, 212] or Kolgomorov complexity [198] as the statistic to rep-resent the distribution of traffic features such as source and destination IP addresses and port numbers. However, we contend that these approaches that only look at one

---

[1] In carrier networks, unmatched SYN packets may not be an anomaly due to the prevalence of asymmetry, especially at the peering links, where packets belonging to the same flow may be routed across different links.

feature at a time lack the important structure and dynamics existing across several features. Thereafter, approaches such as those discussed in refs. [117] and [212] propose techniques based on Principle Component Analysis (PCA) and Residual State Analysis (RSA), respectively, to establish complex relationships between the traffic features using which flows are classified as either legitimate or malicious. Approaches such as PCA and RSA, while robust, are primarily off-line and hence are not effective for worm containment at the high data rate links typical of the Internet core.

15.3.2 Denial of Service

In the context of DoS, three major families of techniques have been proposed in the literature. The first family uses *change-point*-detection algorithms to isolate a traffic statistic's change caused by attacks. These approaches initially filter the target data by address, port or protocol and store the resultant flow as a time series. The time series can be considered as the time-domain representation of a cluster's activity. If a DoS flooding attack begins at time t, the time series will show a statistical change either around t or at time greater than t. One class of change-point-detection algorithms operates on continuously sampled data and requires low amounts of memory and computational resources. An example here is the *Cumulative Sum* (Cusum) algorithm. To identify and localize a DoS attack, the Cusum identifies deviations in the actual versus expected local average in the traffic time series [181, 202]. If the difference exceeds some upper bound, defined either statically or dynamically according to the specified statistical criteria, the Cusum recursive statistic increases for each time series sample. During time intervals containing only normal traffic, the difference is below this bound, and the Cusum statistic decreases until it reaches zero. Using an appropriate threshold against the Cusum statistic, the algorithm identifies an increasing trend in the time series data, which might indicate the onset of a DoS attack. Although most of the researchers have proposed the Cusum algorithm for an effective DoS attack detection, other researchers have extended this detection method to identify the typical scanning activities of computer worms.

The second family uses *wavelet analysis*, which describes an input signal in terms of spectral components. Although Fourier analysis is more common, it provides a global frequency description and no time localization. On the other hand, wavelets provide concurrent time and frequency description, and can thus determine the time at which certain frequency components are present. For detection applications, wavelets separate out time-localized anomalous signals from background noise (the input signal contains both). Ideally, the signal and noise components will dominate in separate spectral windows. Analyzing the energy of each spectral window determines the presence of anomalies. An example of such work can be found in ref. [31], in which the authors decompose the traffic data into distinct time series of average IP/HTTP packet sizes per second, flows per second and bytes per second. They then apply wavelet analysis to each time series, resulting in time-localized high- and mid-band spectral energies. They consider low-frequency content to be daily and weekly activity, and thus not an onset of an abrupt attack. To identify anomalies, they weight a combination of high- and middle-spectral energies, and then they threshold its variability.

The first two families of techniques have been widely used to monitor and profile one single traffic feature at a time. Only recently have researchers come to realize the importance of understanding the internal composition of traffic as a whole due to the fact that, in the event of a network attack, more than one traffic feature is likely to be affected. Thus, extracting and profiling the intrinsic correlation across multiple time series can significantly improve the accuracy of the detection method being used, and thus provide a more prompt response, even in the case of low-rate network attacks. Due to the large amount of data that need to be processed in real time, the research community has focused on how to make this problem more tractable from a computational point of view. Several techniques, such as Multivariate Outlier Detection (MOD) methods [119], PCA [117] and Sketch algorithms [113] have emerged in the literature as dimensionality reduction techniques that enable computationally efficient methods for identifying outliers (or anomalies) in the data set. Common characteristics of all the methods listed above include their effort to fuse multiple aspects of a network's element behavior and quantifying the distance of the current operational state from its normal state. If this distance is larger than a predetermined threshold value, the observed state is considered abnormal and the appropriate recovery actions are invoked. Otherwise it is considered normal.

In this chapter, we introduce a general framework, named DoWitcher, based on the powerful concept of information entropy for NBA-DS that brings together the benefits of the multi-variate time series analysis and the efficiency and reliability of content fingerprinting techniques. The scalability achieved by the proposed algorithm is based on its two-tiered functional architecture. The first tier uses only flow information to perform entropy analysis of the traffic and to identify anomalous activity, hence avoiding inspection of the payload of all packets traversing router interfaces. In order to save resources, e.g. memory and CPU, and for a prompt detection of the ongoing threat, DoWitcher uses two novel metrics, called Pair-wise Marginal Entropy Ratio (PMER) and Pair-wise Joint Entropy Ratio (PJER) which are based on the observation that, during a network anomaly, at least two of the traffic features exhibit diverging behaviors. By looking at the trends of those traffic features, DoWitcher is able to classify network anomalies in different families such as DoS, worms, spam or network scans. For operators interested in knowing more about the threat being detected, DoWitcher automatically pinpoints the malicious flows associated to the ongoing anomaly by generating a flow filter-mask, using which full packet capture is performed over the flows matching the filter. Specialized modules process the new information flowing through the system. In the context of a worm outbreak, we introduce in this chapter an optimized version of the well known Longest Common Subsequence (LCS) algorithm, called *windowed LCS* (LCS^W) that detects high byte similarity in the malicious data-streams, i.e. body or signature of the worm. The novel LCS^W algorithm is memory efficient and able to extract the worm invariant signatures, even from the polymorphic worms, which alter their content during propagation. The remainder of this chapter is structured as follows. In Section 15.4 we introduce the logical architecture of a network behavioral anomaly detection system, and briefly describe the overall functionality of each component. Then we discuss a network deployment of DoWitcher in order for carriers to protect both their core infrastructure and their end customers. Section 15.5 presents, in great detail, step-by-step, the logical operational flow used by DoWitcher, explaining which traffic features are extracted over time, how their histograms are monitored over time, how the structure across the features is captured by profiling one single metric (PMER and PJER), how to profile the behavior of the metric over time and how to detect and classify the network anomaly. Furthermore, we introduce the LCS^W module, able to extract the invariant part of a malicious threat in the context of a worm outbreak. In Section 15.6 we discuss a rich set of results of live experiments that extensively validate the efficiency and performance achieved by the system. Section 15.7 summarizes our findings and concludes the chapter.

15.4   NBA-DS: logical architecture and network deployment

DoWitcher is a highly scalable Network Behavior Anomaly Detection System (NBA-DS). It is able to collect raw packets from network links and router interfaces, locally extract key traffic features and then process all meta-data in one single central location, critical for generating a real-time network-wide view of traffic activity. The system is composed of two modules: *local* and *global* analyzers. The DoWitcher Local Analyzer (DLA), shown in Figure 15.1(a), represents the interface with the network elements that collect raw packets through a *network tap*, e.g. a passive wire tap that collects packets off the wire or a port mirror that collects packets directly from router interfaces (see Chapter 3). The packets are then processed by the *flow reconstruction* module that

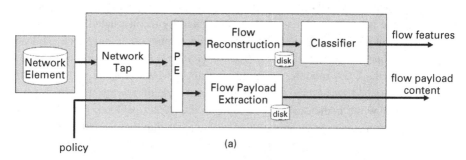

(a)

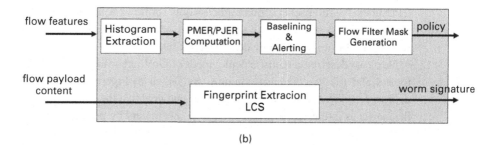

(b)

Fig. 15.1.   Logical architecture of DoWitcher: (a) local and (b) global analyzers.

buffers packets, reorders the out-of-sequence packets and statefully reconstructs TCP and UDP flows in real time. A TCP flow is defined by all the packets identified by the layer-4 5-tuple, in between and including the SYN and FIN packets, while a UDP flow is defined by the set of packets corresponding to a pair of IP addresses and ports until there is no packet arrival within a specified timeout. The flows are then processed by the *Classifier* module that extracts key features from each flow. In each time slot, all DLAs deployed across the network forward this information up to the DoWitcher Global Analyzer (DGA), shown in Figure 15.1(b). The DGA is in charge of several functions: (i) extracting the histograms of the key flow features and summarizing their properties by computing their entropies (Histogram Extraction Module); (ii) grouping all entropies together in one single efficient metric, *PMER* and *PJER* (PMER/PJER computation); (iii) profiling the metric over time and generating alerts when an instantaneous deviation from the historical trend of the current metric is observed (Baseline and Alerting Module); (iv) classifying the anomaly type detected by analyzing the type of divergence/convergence of entropies associated to specific traffic features; (v) composing a policy rule that captures the anomalous activity from a traffic flow perspective (Flow Filter Mask Generation). The policy is then forwarded down to all DLAs. Note that step (v) is used when the end user wants to know more about the specific anomaly being detected. For example, in the case of a worm, a typical policy would request capturing raw packets from a few infected end hosts, using a specific destination or source port and a fixed flow size or range thereof.[2] In the case of email spam, a typical policy would request capturing raw packets from the spammer machine, using a specific destination port and a fixed flow size or a range of flow sizes. As soon as the policy is received, each local analyzer starts *full packet capture* of packets belonging to flows matching the policy rule. After the flows are statefully reconstructed, the *Flow Payload Extraction* module strips out the packet headers and stores only their payloads, generating what is defined as the *flow payload content*. Finally, this information is forwarded up to the DGA module. Different algorithms would be used at the DGA to process this enriched set of data. For example, in the case of a worm, for which we might want to know the malicious body residing in the packet payload, the DGA would extract the worm signature by applying a specific algorithm that we will introduce in this chapter, named LCS^W. We remind the reader that the scalability of DoWitcher comes from the two-phase anomaly detection identification algorithm adopted: we examine the packet headers to infer statistically the existence of an anomaly and initiate the memory- and CPU-intensive full packet capture process only for flows belonging to a limited number of hosts marked as involved in the malicious activity. This architecture resonates well with the multi-tier architecture described in Chapter 3.

Next, we describe a realistic deployment of DoWitcher to protect (i) a carrier's customers and (ii) its core infrastructure. As shown in Figure 15.2, DoWitcher can be deployed in different ways according to the end goal that the carrier wants to achieve. If the carrier is interested in protecting its end customers by providing managed security

---

[2] Polymorphic worms that change their signature during their propagation might alter their flow size by a few bytes.

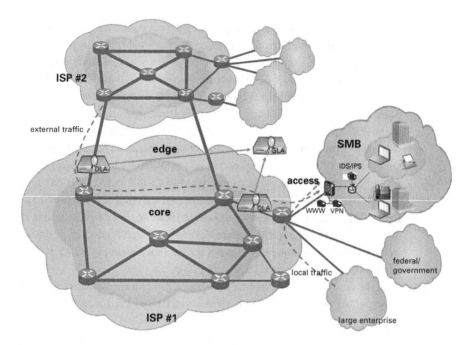

Fig. 15.2. Typical network deployment of DoWitcher in a carrier network: peering and edge links.

services, then DoWitcher can be deployed across the edge links, e.g. links which connect the gateway routers to the backbone routers, thereby protecting a set of customer enterprises while deploying fewer DLAs. If the carrier's desire is to protect its network core from possible roguish behavior of its client networks, then DoWitcher should be deployed at both the edge links as well as at their peering links. This way the carrier can monitor and detect ongoing threats coming from both its peers as well as from its own local customers.

15.5   Algorithm

In this section, we describe in detail the major functions of the system, emphasizing the Classifier (DLA module) as well as all the blocks that comprise the DGA described in Figure 15.1.

15.5.1   Classifier and Histogram Extraction Module

An efficient anomaly detection system must identify and correctly classify the ongoing threat before the real effect becomes visible to the end user, e.g. in unusual high link-bandwidth utilization or large CPU/memory resource consumption. Which traffic features should be monitored and which statistics should be profiled are important questions that we want to answer in this section. In general, a malicious network anomaly tends to disrupt the *normal structure* of Internet traffic due to its intrinsic characteristic of being more uniform or structured than normal traffic in some respects and more random in others. For example, in the case of a worm, a small number of hosts try to find other hosts to infect by attempting to connect to them in a purely random fashion, scanning for a specific vulnerability. Most of the worms connect to other hosts using random source port numbers and the same destination port number. Other worms behave in the opposite way, connecting to other hosts using the same port number and choosing random destination ports. A common characteristic across all Internet worms is their small size in order that they spread as fast as possible. Indeed, a large worm size would prolong infection time and consume bandwidth that could be used for infecting other targets. The typical flow size of a worm ranges from a few hundred to a few thousand bytes. The communication patterns fall into the two types represented in Figures 15.3(a) and (b). Note that the traffic feature denoted as *Fsize* is adopted in this context with an interchangeable meaning of bytes per packet (such as in Figure 15.3(a) – (d)) and bytes per flow (such as in Figure 15.3(e) and (f)).

In the case of scan activity, targets can be either a specific host, i.e. explorative phase in search of open ports to be exploited later (also known as *port scan*), or a broader

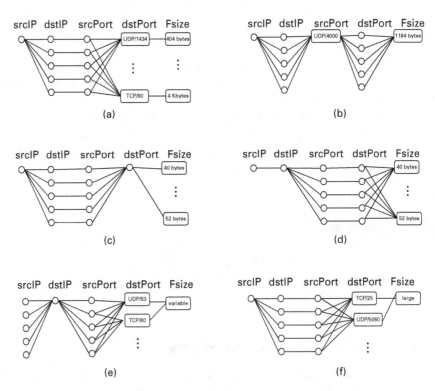

Fig. 15.3.   Communication behavior of common Internet threats. (a) Worm 1 attack (Sasser, CodeRed); (b) Worm 2 attack (Witty); (c) network scan attack; (d) port scan attack; (e) DDoS attack; (f) spam attack. Note srcIP = source IP; dstIP = destination IP; srcPort = source port; dstPort = destination port.

pool of hosts, i.e. explorative phase in search of hosts having a pre-selected port open (also known as *network scan*). During such attacks, usually lasting no longer than a few minutes, a large number of flows are generated by the same host, directed to one or more hosts and using either a large number of distinct source ports or destination ports. Such flows have the characteristic of containing very few packets (typically 1 to 3 packets per flow with 40, 44 and 52 Bytes per packet). The scan of the IP space or the port range can be executed either sequentially or in a purely random fashion. Figures 15.3(c) and (d) summarize common patterns of network scan and port scan attacks.

In the case of a DDoS attack, many machines open many sessions toward the same target host in a very synchronized fashion and exploit the same vulnerability on the target. As a consequence, a DDoS attack is characterized by many flows directed to the same destination host and destination port and originated by a large number of different source IPs. Due to the fact that a DDoS attack can manifest itself in a variety of different ways, targeting both the data or the control plane of an Internet communication, its flow size does not represent an important traffic feature for its detection (see Figure 15.3(e)). On the other hand, the strong coordination among the attacker machines can be fundamental to distinguish efficiently a real DDoS from a flash-crowd event (many legitimate hosts contacting the same destination host to search for the same thing, e.g. the latest movie trailer or video clip on a website).

Last, but not least, an important network attack is spam, historically being associated with email traffic only, but today covering more services such as VoIP, SMS, etc. The pattern behavior of a spammer can often be confused either with a worm or with a network scan attack due to its similar behavior; i.e. a spammer tries to contact many distinct hosts using a specific service such as email (TCP/25) or VoIP (UDP/5060 or UDP/5061). In order to distinguish the spam activity from other forms of attacks, it is important to monitor the destination port being used on such flows, the size of the communication (usually spam flows are much bigger than worm flows and network scan flows) and the diversification of the destination IPs being selected (usually worms and network scans target hosts belonging to the same subnet first, while a spammer usually selects victims much more randomly). A common communication pattern of a spam attack is shown in Figure 15.3(f).

An important commonality across all the attacks is related to the small set of *flow* traffic features whose histograms tend to be affected during the spread of those anomalies. More specifically: (i) source IP, (ii) source port, (iii) destination IP, (iv) destination port, and (v) flow size.

Indeed, in the case of a worm, a shift in the source-IP histogram will be expected whenever the number of flows generated by the infected hosts grows to be a significant part of the total observed flows; the source IP addresses of the scanning hosts will be seen in many flows, and the distribution of the source IP address will become more skewed around the few infected hosts. A similar thing happens on the source-port and destination-port histograms. If any attacker scans for a specific vulnerability, these scans often have to go to a specific target destination port. The source ports of these connections are usually selected in some weakly random fashion from a range of possible source ports. Examples of these behaviors are visible in worms seen in the past, such as Sapphire (destination-port 1434), CodeRed (destination-port 80), Welchia (destination-port 135) and many others. A few worms, such as Witty, behave unexpectedly using a fixed source port and variable destination ports. Similar consideration holds for the flow-size histogram. If a specific flow size becomes a significant component of the overall traffic, the distribution of the flow size will become more skewed around the flow size used by the worm. On the other hand, the destination-IP histogram is expected to flatten due to the inherent random scanning activity of the destination hosts.

Similarly, port scans and DDoS attacks would affect the same flow traffic features. For example, in the context of a port scan, the scanner machine initiates the scan activity toward the victim machine by opening many TCP or UDP flows and scanning the entire port space either randomly or in a sequential fashion. As a consequence, we expect the source IP and destination-IP histograms to become more centered around the IP address of the scanner and the victim host, respectively, while the destination-port histogram is expected to be flatter as many more destination ports will be hit than normal. On the other hand, when a victim is under a DDoS attack, we would expect to see the source-IP histogram flatter, while the destination-IP and destination-port histograms will be more centered around the IP address of the victim and the specific vulnerability exploited by the attack, respectively.

In the following, we exploit the above observations in order to provide a proper classification of malicious threats. We remind the reader that although those flow traffic features prove to be very efficient in detecting the threats mentioned above, they should not be considered as an exhaustive solution to the problem, but that they could work in concert with other features, previously proposed in the literature.

A typical Internet link may serve traffic to hundreds of thousands of distinct IP addresses at any given time. Hence, it may be prohibitive to construct histograms over distinct IP addresses. DoWitcher's histogram extraction is designed with the goal of providing memory savings by aggregating flows over clusters of IP addresses. Clusters might be defined either *statically*, as pre-defined blocks of IP addresses, i.e. subnets (/16, /24), or *dynamically*, as groups of hosts behaving similarly over time in terms of traffic.[3] In the following, we refer to *clusters* as subnets.

15.5.2   PMER/PJER Computation Module

In order to track changes in the shape of the featured histograms, we introduce the concept of *entropy*, which measures the degree of randomness in a data set; the more random the data, the higher its entropy. In the following, we mathematically introduce the concept of entropy.

Let's assume we are monitoring a generic key feature X over time for a specific cluster A and let $M^X(x)$ be its frequency distribution, i.e. the number of times we see

---

[3] Having the capability of identifying client vs server activity helps DoWitcher to scale even further as it will process less traffic, e.g. only client traffic, while removing server traffic, based on the assumption that an attack is not very likely to be launched from servers.

an element $x \in X$. From the frequency distribution $M_i^X(x) = \{x_i\}$ in time-window i, we can derive the empirical probability distributions $P_i^X(x)$ as follows:

$$P_i^X(x) = \left\{ p_i^X | p_i^X = \frac{x_i}{m^X} \right\},$$

where $m^X = \sum x_i$ is the overall number of flows that contributed to the distribution during time-window i.

From this probability distribution P_i^X, we calculate the information entropy H_i^X as

$$H_i^X = - \sum_{p \in P_i^X} p \log_2 p,$$

where, by convention, $0 \log_2 0 = 0$. As known, the measure of entropy provides a means of quantifying the uniformity of a distribution [125]: low entropy indicates high probability in a few elements (concentrated usage of the same port, high traffic from the same source), while high entropy indicates a more uniform usage (random scan of destination IP, variable source port). Since we are using $\log_2(\cdot)$ in our definition, each H_i^X will assume values in a range between zero and $\lceil \log_2(N^X) \rceil$, N^X being the maximum number of distinct values X can assume in the time window. In order to have a metric H_i^X independent of its support size, we normalize the entropy by the size of its support, i.e. $Support(N^X) = \lceil \log_2(N^X) \rceil$. This is also known in the literature as Relative Uncertainty (RU) and is formally described as:

$$RU_i^X = \frac{H_i^X}{Support(N^X)}.$$

In the remainder of this chapter we refer to H^X with the meaning of RU^X.

Single global metric

As discussed before, methods based on PCA and RSA outperform previous approaches since they capture the structure across features. Similarly, DoWitcher is based on the hypothesis that, during an anomaly outbreak, the RUs of at least two of the five features will diverge [198]. In order to capture such dynamics, we propose in this chapter two novel metrics, namely the *pairwise joint entropy ratio* and the *pairwise marginal entropy ratio*, which exhibit a stable behavior under normal conditions while showing a sharp increase even during a low-volume anomaly or scan activity.

The Pairwise Joint Entropy Ratio (PJER) is defined as the maximum over all feature-pairs $((x, y))$ of the ratio between S_i^{xy} and its average $Avg\left(S_i^{xy}\right)$ computed using the last N_S time-windows:

$$S_i = max_{(x,y) \in K} \left| \frac{S_i^{xy}}{Avg(S_i^{xy})} - 1 \right|, \tag{15.1}$$

where $S_i^{xy} = H_i^{xy} / \left(H_i^x + H_i^y\right)$ represents the instantaneous ratio between the joint RU (H^{xy}) and the sum of the two marginal RUs ($H^x + H^y$) at time i, also called the *joint-entropy-ratio*; $Avg(S_i^{xy}) = 1/N_S \sum_{k=i-N_S}^{i-1} \left[H_i^{xy} / \left(H_i^x + H_i^y\right)\right]$ represents the average value of S^{xy} over the last N_S time-windows. At each point in time, the metric computes the shift in the slope between the instantaneous value of S^{xy} and its averaged value $Avg(S^{xy})$ for all feature-pairs. The absolute value is considered such that it is an always-positive process. The maximum is used to consider the largest instantaneous deviation of this metric at each point in time. The reason why we decided to normalize the joint RU, H^{xy}, by the sum of the marginal RUs, H^x and H^y, is that the single joint RU does not capture the deviation of the two associated features, but only the fact that specific pairs (x, y) of the joint distribution become more used than others, e.g. the joint distribution becomes more skewed around a few (x, y) values. Note that, since $H^{xy} \leq H^x + H^y$, the S^{xy} can only assume values between 0.5 and 1. By normalizing the joint RU by the sum of the marginals, we track over time the dependency of the feature-pair, e.g. the reduction in uncertainty X due to the knowledge of Y. Indeed, under normal traffic conditions, the two features exhibit a very structured dependency, while under anomalous activity the two features X and Y become more independent from each other, leading to an increase of the joint entropy ratio S^{xy} value. In the limit case, when the two features will be completely independent, $S^{xy} = 1$. Thus, the *PJER* metric is very accurate in detecting the feature-pair that is deviating the most and it is able to identify the most frequent pair values (x, y) observed. Unfortunately, the computation of this metric is space- and memory-intensive, since we must build and store in memory the histograms of both marginal and joint distributions of all the key features monitored. In order to overcome this issue, we introduce a second metric that uses only marginal distributions.

The Pairwise Marginal Entropy Ratio (PMER) is defined as the maximum over all feature-pairs $((x, y))$ of the ratio between the marginal RUs (H^x and H^y) and its average computed using the last N_S time-windows:

$$R_i = max_{(x,y)\in K} \left| \frac{R_i^{xy}}{Avg(R_i^{xy})} - 1 \right|, \qquad (15.2)$$

where $R_i^{xy} = H_i^x / H_i^y$ represents the instantaneous ratio between the two marginal RUs H^x and H^y; $Avg(R_i^{xy}) = 1/N_S \sum_{k=i-N_S}^{i-1} H_k^x / H_k^y$ represents the average value of R^{xy} over the last N_S time-windows. At each point in time i, the metric captures the shift in the slope between the instantaneous value of R^{xy} and its average value $Avg(R^{xy})$ for all feature-pairs. As for *PJER*, *PMER* monitors the maximum divergence from normal behavior across all possible feature-pairs. As the reader will note, the *PMER* metric is less space- and memory-intensive than *PJER* since it uses only marginal RUs for its computation. On the other hand, since *PMER* does not contain information about joint entropies, it is not able to identify accurately pair values (x, y) that are characteristic of a worm outbreak, but can only infer the most likely ones.

15.5.3   Baselining and Alerting Module

Our anomaly detection consists broadly of two phases: off-line baselining and on-line detection. First, we perform an off-line characterization of traffic from system logs, assuming that the traffic consists of legitimate flows solely, i.e. it is not influenced by any worms or other anomalies propagating through the network. Next, in an on-line phase, we compare the ongoing traffic statistics with the legitimate profiles obtained previously. Using a weighted moving average of the traffic statistics in the past observations, we forecast their value in the next time interval and flag traffic as suspicious if there are significant deviations from the forecasted value. In the following, we mathematically introduce the forecasting profiling algorithm for the PMER metric; the same algorithm can be applied for the PJER metric. The algorithm is as follows.

Step 1 Define an averaging set $\mathcal{R}_i$ with $|\mathcal{R}_i| = W$ containing the last $W > N_S\ R_j$ samples considered being *in profile*. During baselining, all the R_j are considered *in profile* and are included in $\mathcal{R}_i$; $\mathcal{R}_i$ is valid only for $i \geq W$, so we need to collect data for at least W time-windows before being able to run the algorithm.

Step 2 Maintain a running average of R_i over $\mathcal{R}_i$:

$$\overline{R}_i = \frac{1}{W} \sum_{R_n \in \mathcal{R}_i} R_n.$$

During baselining, since all the last $W\ R_i$ samples are in $\mathcal{R}_i$, this becomes

$$\overline{R}_i = \frac{1}{W} \sum_{n=i-W+1}^{i} R_n.$$

Step 3 Define control coefficients $\hat{\alpha}_i^{\max}$ and let $\hat{\alpha}_i^{\max} = \hat{\alpha}_k^{\max} = 1$ for $0 < k < W$.

Step 4 For $W < i \leq T_w$, T_w being the length of the baselining period, we have the following:

$$\text{if}\quad R_i > \hat{\alpha}_{i-1}^{\max}\overline{R}_{i-1} \quad \text{then}\quad \hat{\alpha}_i^{\max} = \frac{R_i}{\overline{R}_{i-1}}$$

$$\text{else}\quad \hat{\alpha}_i^{\max} = \hat{\alpha}_{i-1}^{\max}. \tag{15.3}$$

At the end of the baselining period, $\hat{\alpha}_{T_w}^{\max}$ will contain the largest measured excursion between one sample and the average $\overline{R}$ value in the near past. We freeze this value, defining $\alpha^{\max} = \hat{\alpha}_{T_w}^{\max}$.

After the baselining period, we will report an anomaly whenever $R_i > \alpha^{\max}\overline{R}_{i-1}$. If no anomaly is revealed, the sample R_i is added to $\mathcal{R}_i$, dropping the oldest value in the set. If we detect an anomaly, the sample is discarded (i.e. not included in the set used to compute the baseline for the next sample) and the average is not updated.

The set for the calculation of the running average $\overline{R}$ is therefore detached from the samples coming from the measurements and it will contain only *good* measures.

As soon as an anomaly is detected, the system classifies the anomaly by analyzing which flow traffic features have experienced a change in their distribution compared to normal behavior, either as an increase or a drop in RU values. In order to classify anomalies that show similar communication patterns correctly, such as worm, network scan and spam, a combination of temporal trends in the relative uncertainties, pre-determined ranges of flow size (number of bytes per flow or number of packets per flow), packet size (number of bytes per packet) and destination ports being used are simultaneously considered. A high-level description of such a heuristic is shown in Table 15.1.

Table 15.1. Simple heuristic for attack classification: characteristic temporal trends of relative uncertainty (RU), range of packet sizes, flow sizes and destination ports. The symbol – is used with the meaning of "any."

Attack type	Worm 1	Worm 2	Network scan	Port scan	DDoS	SPAM
RU(srcIP)	↓	↓	↓	↓	↑	↓
RU(dstIP)	↑	↑	↑	↓	↓	↑
RU(srcPort)	↑	↓	↑	↑	–	–
RU(dstPort)	↓	↑	↓	↓	↓	↓
RU(Fsize)	↓	↓	↓	↓	↓	↓
Bytes/Pkt	[300–600]	[300–600]	[40–60]	[40–60]	–	–
Pkts/Flow	–	–	{1, 2}	{1, 2}	–	–
dstPorts	$\neq$ {25, 5060, 5061}	$\neq$ {25, 5060, 5061}	–	–	–	{25, 5060, 5061}

15.5.4   Selecting the flows that matter: Flow-Filter Mask Generation Module

As soon as an alert is raised, DoWitcher analyzes the cause of the deviation and identifies which features were involved in the anomaly, i.e. sudden change in their marginal RUs. At this point, if the end user wants to know specific details of the anomaly by requesting to process packet payload information, then the algorithm focuses on the features for which it notices a decrease in their marginal RU values between the previous and the current time-window, i.e. features whose histograms suddenly become concentrated around specific elements. The elements contributing the most to the decrease of the RU values are identified using the concept of *relative entropy*, defined in the following. Assume that, at a specific point in time i, we detect an anomaly from one of the monitored clusters, according to the forecasting approach. We remind the reader that this condition is an alert if, and only if, two of the five marginal RUs deviate in the opposite direction. Assume X is one of the key features that exhibits a decrease in its RU value from time-windows $i - 1$ to i, i.e. $H_i^X < H_{i-1}^X$. At this point we need to identify the set $L_{X_i} = \{x_i \in M_i^X(x)\}$ that is contributing the most to the decrease. The cardinality of the set L_{X_i} is defined as an input of the algorithm, and it is represented by $|L_{X_i}|$. Given the empirical probability distributions $P_i^X(x)$ and $P_{i-1}^X(x)$, we compute the relative entropy RE_i^x for each of the elements $\{x_i \in M_i^X(x)\}$ as follows:

$$RE_i^x = \left(p_i^X / Support(N_i^X) \right) \log_2 \frac{p_i^X / Support(N_i^X)}{p_{i-1}^X / Support(N_{i-1}^X)}.$$

Then we sort the $\{x_i\}$ according to their relative entropy value RE_i^x and we store in the set L_{X_i} the elements contributing the most to RE_i^x. The cardinality of L_{X_i} depends on the amount of information required for further analysis.

By applying the above procedure to all key features experiencing a drop in their entropies from $i - 1$ to i, we generate a flow-filter mask that will be used in the next time-window to collect per-host flow information.

15.5.5 Example of Packet Payload Module: worm fingerprinting

In this section, we introduce a new module to be used in the context of worm attacks with the final goal of extracting the worm malicious code, i.e. worm fingerprinting. We note that other modules can be developed within the framework presented in this chapter to serve other purposes.

After extracting the payloads of the flows that match the flow-filter mask, the payloads are compared to obtain the worm fingerprint. DoWitcher provides two different algorithms for fingerprint extraction: (i) Rabin Fingerprints (*RF*) and (ii) Longest Common Subsequence (LCS) Windowed Fingerprints (referred to as LCS^W). The two algorithms differ in terms of computational complexity and their ability to extract complex worm signatures.

When speed matters the most, DoWitcher uses the *RF algorithm*. As shown in refs. [112] and [180], high-speed calculation of fingerprints (or checksums) for the packet payload is a requirement when it comes to identifying repeating content at line speed. While any hashing algorithm could be used to generate the fingerprint, Rabin fingerprints [165] are among the most used hash functions due to their good hashing properties. DoWitcher currently considers a sliding window of 150 bytes over the packet content and stores the number of times each fingerprint is seen in all traffic under investigation. When a fingerprint counter reaches a chosen threshold T_F, DoWitcher assumes that the analyzed flow is a worm due to its high presence in the suspect traffic. We point out that the size of the window used by DoWitcher can be changed to achieve the best tradeoff between the algorithm's efficiency (by using large windows) and its effectiveness in identifying worms containing small variations in the payload content (by using smaller windows).

DoWitcher also provides the ability to detect and extract complex signatures associated with the obfuscating strategies employed by *polymorphic worms*. These worms have the ability to change their signatures during their propagation into the network by introducing wild-cards or completely random characters, or by encrypting the payload while hiding the decryption key somewhere around the payload. However, even polymorphic worms have an invariant across different flows. We pose this problem as determining the longest common subsequence between two strings, which has been extensively studied in the past. A string s is said to be a subsequence of string S if s can be obtained by deleting zero or more characters from string S. Thus, a string s is the longest common subsequence of strings S and T if s is a common subsequence of S and T and there is no other common subsequence of S and T of greater length. For example, if two packet payloads contain the strings houseboat and computer, the longest common subsequence that the LCS reports is out. In contrast, since the *RF* algorithm looks for the common substring, it will not report any commonality between the two packet payloads.

We first discuss a simple dynamic programming approach toward solving the LCS problem [8], which has polynomial complexity $O(nm)$, where n and m represent the length of the two strings A and B in tokens. A token is a contiguous set of characters, i.e. a substring that is, by default, set to one character. As shown in Box 15.1, for each of the m starting points of A, the algorithm checks for the LCS of tokens starting at each of the n starting points of B. Thus, the total number of checks is $O(mn)$. At the end of the algorithm, the LCS is obtained as $s(m, n)$ with length $L(m, n)$.

Box 15.1. Longest Common Subsequence Algorithm

(1) Parameters: Input1 A, Input2 B;
(2) Parameters: Length of input1 m tokens, Length of input2 n tokens;
(3) LCS(A,B)
(4) Allocate memory L[m][n]
(5) **for** $i := 0$ to m **do**
(6)   L($i, 0$) := 0;
(7) **for** $j := 0$ to n **do**
(8)   L($0, j$) := 0;
(9) **for** $i := 1$ to m **do**
(10)   **for** $j := 1$ to n **do**
(11)   **if** $A[i] == B[j]$ **then**
(12)   L(i,j) := 1 + L(i-1,j-1);
(13)   **else**
(14)   L(i,j) := max(L(i-1,j),L(i,j-1))
(15) return(L(m,n));

The space complexity of the dynamic algorithm, $O(mn)$, could quickly become prohibitive when the size of the two payloads being compared increases. For instance, if the two flows being compared have payloads of 10 Kbytes each, (the CodeRed worm had a payload of 4156 bytes), then the required memory is quite high at 100 Mbytes. Hence, in this chapter, we also introduce a windowed version of the LCS algorithm namely, LCS^W, which trades off memory utilization for a small increase in the CPU utilization and a small reduction in accuracy of the signature extraction (see Algorithm in Box 15.2).

The motivation for the LCS^W algorithm is to compare smaller chunks across flows, thereby requiring smaller tables in memory and still be able to construct the complete signature without a significant loss in accuracy. We motivate this via a naive windowing algorithm. Suppose there are two flow payloads containing the CodeRed signature, each split across two windows, GET./default.ida?|..........| and GET./def........|..ault.ida?|, where the "|" is the window delimiter and "." denotes non-matching characters. Say the naive algorithm compares the first window of the two flows and obtains the signature as GET/def, and, on comparing the first window of the first flow with the second window of the second flow, it obtains ault.ida?. Further, say this algorithm concatenates the outputs from the individual window-by-window comparisons to obtain the seemingly correct signature: GETdefaultida?. Now suppose the second flow was instead GET./def........|..GET./def.|, then this naive algorithm would yield the wrong signature. In summary, the windowing algorithm must ensure that the signatures obtained are correct, once per-window signatures are combined.

Thus, our windowed version of LCS, LCS^W, employs a divide-and-conquer approach and begins by first dividing each input string into a number of smaller windows, depending on the initial window size. At each iteration, a window of the first string is then compared with all the windows of the other string; if the number of matching tokens is less than a minimum value, the two windows are dropped. Otherwise, the two windows are added to their respective flow buffers for further analysis, while the window size is cut in half. Thus, beginning the content analysis over large windows allows us to remove the non-matching portions quickly, while progressively reducing the window size over every iteration still allows us to zoom into the actual matching content.

Box 15.2. Longest Common Subsequence Window (LCS^W) Algorithm

(1) Parameters: Initial window size i, Minimum window size min;
(2) Parameters: Token size b bytes, Minimum matched tokens per window t;
(3) Inputs: Flow1 f_1 , Flow2 f_2;
(4) Initialization: Window size $w := i$, $pos_1 := 0$, $pos_2 := 0$;
(5) **while** $w > min$ **do**
(6)   **while** $pos_1 < len(f_1)$ **do**
(7)   **while** $pos_2 < len(f_2)$ **do**
(8)   Number of tokens matched k:=LCS(A,B);
(9)   where, A $:= f_1[pos_1\text{-}(pos_1 + w)]$ and
(10)   where, B $:= f_2[pos_2\text{-}(pos_2 + w)]$;
(11)   **if** $k > t$ **then**
(12) $newf_1 := newf_1 + A$;
(13) $newf_2 := newf_2 + B$;
(14) $pos_2 := pos_2 + w$;
(15) $pos_1 := pos_1 + w$;
(16) $w := w/2$;
(17) $f_1 := newf_1$;
(18) $f_2 := newf_2$;
(19)   Reset $newf_1$ and $newf_2$;
(20)   return $(k)_i$

15.6   Case study: worm attacks

In this section, we present details of a testbed implementation and results that justify the efficiency and efficacy of DoWitcher in detecting low-rate anomalies. Although the algorithm is a general framework for network anomaly detection, in this section we focus our attention on applying DoWitcher in the context of low-rate worms as well as polymorphic worms.

In our experiments, we use real-world traces obtained from a large wireless ISP and a large South American carrier network.

15.6.1   Testbed

Recall from Section 15.4 that we introduced the DLA and DGA modules as comprising the DoWitcher system. These modules are co-located on the same physical machine, which we henceforth call the packet analyzer. To validate the efficacy of DoWitcher, we use a testbed consisting of two machines, connected via an OC-48 link (2.4 Gbps). Each machine has two Intel$^{(R)}$ Xeon$^{(TM)}$ CPU 3.40 GHz processors and 4 Gbyte memory available to a process running on the Linux kernel 2.4.21. We replay traces by using *tcpreplay* from one machine, while the other one is configured to sniff packets off-the-wire and runs our packet analyzer. The packet analyzer passively reads packets off-the-wire and in its layer-4 mode is configured to parse packets, reconstruct flows and provide annotated vectors that describe the flow's key traffic features. In its layer-7 mode, the packet analyzer extracts payloads of packets that match a flow-filter mask. The output of traffic features and reconstructed flow payloads from the packet analyzer are then processed by an anomaly detection daemon and a fingerprint extraction daemon, respectively, each of which are running on a different Intel(R) Xeon(TM) CPU 3.40 GHz machine. The capability of our DoWitcher system while parsing the layer-4 information is pegged at a line-rate of OC-192, while processing up to 3.7 million packets per second, where the average packet size is 376 bytes.

15.6.2   PMER versus PJER metric

This experiment refers to a 320 hour packet trace captured from a large wireless ISP in which two outbreaks of the Sasser worm are present. The Sasser worm was able to propagate by leveraging a flaw in Microsoft Windows LSA (Local Security Authority) Service and was able to execute without requiring any action on the part of the user. Data were collected and aggregated every hour, so the time-window duration for the algorithm is set to one hour. Due to the low number of hosts in the single subnet, data are analyzed only at the network level, i.e. considering only traffic being generated inside the network and matching the network 16-bit mask. No payload was available for the flows, so no fingerprinting was possible for this data set. The baselining period lasts for 180 hours and then the algorithm enters the detection mode.

In Figure 15.4(a), we show the evolution over time of the RUs of the four marginal distributions, e.g. srcIP, dstIP, srcPort, dstPort, and in Figure 15.4(c) we show the alerts generated by DoWitcher when using the PMER with forecasting approach. A clear daily traffic pattern can be identified, a pattern that is broken when some hosts in the network became infected by the Sasser worm (TCP connection to port 5554). Two outbreak periods are clearly visible in these plots. The first sign of activity happens between hour 270 and 275. At this time, only one infected host was present in the network, but it exhibited very strong activity contacting a large number of distinct destination IP addresses (note the decrease in the marginal entropy of source IP distribution and

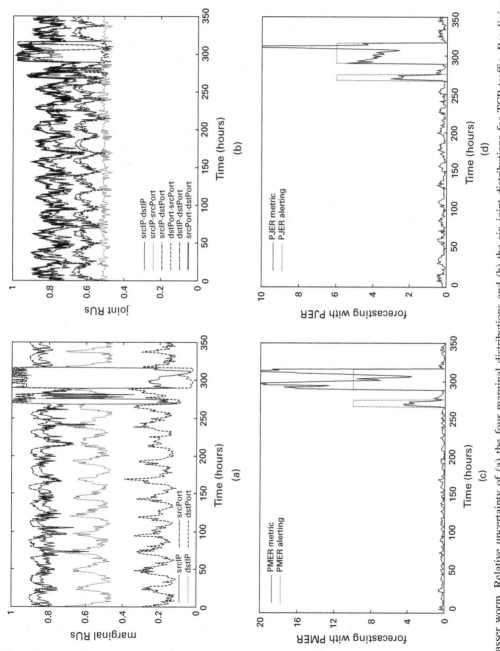

Fig. 15.4. Sasser worm. Relative uncertainty of (a) the four marginal distributions and (b) the six joint distributions for TCP traffic. Baselining using the forecasting method and alerting using (c) PMER and (d) PJER metrics.

the increase in the marginal entropy value of the destination IP distribution) by using random source ports and a specific destination port (note the decrease in the marginal entropy of destination port distribution and the increase in the marginal entropy value of the source port distribution). A general comment that can be made is that, due to the fact that most of the legitimate traffic of the network was already destined to TCP port 5554, the destination port distribution was not altered in a significant way by the worm, with only a limited decrease. At this point, no bandwidth consumption was observed by the service provider and any technique based on volume, e.g. volume of traffic generated by the infected host, would have missed this preliminary worm activity. A second sign of activity was detected 16 hours later (hour 291). At this time, five new hosts were infected and the worm started to have a considerable impact on the network bandwidth by increasing the bandwidth consumption by around three times more than during non-worm periods. From this time on, the population of the infected hosts grows, with an average infection rate of 20 new hosts every hour, leading to a bandwidth consumption of 30 times that of normal behavior 30 hours after the first sign of activity. The worm was then successfully mitigated, around hour 320, and the network returned to normality. In Figure 15.4(c) we show DoWitcher in action using the PMER metric and a forecasting method for baselining and alerting. Note how efficiently (no false positive was generated by DoWitcher) and quickly (in one time-window) DoWitcher is able to detect the worm since its first outbreak, when no bandwidth impact was visible to the operator.

In Figure 15.4(b), we show the RUs for the six pairwise joint distributions and in Figure 15.4(d) we show the forecasting method. As you can see, DoWitcher was again able to detect the two worm outbreaks with no false positives.

One important conclusion that we draw from this experiment is that the PMER metric is as effective as the PJER metric, in that the alerts via PMER are detected exactly when PJER metric alerts would be detected. This observation affirms our hypothesis that worm traffic alters the ratios of marginal entropies such that a worm can be detected by just profiling their divergence without the need for constructing joint distributions, as achieved in the PJER metric.

15.6.3   Low-rate worm

Next, we present results which highlight the efficacy of DoWitcher in detecting even the low-propagation-rate worms, which are undetectable via volume metrics of total bandwidth, total packets or number of new flows.

In this experiment, we use a 20 minute trace from a large South American carrier network and use a time slot of 30 s for calculation of the histograms and the PMER metric. The baselining period T_w of the PMER metric lasts until slot number 24, while the PMER is calculated by using the following values for the sliding windows: $W = 5$ and $N_S = 5$. Observe from Figure 15.5(a) that the traffic characteristics in the trace are stable over time, i.e. the histogram for the source network addresses obtained over two consecutive time slots (at generic times t and $t + 1$) does not vary much. Figures 15.5(b) and (c) show the histograms for destination port and flow size.

We inject a varying proportion of synthetic CodeRed worm traffic into the trace around time slot 35 according to the following worm propagation model. One IP

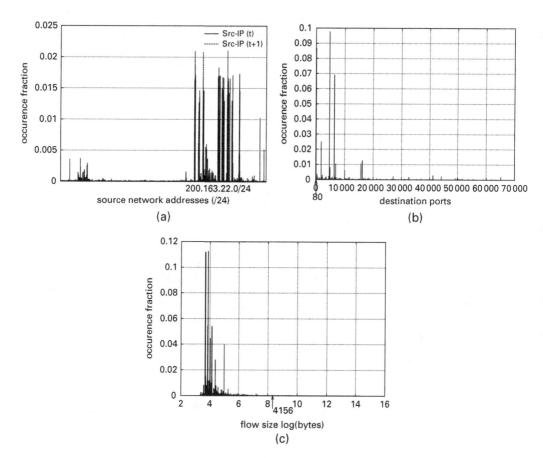

Fig. 15.5. Histograms of features in the carrier trace over a time slot of 30 s. (a) Source network address (/24); (b) destination port; (c) flow size.

address, belonging to one of the most widely used subnets, is assumed to be infected. This infected machine then generates x worm flows (TCP) per slot destined to port 80. The infected host performs a network scan and initiates one TCP flow per destination IP address, where the destination IP addresses are chosen as follows: 50% of the addresses contacted did not occur in the last time-window and the rest were seen in the last window. The source port is chosen randomly and each worm flow consists of a total of 15 packets and a total flow size of 4156 bytes, including the TCP handshake and teardown packets. In our experiments, we increase the number of worm flows generated in time slot 35 as $x = [0{-}1000]$ until the worm is detected via our PMER-metric-based alerting. Observe from Figure 15.6 that the RUs for source network address, flow size and destination port decrease on increasing the proportion of worm flows, since the infected host, flow size of 4156 and destination port of 80 begin to dominate the histograms. In contrast, the RUs for destination IP address and source port increase with increasing proportion of worm flows, given that the worm introduces greater randomness in these histograms.

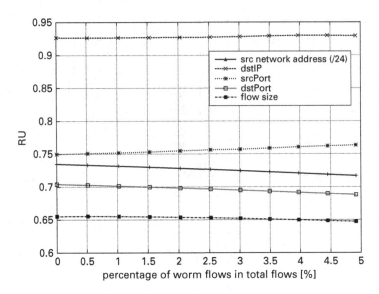

Fig. 15.6.   Changes in RU on introduction of worm flows. At time slot $t = 35$.

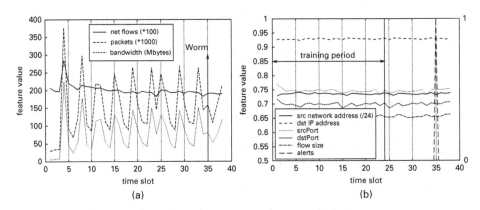

Fig. 15.7.   (a) Volume behavior metrics on a low-rate worm which comprises 1.0% flows, 2.0% packets and 1.1% bytes of normal traffic. (b) RUs of the five traffic features monitored. The worm occurs at time slot 35 in the trace and is detected via the PMER metric while indistinguishable via any volume metric.

The worm becomes detectable via the PMER metric when there are 200 worm flows per slot (see Figure 15.7(b)). This is indeed a low-propagation-rate worm, since it comprises 1.0% flows, 2.0% packets and 1.1% bytes of total traffic in that time slot, and is hidden within the clean traffic completely (Figure 15.7(a)).

15.6.4   Fingerprinting

We present some indicative results reflecting the benefits of using LCS^W in contrast with LCS. We run several experiments aimed at understanding the performance achieved by LCS^W in (i) detecting, isolating and correctly reporting the invariant part

Table 15.2. LCS^W memory and CPU requirements as a function of the size of the worm signature

Signature size (Kbytes)	Total flow (%)	Memory (Mbytes)	CPU (%)
1	13.33	5.0	57.2
2	26.66	7.1	58.1
3	40	16.0	57.5
4	53.33	28.1	57.6
5	66.67	33.0	57.3
6	80	35.0	57.1

GET /default.ida?xx..[220*x]x		
GET /default.ida?xx..[220* x]x		

	GET /default.ida?xx..[220 *x]x	
GET /default.ida?xx..[110 *x]x	xx..[110 *x]x	%u9090%u6858

GET /default.ida?xx..[110 * x]x	xx..[50 * x]x	xx..[60 * x]x%u9090
GET /default.ida?xx..[50 *x]x	xx..[70 * x]x	xx..[100 * x]x%u9090

Fig. 15.8. Spread of worm invariant signature across flows in different experiments.

of a worm, i.e. worm fingerprint or signature, when the signature of the worm is fragmented in several chunks and distributed across the flow; and (ii) the memory and CPU utilization required to execute the algorithm.

In our experiments, we start considering two flows of size 7.5 Kbytes characterized by exactly the same invariant part (emulating the invariant part of a worm flow, and called, in this context, "signature"). In each experiment we operate on the second flow by increasing its signature size and spreading it across its payload. The first flow is kept in its original form across all experiments. We increase the signature size at incremental steps of 1 Kbyte (see the first column of Table 15.2), which will cause the size of the flow to increase accordingly (see the second column of Table 15.2). Each new chunk of 1 Kbyte being added to the flow is placed after the previous chunk, but not immediately next to it, i.e. the two chunks will be interleaved with bytes associated with the original content of the flow (see Figure 15.8).

Interestingly, we have found that the LCS algorithm is completely insensitive to the size of the worm signature, showing a constant utilization both in terms of memory (56.25 Mbytes) and CPU (55%). In contrast, LCS^W, which is configured with the following initial parameters: (i) initial window size $w = 4000$ bytes, (ii) minimum window size $min = 200$ bytes and (iii) token size $b = 4$ bytes, utilizes proportionately less memory depending on the signature length and its spread across the flow payload (see Table 15.2), at the cost of only an approximately 2% increase in CPU utilization. Moreover, LCS^W always returns the right signature length. This justifies the performance tradeoff provided by LCS^W over a regular LCS algorithm.

15.7   Lessons learned

In this chapter, we presented DoWitcher, a novel system for network anomaly detection at the carrier network link speeds (OC-48, OC-192 and upwards). DoWitcher provides the capability of detecting zero-day network threats by analyzing the key features of traffic flows, classifying the network anomaly according to specific pattern dynamics and finally isolating the suspect malicious flows via the generation of a flow filter-mask. DoWitcher further comprises specialized modules for gathering and processing packet payload information related to malicious flows. We presented a specific module for the extraction of the worm content signature via a windowed LCS (LCS^W) algorithm applied over the flow payload content of isolated flows. Thus, in contrast to previous approaches, DoWitcher derives its scalability by this two-tiered approach, where only the layer-4 traffic features are used for the detection and classification of network anomalies, while anomaly details are deeply analyzed only on a few suspect flows. Through testbed experiments, we established the efficacy and efficiency of DoWitcher in detecting the more sophisticated stealth worms: low-propagation-rate worms, which utilize as little as 4.4% of the total network bandwidth, and polymorphic worms, which spread their signature across multiple packets. Similar results were obtained for other anomalies such as DoS/DDoS, port scans, network scans and spam.

16 Detection of control-plane anomalies: beyond prefix hijacking

As already presented, the Internet routing system is partitioned into tens of thousands of independently administered Autonomous Systems (ASs). The Border Gateway Protocol (BGP) [170] is the de facto inter-domain routing protocol that maintains and exchanges routing information between ASs. However, the BGP was designed based on the implicit trust between all participants and does not employ any measure to authenticate the routes injected into or propagated through the system. Therefore, virtually any AS can announce any route into the routing system, and sometimes the *bogus routes* can trigger large-scale anomalies in the Internet. A canonical example occurred on April 25, 1997, when a misconfigured router maintained by a small service provider (AS7007) in Virginia, USA, injected incorrect routing information into the global Internet and claimed to have optimal connectivity to all Internet destinations. As a result, most Internet traffic was routed to this small ISP. The traffic overwhelmed the misconfigured and intermediate routers, and effectively crippled the Internet for almost two hours [16]. Since then, many such events have been reported, some of them due to human mistakes, others due to malicious activities that exploited vulnerabilities in the BGP in order to cause large-scale damage. For example, it is common for spammers to announce an arbitrary prefix and then use that prefix to send spam from the hijacked address space, making the trace back and the spammer identity discovery much more difficult [167]. Similar behavior was noticed for other types of attacks, such as application-layer DDoS attacks [168].

There are a myriad of different ways that attacks can be used to break the stability of the routing infrastructure. Examples of such are attacks aimed at (i) directly compromising the routers, (ii) disrupting the peering relationships or (iii) compromising the routing topology and reachability information by falsely injecting information. The latter two families of attacks are known as "bogus route creation."

The first family of attacks focuses on gaining unauthorized access to the router element and full control over its functions. At this point, the attacker has full capability to modify the router configuration and thus manipulate the routing information. Thus, an operator should deny external access to routers at every opportunity with security-configured passwords, privilege levels, logging-in and protocols (i.e. usage of SSH

Portions reprinted, with permission, from Qiu, J., Gao, L., Ranjan, S. and Nucci, A. (2007). "Detecting bogus route information: going beyond prefix hijacking", *IEEE SecureComm, 3rd International Conference on Security and Privacy in Communication Networks*, Nice, France September 2007.

rather than Telnet). The second family of attacks is focused on de-stabilizing the peering relationships via DoS/DDoS attacks or more subtle attacks using malformed packets exploiting state machine vulnerabilities. In this case, the attacker has to find the address of a router in a critical location (i.e. the entry point into the network) and then attack the router using normal DoS/DDoS methods or sending TCP reset packets to cause the sessions with its peers to fail. The third family of attacks is focused on misdirecting traffic into a black-hole or along a monitored path, or on claiming false ownership of a block of prefixes, so that the attacker can gain access to the information in the data stream or use the block of prefixes for malicious purposes. The attacker might even force a routing loop in the system, causing wide-scale network outages. These types of attacks require the attacker to compromise/access a router attached to the Internet and cause it to inject false routing information. This requires manipulation of the configuration of the router, possibly gaining access to a console, or through SNMP, etc. At this point, the attacker is said officially to participate in the routing system. In this chapter we focus on the latter two families of routing attacks and present the most common techniques and operational countermeasures; we leave the first family to be handled by operators through correct security practices.

There are several ways that can be used by an attacker to fabricate "bogus routes" to gain control of an arbitrary address space. In this chapter we describe the two most common approaches, known as *prefix hijacking* and *path spoofing*. Prefix hijacking is one kind of bogus-routing announcement in which an attacker AS, also referred as a *rogue AS*, announces itself as the origin of a prefix which it does not own. Path spoofing is an even stealthier approach used to inject bogus-routing information into the routing infrastructure. Under normal circumstances, a route must be propagated along the valid AS-path, i.e. a chain of legitimate ASs, in which each pair of consecutive ASs must have real peering sessions established. In the case of path spoofing, the malicious AS spoofs the valid AS-path and alters its legitimate nature by inserting fake AS(s) and/or deleting some of the legitimate ASs.

Despite the criticality of the problem, to date no solution has been widely accepted and globally deployed in the Internet. From a commercial stand point, most security vendors have designed their products completely ignoring the *control-plane* (routing data) aspect, focusing only on the Internet *data plane* (packet headers and payloads). Internet security systems typically operate on the simple principle of detecting malicious data-plane behavior and erecting data-plane filters. On the other hand, the research community has been remarkably active over the last few years to try to fix, or at least patch, the control-plane problem. Many approaches have been proposed to address several issues related to the integrity, authentication, confidentiality, authorization and validation of BGP data (see S-BGP [109] and soBGP [205]), but their comprehensive deployment is still unforeseeable [51]. Thus, ISPs are left to face any security incidents on their own, largely relying on ad hoc filters to remove bogus routes from their routing tables. As more bogus routes emerge, it is imperative to provide ISPs with a practical and reliable system to identify the bogus-routing information and thereby detect malicious activities associated with them. Due to the importance of the problem, and the lack of a definitive answer, we reserved this last chapter to cover this thematic.

We present a real-time bogus BGP route detection system to protect ISP networks against prefix-hijacking and path-spoofing episodes. The system extracts and learns the route properties from the historical BGP routing data and then uses the knowledge to detect bogus routes. Although BGP routes are very dynamic (i.e. the routes change quite frequently), two basic metrics that constitute the BGP routes are relatively persistent over time: (i) binding of the *prefixes and their origin ASs* and (ii) the *neighboring relationship between ASs* (i.e. the peering status between ASs). In the remainder of this chapter, we refer to the above components by the general term *routing objects* and use them in our methodology. As the system continues to track the history of the objects over a long enough period of time, it is able to capture the majority of the legitimate relationships that have shown stable histories. Accordingly, the never-seen or short-lived objects are deemed illegitimate and the routes composed of them are considered bogus.

The objects obtained in such a simplistic learning process, however, are neither clean nor complete. We further enhance our methodology by providing (i) a simple method for the creation of a robust and clean history by identifying and filtering objects likely to be associated to bogus routes (i.e. used during the learning phase) and (ii) several heuristics to enhance the performance of the bogus-route detection that rely upon the characteristics of attacker behavior and common network operational practices. We then explore the correlation among the routes triggered by identical events to calibrate the detection results. These efforts yield low false-positive rates and high detection rates. Through a series of experiments, we prove the efficacy of our system in bounding the false-positive rate to 0.2%, which translates to around 20 alarms daily. By evaluating our system performance against several documented incidents, we also show that the system can achieve false-negative rates as low as 0%, with false-positive rates no greater than 0.02%.

The rest of the chapter is organized as follows. Section 16.1 briefly describes the fundamental aspects of the BGP routing protocol that are relevant to this chapter. More details on BGP are provided in Chapters 2 and 10. Section 16.2 presents the common practices being adopted by ISPs to deal with this problem and highlights their major associated limitations. After reviewing the related work in Section 16.3, in Section 16.4 we introduce a novel system architecture to detect bogus routes. Section 16.5 introduces the concept of routing information objects and the basic detection algorithms. We explore various heuristics to improve the system performance in Section 16.6. The experimental evaluation of the proposed solution is described in Section 16.7, and Section 16.8 concludes the chapter.

16.1   Brief introduction to the Border Gateway Protocol

The Internet is a network of networks that share information via routers. A group of routers under the same administrative control is considered to be an AS. As described in Chapter 2, there are three types of ASs: stub, multi-homed and transit. Stub ASs are communication end points, with connections to the rest of the Internet only through a single upstream provider. Multi-homed ASs are similar to stub ASs, but possess multiple upstream providers. Transit ASs have connections to multiple ASs and allow traffic to flow through to other ASs, even if the traffic does not originate or terminate within them. While the communication within an AS is accomplished using an Interior Gateway Protocol (IGP) such as OSPF and IS–IS, the communication between ASs is accomplished via an Exterior Gateway Protocol (EGP). The de-facto standard EGP in use on the Internet is the Border Gateway Protocol (BGP). A router running the BGP protocol is known as a *BGP speaker*; BGP speakers communicate using TCP and become *peers* or *neighbors*. Each pair of BGP neighbors maintains a *session*, over which information is communicated. BGP peers within the same AS (internal peers) communicate via an internal BGP (iBGP). External BGP (eBGP) is used between speakers in different ASs (external peers). Each AS originates one or more address *prefixes*. A prefix is a representation for a block of IP addresses. Prefixes are expressed as "prefix/# most significant bits." For example, the prefix 192.68.0.0/16 has 16 significant bits and thus represents all of the IP addresses between 192.68.0.0 and 192.68.255.255 inclusive. Each AS establishes a path for the prefixes advertised by BGP. To simplify things, the paths are vectors of ASs that any packet must traverse to reach the IP address. The last AS in the path is the *origin* of that address and its parent prefix. These vectors are stored in a routing table and shared with neighbors via BGP.

BGP peers constantly exchange *Network Layer Reachability Information* (NLRI), e.g. known paths and prefixes, via UPDATE messages. Each peer updates its routing tables based on its neighbors' NLRI, and forwards that information to its other neighbors. This flooding process ensures that all ASs are informed of the reachability of all prefixes. As long as the session is active, peers use UPDATE messages to inform each other of routing table changes, which include the addition of new routes and withdrawals of old ones.

An AS may, and often should, receive multiple paths to a single prefix. BGP uses a complex algorithm to select which of these paths to use to forward and advertise to its neighbors. Policy communicated in UPDATE messages, as well as local configuration, may influence this process, as described in Chapter 10. However, in the absence of mitigating policy, and subject to several other factors, BGP will select the shortest path (as measured in AS hops).

Policies configured in a BGP router allow it to filter the routes received from each of its peers (import policy), filter the routes advertised to its peers (export policy), select the routes based on desired criteria and forward traffic based on those routes. BGP routers can be configured with route preferences, selective destination reporting (i.e. report a destination to some neighbors and not others) and rules concerning path editing. Setting path preferences usually involves path editing, such as adding AS numbers to a path to discourage its use (a technique known as AS-path prepending). These aspects of the protocol enable BGP to adhere to desired policies.

Although BGP has had success as a policy-based inter-domain routing protocol, there are a number of issues that suggest that the Internet may have evolved beyond BGP's current incarnation. In Section 16.2, we discuss a few of the known security issues that, if exploited by attackers, can be used to cause global network damage to the entire Internet infrastructure and all entities attached to it.

16.2 Vulnerabilities, consequences and common practices

The fundamental weakness of BGP that makes it vulnerable to attacks is its distributed nature, being run by hundreds of thousands of routers. Each AS is indirectly connected to every other AS in the Internet and thus adversaries can affect routers and networks far removed from their peers by exploiting this scale and inter-connectedness in a variety of different forms. As mentioned above, two of the most dangerous attacks that can be launched against the routing infrastructure at large scale are known as prefix hijacking and path spoofing. In what follows we describe these two types of attacks in more detail.

A prefix-hikacking attack occurs when an AS advertises incorrect information through BGP UPDATE messages to routers in neighboring ASs. A malicious AS can advertise a prefix originated from another AS and claim that it is the originator. Neighboring ASs receiving this announcement will believe that the malicious AS is the prefix owner and route packets to it. The real originator of the AS will not receive the traffic that is supposed to be bound for it. If the malicious AS chooses to drop all the packets destined to the hijacked addresses, the effect is called *black-holing*. This attack makes the hijacked addresses unavailable. Note that the outage outwardly looks like any other kind of outage, and is often difficult to diagnose. If the malicious AS chooses to forge all addresses in a block, using hosts and devices within its control, the effect might be much more severe. Unless properly authenticated using some other security service, one can impersonate all of the services and resources of the hijacked address space. The malicious AS can then analyze the traffic it receives, possibly retrieving virulent sensitive information such as passwords or confidential information.

Prefix hijacking can be further classified into the following four subclasses: (i) *duplicate-prefix hijacking*, in which other ASs own the exact same prefix as the hijacked prefix; (ii) *sub-prefix hijacking*, in which the hijacked prefix is a subnet of the legitimate prefixes of other ASs; (iii) *super-prefix hijacking*, in which the hijacked prefix is the supernet of the legitimate prefixes of some ASs; and (iv) *independent-prefix hijacking*, in which there is no overlap between the hijacked prefix and the legitimate prefixes of any AS. If the origin AS is legitimate to originate this prefix, but the AS-path is not valid, namely it contains non-existent AS-links, the route is deemed as path-spoofing.

Path spoofing is another method used by a malicious AS to spread misinformation by tampering the path attributes of an UPDATE message. As previously mentioned, BGP is a path vector protocol; reaching a final destination implies the transmission of packets across a series of ASs, as they appear in the BGP path string. An AS can modify the path it receives from other ASs by inserting or deleting ASs from the path vector, or by changing the order of the ASs, in order to create routing delays or to allow the malicious AS to alter network traffic patterns. By altering attributes in an UPDATE message, such as the multi-exit discriminator (used to suggest a preferred route into an AS to an external AS) or the community attribute (used to group routes with common routing policies), traffic engineering and routing policy can be severely undermined. Path spoofing can be further classified into the following two subclasses: (i) *spoofing with forged AS-link* and (ii) *spoofing with route redistribution*.

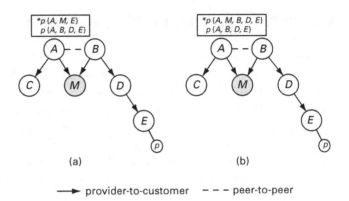

Fig. 16.1.   Examples of path spoofing. (a) Spoofing with forged AS link *ME*: attacker as *M* fakes the AS path (A, B, D, E) to prefix p by announcing the existence of direct link *ME* to E (A, M, E) (b) Spoofing with route redistribution: attacker as *M* places himself between AS *A* and AS *B*. Traffic to prefix p is routed from AS A through AS *M* to AS B (A, M, B, D, E).

In Figure 16.1 we provide two examples of the two path-spoofing attacks. In Figure 16.1(a) we show an attacker AS *M* spoofing the path by bluntly faking a nonexistent path, e.g. *ME* to *E* (called *fake-link path spoofing*). Alternatively, *M* can deliberately redistribute routes between its providers or peers, e.g. from its provider *B* to another provider *A*, as shown in Figure 16.1(b). This way the malicious AS *M* appears as a transit AS. Such routes are illegitimate since they break the commercial agreements that define the route import/export policies between *M* and the relevant neighbors and enable *M* to access traffic that it is not supposed to (called *route-distribution path spoofing*).

The consequences of these attacks are as diverse as their approaches. BGP sessions can be prematurely severed, networks and ASs can be made unreachable, the address space can become highly fragmented, and other undesirable outcomes can result from such an attack. Attacks can be used in concert to amplify their effect or to enable further malicious activity. Examples are disclosure of confidential information, deceptive or incorrect information introduced into the network through message modification, the disruption of network activity through denial of service (black-hole a route, for example, causes denial of service for that prefix), and the usurping of router services and functions. Consider the ramifications of a dysfunctional routing system under attack. An individual router is subject to being overloaded with information, knocked off-line or taken over by an attacker. An AS can have its traffic black-holed or otherwise misrouted, and packets to or from it can be grossly delayed or dropped altogether. Malfunctioning ASs harm their peers by forcing them to recalculate routes and alter their routing tables. These events can disrupt international backbone networks and have the potential to bring a large part of the Internet to a standstill. As a consequence, the threat against BGP is a matter of grave concern to anybody reliant on the Internet.

Due to the lack of a viable, globally accepted and deployed commercial solution,[1] ISPs are implementing local countermeasures. Examples are based on (i) *generalized*

---

[1] S-BGP [109] and soBGP [205] are prohibitively expensive to be deployed in commercial networks and thus remain, as of today, interesting architectures under consideration for standardization by the Internet Task Force (IETF).

time-to-live (TTL) security mechanisms and (ii) *defensive routing policies.* The first mechanism represents a simple way for protecting peers from remote attacks. This approach builds on the premise that, in the vast majority of BGP peering sessions, the two peers are adjacent to each other. The TTL attribute in an IP packet is set to a value that is decremented at every hop. For example, if a packet traverses four hops from source to destination, the TTL decreases by four. Routers using this method set the TTL of an IP packet to its maximum value of 255. When a BGP peer receives a packet, it checks the TTL and, if this value is lower than 254 (decremented by one), the packet is flagged or discarded outright. This prevents remote attacks which come from more than one hop away, as those packets will have TTLs lower than the threshold value of 254. The second mechanism is used to filter bad and potentially malicious announcements, and to manipulate potentially dangerous attributes of received routes. BGP speakers commonly filter prefixes that are "Documented Special Use Address" (DSUA) prefixes (e.g. loop-back addresses) and bogus (advertisements of address blocks and AS numbers with no matching allocation data). Several organizations maintain the list of bogus prefixes (such as CIDR), and ISPs use these lists to filter "suspicious" announcements, like the ones associated with conflicting prefixes, or those announcements containing private ASs, or those coming from unexpected downstream ASs. A policy of careful ingress and egress filtering aids in maintaining security for both the local AS and its neighbors, and is widely held to be the most widely deployed and effective BGP security measure.

Unfortunately, TTL mechanisms or smart filtering are not a replacement for a strong security architecture. For example, the TTL approach protects peers communicating via single-hop BGP sessions, except in those cases where the directly connected peer has been compromised. While the method is less effective for multi-hop BGP sessions, it still closes the window on several forms of attacks (i.e. it lessens vulnerability to attacks launched by attackers sitting between the BGP speakers). Similarly, the filtering rules are fundamentally limited by the heuristics they reflect, and can only remove announcements which are overtly bad.

This leads us to investigate new techniques that can assist ISPs in their daily network operation. Before introducing our new architecture to detect bogus route information, we present the results of the most active research being carried out on this topic.

16.3   Related work

The Pretty Good BGP (pgBGP) [108] and the Prefix Hijacking Alerting System (PHAS) [116] are recent solutions in the prevention of BGP prefix hijacking based on historical BGP routing data. Both systems keep track of the origin ASs of every prefix over time and identify the suspicious routes whose prefixes are originated from unknown ASs. However, they can detect duplicate- and sub-prefix hijacking routes only. Meanwhile, there is increasing evidence pointing to the use of super-prefix hijacking to send email spam [167,175]. A savvy attacker may be expected to employ not only prefix hijacking, but also stealthier path-spoofing attacks. The system presented in this chapter provides comprehensive detection of both prefix hijacking and path spoofing. Krügel *et al.* have proposed to gather route validation information through passive monitoring of BGP routing traffic to identify BGP anomalies [114], in which they exploit AS clustering based on hierarchies, i.e. either core or peripheral, and geographic locations. In a valid path, two neighboring ASs should be in the same geographic cluster and the path should traverse the core at most once. This method is computationally expensive, while the obtained AS topology model is coarse. In contrast, we propose a lightweight, yet sophisticated, directed AS-link topology model to identify path spoofing. Meanwhile, because the information learned from history is inevitably limited, all above work suffers from high false positives. Hu *et al.* [211] proposed to use active probes on the data plane to improve detection accuracy. Instead, we explore heuristics to reduce the false-positive rate based on the control-plane information alone.

16.4   Detection system architecture

In this chapter, we present a real-time bogus BGP route detection system based on information learned from historical BGP routing data. As a path-vector routing protocol, a route in the BGP system mainly consists of a prefix, which represents the destination, and an AS-path, which is a sequence of ASs along which the route is propagated from the origin to the local AS. Although BGP routes change dynamically, they comprise two basic *routing information objects*: the association of prefixes and their origin ASs and the AS-links that represent the neighboring status between ASs. The objects represent the structure of the inter-domain routing system and are found to be relatively stable over time; the reason being that ASs have to undergo lengthy and costly procedures to request address spaces from their providers or Regional Internet Registries (RIRs), coordinate connections and negotiate commercial agreements. Hence, an AS typically maintains its existing prefixes and its connectivity with neighboring ASs as long as possible. Thus, our detection hypothesis is that, since the majority of BGP routes are believed to be legitimate and stable, we can learn the majority of the legitimate routing objects over time based on received routes and then use the learned information to identify the bogus routes.

In Figure 16.2, we show the architecture of the system we propose. By peering with several BGP routers, the system (i) passively receives routing data, (ii) extracts and consequently stores routing information objects from the received routes and (iii) examines whether the routes are bogus or not. Note that, due to the learning-based approach, the system needs an initialization phase to accumulate enough route information, i.e. history, to create its internal knowledge base. After that, every route is examined as soon as it arrives and the results are presented to a network operator in real time.

The detection system can be deployed in two scenarios. First, it can be deployed by a service provider, typically Tier-1, to protect its routing system. Given that routers within an AS usually have similar views, to diversify the received information the detection system can peer with not only the routers within the deployed AS, but also those in neighboring ASs. However, the routes received from the neighboring ASs should be consistent with those from the local AS. For example, routes from customer ASs should

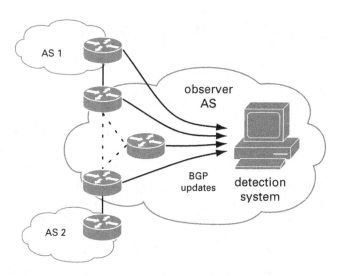

Fig. 16.2.   System architecture.

be customer routes only. Second, the system can be deployed as a bogus route monitoring system for the global Internet, which analyzes the BGP routing data retrieved from several public BGP data repositories, such as ROUTEVIEWS [14] or RIPE RIS [13].

16.5   Basic detection algorithm

In this section, we present a basic detection algorithm based on building a historical database of routing information objects to detect bogus routes.

16.5.1   Collecting routing information objects

A BGP route mainly consists of a prefix p and an AS-path $\{a_k, \ldots, a_0\}$. a_k is the *observer AS*[2] and a_0 is the *origin AS*. The *direction* of an AS-path is defined as from the observer AS to the origin AS. From a received BGP route of prefix p with AS-path $\{a_k, \ldots, a_0\}$, we extract (1) the *prefix-originAS association*, which is the tuple (p, a_0) and (2) the *directed AS-links*, which are *directional* AS-pairs $a_i \rightarrow a_{i-1}, i = k, \ldots, 1$ with the same direction as the AS-path. Note that a_i is said to be the *upstream* of a_{i-1} and a_{i-1} is the *downstream* of a_i. A prefix-originAS association records the binding between a prefix and one of its origin ASs. A directed AS-link indicates that the two ASs are neighbors. More importantly, the direction encodes the import/export routing policies of the two ASs from the viewpoint of the observer AS (i.e. the downstream AS allows routes to be *exported* to the upstream AS, while the upstream AS *imports* the

---

[2] Depending on the type of peering session between the detection system peer and the BGP routers, the observer AS number might not appear in the AS-path. In this case, the detection system should add the number.

routes from the downstream AS). At time t, the extracted prefix-originAS associations and directed AS-links during the observation window with length T, which starts at $t - T$ and ends at t, compose the sets $\mathbb{A}[t - T, t)$ and $\mathbb{L}[t - T, t)$, respectively.

16.5.2   Bogus routes detection algorithm

Given, a route $(p, \{a_k, \ldots, a_0\})$, we use the procedure ISBOGUSROUTE shown in Figure 16.3 to verify its legitimacy. The algorithm first verifies the AS-links sequentially in the direction of the AS-path and then the prefix-originAS association. It stops when it finds the first illegitimate object and returns the illegitimate object. As routes are propagated in the reverse direction of its AS-path, i.e. from the origin AS to the observer AS, an AS can forge anything downstream but nothing upstream. Thus, the detection algorithm qualifies the upstream AS of a suspicious directed AS-link or the origin AS of an illegitimate prefix-originAS association as the potential attacker. Note that this procedure returns the first encountered suspicious object only, even though a route might be manipulated by several attackers and contain multiple suspicious objects.

The detection procedure is based on the *legitimate* routing information objects, which are precisely determined with procedures ISLEGITIMATELINK and ISLEGITIMATE-ASSOCIATION shown in Figure 16.3. Based on the overly simplistic assumption that anything observed in the past is considered valid at present, the routing information objects in $\mathbb{A}[t - T, t)$ and $\mathbb{L}[t - T, t)$ are considered legitimate. Moreover, as an AS can aggregate or de-aggregate its prefixes, the prefix-originAS associations derived from

Algorithm : ISBOGUSROUTE$(p, \{a_k, \ldots, a_0\})$

for $i \leftarrow k$ **to** 1
    if $\neg$ISLEGITIMATELINK$(a_i \rightarrow a_{i-1}, t)$
        return (True, $a_i \rightarrow a_{i-1}$)**if** $\neg$ISLEGITIMATEASSOCIATION$((p, a_0), t)$
    return (True, (p, a_0))**return** (False)

Algorithm : ISLEGITIMATELINK$(a_i \rightarrow a_{i-1}, t)$

if $a_i \rightarrow a_{i-1} \in \mathbb{L}[t - T, t)$
    return (True)
else return (False)

Algorithm : ISLEGITIMATEASSOCIATION$((p, a_0), t)$

if $(p, a_0) \in \mathbb{A}[t - T, t)$
    return (True)
    else if $(P, a_0) \in \mathbb{A}[t - T, t) \wedge p \Subset P$
        return (True)$/ * de - aggregation * /$
    else if $(p_i, a_0) \in \mathbb{A}[t - T, t) \wedge p_i \Subset p \wedge p = \cup_i p_i$
        return (True)$/ * aggregation * /$
    else return (False)

Fig. 16.3.   Pseudo-code of the detection algorithm.

prefix aggregation or de-aggregation can also be considered as legitimate. However, the caveat is that, if a prefix P and its subnet p are assigned to two different ASs, say A and B, then the associations that are derived from the de-aggregation of (P, A) might not be legitimate since the relevant prefixes might be the subnet of p, which is in the address space of B. Thus, we introduce the term immediate subnet. A prefix p is the *immediate subnet* of a prefix P, denoted by $p \Subset P$, if no legitimate prefix-originAS association has prefix that is the subnet of P and the supernet of p.

16.5.3 Classification of bogus routes

Bogus routes identified by the detection algorithm presented in Section 16.5.2 are then further classified according to the type of illegitimate object being returned and according to the following rules.

A route is identified as *path spoofing* if the illegitimate object found by the procedure ISBOGUSROUTE is an AS-link. We further characterize it as either *redistribution* or *fake-link* path spoofing. Since the direction of a directed AS-link implies the import/export policy of the relevant ASs, if a directed AS-link is not legitimate, but its reversed counterpart is legitimate, the reversal of the AS-link direction might indicate a policy violation. Further, redistribution path-spoofing routes might come along with not only reversed AS-links, but also some hidden AS-links that are invisible in both directions under normal configurations. Therefore, given a spoofed path, we check all the illegitimate links we find within it. If one of them reverses direction, the route is classified as redistribution path spoofing. Otherwise, the path is fake-link path spoofing.

If the AS-path of a route is valid, but the prefix-originAS association, say (p, a_0), is found to be illegitimate, the route is deemed as *prefix hijacking*. In addition, if there exists a legitimate association (p, x) with identical prefix but different origin AS, (p, a_0) is duplicate-prefix hijacking. Otherwise, if there exists a legitimate association (P, x), $x \neq a_0$, and p is the immediate subnet of P, (p, a_0) is sub-prefix hijacking. Otherwise, if there exists a legitimate association (q, x), $x \neq a_0$, and q is the immediate subnet of p, (p, a_0) is super-prefix hijacking. In any other cases, (p, a_0) is deemed as independent-prefix hijacking.

16.5.4 Justifying the detection algorithm

Next, we attempt to justify the correctness of our detection algorithm. In order to do so, we assume that the algorithm is based on "perfect" routing information objects, which are perfect in the sense that they consist of all the legitimate objects that should be visible to the observer AS when the detection is performed. This is an extremely idealistic scenario. However, it helps in isolating the errors caused by the imperfections of the routing information objects, which are the input of the algorithm. By using the ideal prefix-originAS association set thus obtained, our detection algorithm, which classifies a route as bogus if it fails the presence test, must have 100% detection accuracy. However, the correctness of our detection algorithm for path-spoofing routes is not as straightforward, as we discuss in the remainder of this section.

In theory, we should use a complete Internet AS topology to detect path spoofing. In particular, to detect redistribution path spoofing, we have to know at least the AS relationships between ASs. In the Internet, the import/export routing policies are dictated by the commercial agreements between ASs. Although the commercial agreements and the corresponding policies appear in diverse forms and are usually very complex, the majority of them define two typical relationships, i.e. provider-to-customer and peer-to-peer relationships [82]. In the former, the provider charges the customer for the data delivery that allows the customer to access the rest of the Internet. In the latter, the two ASs reciprocate free data delivery between each other's customers. Thus, an AS in the path should not redistribute routes between its providers or peers. Namely, the path is said to follow the valley-free rule, described as follows.

Rule 1 (valley-free). In a valid path in a relationship annotated AS topology, a provider-to-customer or peer-to-peer link should be followed by a provider-to-customer link.

Thus, given such a topology, any route whose AS-path contains AS-links that are not present in the topology is classified as a fake-link path spoofing, whereas any route containing links not conforming to the valley-free rule is classified as redistribution path spoofing.

It is difficult, however, to obtain such a complete AS topology annotated with AS relationships from the real-time route announcements supplied by the observer AS. This is because, first, there is no way to obtain a complete Internet AS topology based on the BGP route announcements from a few vantage points [54]. For example, certain private peering links would never appear due to policy constraints. Second, although many algorithms are available to infer the AS relationships, such as those given in refs. [82], [192] and [34], any inaccuracy about the inferred AS relationship might severely affect the performance of the overall detection algorithm. As a consequence, we decided to take the AS topology as a direct input to our detection algorithm.

As a result, in this chapter, we use an Internet AS topology annotated with directed AS-links to detect path spoofing. According to the procedure ISBOGUSROUTE shown in Figure 16.3, if a path is valid, all links in the path have to be (1) present and (2) in the same direction as the path. We refer to the second condition as the direction-conforming rule.

Rule 2 (direction-conforming). In a valid path in an AS topology annotated with directed AS-links, all AS-links must be in the same direction as the path.

In Appendix D we show that the direction-conforming rule is roughly equivalent to the valley-free rule when used to detect redistribution path spoofing. On the one hand, any path that does not follow the direction-conforming rule is not valley-free, which implies zero false positives. On the other hand, for Tier-1 ASs, any direction-conforming rule must be valley-free; for non-Tier-1 ASs, except the redistribution path spoofing routes launched by their provider ASs, which might be direction-conforming, other direction-conforming paths must be valley-free. Nonetheless, such undetectable spoofing paths are very unlikely to exist in the Internet since the number of Tier-1 ASs is limited. (Indeed, by using the ROUTEVIEWS tables dated December 20, 2006, and inferring the AS relationships using the approach proposed in ref. [82], we have noted that the total number of Tier-1 ASs in the Internet is 27, representing only 0.1% of all the ASs composing the Internet.) Meanwhile, most of them are backbone ISPs and have little incentive to do path spoofing. Therefore, the detection algorithm is very accurate. More importantly, the directed AS-links can be derived from the BGP route announcements directly, which makes the real-time detection feasible.

16.6   Refining the detection algorithm

The detection algorithm presented in Section 16.5.2 relies upon the legitimacy of the routing information objects learned from BGP routing data during the sliding window $[t - T, t)$. The quality of the objects learned determines its detection accuracy. However, the sets $\mathbb{A}[t - T, t)$ and $\mathbb{L}[t - T, t)$ used in our basic detection systems are directly obtained from the BGP data and may not be accurate. On the one hand, the sets are not clean. Beside legitimate objects, these sets may also contain the illegitimate ones carried by the bogus routes, which would make the future announcements of the bogus routes that contain the same objects undetectable. To eliminate these illegitimate objects, we strengthen the criteria to determine legitimacy in objects. On the other hand, the sets are not complete. It is impossible to obtain all routing information objects in the Internet from the history, because first, the observation window and the views of neighboring routers limit the available routing information objects, and second, the Internet keeps growing while the objects are learned from the history. The new arrivals are naturally missing. To address these limitations, we properly lengthen the observation window and increase the number of views to obtain more objects with moderate cost. Moreover, we explore various heuristics to infer additional possibly hidden and new objects.

16.6.1   Removing transient routing information objects

Intuitively, bogus routes are naturally stealthy and thus can not be expected to last long. Accordingly, the routing objects carried by the bogus routes are very likely to be transient. Therefore, we remove these transient objects from the BGP routes collected during the learning phase. We refer to this process as *baseline cleaning*.

   We first identify the metrics that measure the stability of the routing information objects. The detection system maintains a routing table $\mathbf{R}(t)$ that stores all routes from its peering BGP routers at time t; $\mathbf{R}(t)$ is continually updated with the routing updates from the neighboring routers. We say a routing object o exists at time t if there is at least one route in $\mathbf{R}(t)$ having o. Otherwise, o does not exist. Given an observation window $[t - T, t)$, the *(cumulative) uptime* of o, denoted by $u_o[t - T, t)$, is the sum of the durations of all the periods that o exists. Further, the *lifespan* of o during the window, denoted by $l_o[t - T, t)$, is the timespan when o first and last exists in the window.

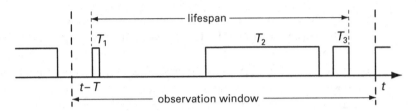

Fig. 16.4.   Illustration of $H_o(t)$.

For example, in Figure 16.4, during the observation window $[t - T, t)$, the uptime is $u = T_1 + T_2 + T_3$ and the lifespan is the length of the shown interval.

We can use the uptime to redefine legitimate objects by requiring that a legitimate object should have uptime longer than a threshold θ_u in the observation window $[t - T, t)$. Similarly, we can use the lifespan metric as a further discriminator by requiring that a legitimate object should have a lifespan longer than θ_l.

We apply the two criteria, respectively, to the prefix-originAS associations and the directed AS-links. As the uptime of an object is never longer than its lifespan, the uptime criterion is more stringent than that of lifespan. Compared with prefix-originAS associations, directed AS-links have less visibility because the network topology and routing policies can limit the visibility of an AS-link to the observer AS. For example, a multi-homed stub AS announces its prefixes through its primary and backup links alternately. From the viewpoint of the observer AS, the links show up intermittently, while the prefix-originAS associations of the AS appear continuously. Therefore, prefix-originAS associations are more likely persistent over time compared with directed AS-links. Thus, we apply the uptime to the former and the lifespan to the latter.

Accordingly, in the procedures in Figure 16.3, we should replace the set $\mathbb{A}[t - T, t)$ by the refined set $\mathbb{A}'[t - T, t) = \{o | o \in \mathbb{A}[t - T, t), u_o[t - T, t) > \theta_u\}$ and $\mathbb{L}[t - T, t)$ by $\mathbb{L}'[t - T, t) = \{o | o \in \mathbb{L}[t - T, t), l_o[t - T, t) > \theta_l\}$.

16.6.2   Inferring potentially legitimate objects

By analyzing the behavior of attackers and the common practices in prefix assignment/allocation and AS peering in the Internet, we use the following heuristics to explore those possibly hidden or new routing information objects. These heuristics would supplement the procedures in Figure 16.3 to justify further the legitimacy of objects. The objects that are not legitimate based on the procedures, but comply with some of these heuristics, are said to be *potentially legitimate*.

Attacker behavior heuristics

The heuristics presented in this section rely on the attacker behavior and the likelihood to achieve his/her final goal. As an attacker announces bogus routes to gain control of address spaces, if a suspicious route cannot help the attacker achieve the goal, it should not be considered as a prefix-hijacking or path spoofing-route, but it is more likely legitimate. Accordingly, the relevant objects might be legitimate.

Path Extension (PE) heuristic

Suppose that the AS-path of a prefix p is extended from the origin AS to a new AS, e.g. the AS-path changes from $\{A, B, C\}$ to $\{A, B, C, D, E\}$. The route cannot let D and E access p's traffic as long as C is the legitimate origin because any traffic from A to p would stop at AS C. Therefore, this kind of route cannot be announced for malicious purposes, but is more likely caused by legitimate operations, such as address suballocation. Therefore, we consider this new route to be valid and the relevant new objects, e.g. $(C \rightarrow D)$, $(D \rightarrow E)$ and (p, E), as legitimate. Note that this kind of route might also result from misconfiguration. We find that typos in AS-path prepending can cause this kind of route. An AS, say A, types its own number as A' when configuring prepending lists. If its own AS number is added after the prepended list, the resulting AS-path will resemble $\{\ldots, A, A'\}$.

En-route AS (EA) heuristic

The ASs in the path to a prefix are called the *en-route ASs* of the prefix. Since the en-route ASs of a prefix have already had access to the traffic of this prefix, they have no further motivation to hijack or spoof routes to this prefix. Therefore, if the AS-path to a prefix contains a new directed AS-link whose upstream AS is an en-route AS of the prefix, the link should be legitimate. Similar to the previous heuristic, misconfiguration such as typos in AS prepending could result in this kind of route.

In order to capture the legitimate en-route ASs of prefixes, we introduce a new type of routing information object called *prefix-enrouteAS association*, which is a tuple $\langle p, a_i \rangle$ of prefix p and one of its en-route ASs a_i. We also use the lifespan of the associations with threshold θ_e to identify the legitimate ones because the prefix-enrouteAS associations also have limited visibility similar to the directed AS-links.

Common-practice heuristics

Further, we explore several common practices that are widely adopted in the Internet to infer some reasonably hidden or new routing information objects. We use these observations and characteristics to propose new heuristics.

Address Expansion (AE) heuristic

In order to optimize routing table size, RIRs try to assign ISPs new address spaces that can be aggregated with their existing prefixes [18]. Meanwhile, after ISPs obtain a large block of addresses, they may initially announce part of them and then gradually announce others. Thus, an AS is likely to announce new prefixes that can be aggregated with their existing prefixes in the same "virtual" supernet. We allow an AS to expand its existing prefixes to a virtual supernet by at most 2^δ times, where δ is called the *expansion factor*. New prefix-originAS associations in the expanded space are deemed legitimate.

Neighboring (NB) heuristic

For two neighboring ASs, either of them can originate routes to the co-located prefixes. Meanwhile, an AS can suballocate its address space to its customers. Therefore, two neighboring ASs might be able to originate the prefixes of each other. Therefore, given two neighboring ASs A and B, if A has prefix p, B might also be legitimate to announce the route of prefix p.

Address Sharing (SH) heuristic

Similarly, if two ASs share prefixes, they might be neighbors. The heuristic can help find hidden links. For example, a customer AS has a subnet of the provider AS while the AS-links between them are invisible since the provider AS announces the aggregated route instead of the more specific route originated from the customer. Further, since two ASs that share address space might be neighbors, according to the neighboring heuristic the two ASs can further share other address spaces.

Backbone (BA) AS heuristic

The Internet backbone ASs have worldwide presence and can virtually peer with any AS. Thus, any new directed AS-link from a backbone AS to another AS might be legitimate. The key to this heuristic is to identify the backbone ASs. Usually, the backbone ASs have dense connectivity. If the in-degree of an AS, i.e. the number of its upstream ASs, is more than a threshold G, it is considered as a backbone AS. We decide to choose the in-degree of an AS instead of its out-degree, because the in-degree of an AS is harder to forge.

16.6.3   Event-base clustering and calibration

As the introduction of heuristics can make some actual bogus routes undetectable and introduce false negatives, we further utilize the concept of event clusters to calibrate our detection. Because the routes triggered by the same cause are likely to share similar characteristics, if some of them are found bogus, others are also likely bogus, even if they have been identified as potentially legitimate with certain heuristics. We use this *Event Calibration* (EC) to correct the mistakes because of heuristics overuse. We use the following clustering process to group routes into event clusters. First, the routes in the same cluster must be temporally correlated. Suppose the routes in a cluster are announced at time $t_1 \leq t_2 \leq \cdots \leq t_n$, then $t_{i+1} - t_i \leq d$ and $t_n - t_1 \leq D$, i.e. the two consecutive routes should not be spaced out by more than d and the whole cluster should not span a period longer than D. The temporal clustering of BGP routing updates has been intensively studied in ref. [53]. We use the typical value of $d = 70\,\mathrm{s}$ and $D = 600\,\mathrm{s}$ as suggested in ref. [210]. Second, routes in the same cluster share the same cause. Since the detection algorithm (see Figure 16.3) can pinpoint the possible attacker AS behind a bogus route, which is either the upstream AS of the suspicious link or the origin AS of the suspicious association, the routes in the same cluster should also share the same attacker AS.

16.7 Experiments

In this section, we first use BGP routing data to investigate the values of the various parameters used in the detection algorithm and to validate the key ideas captured by our heuristics. Then we evaluate the performance of the detection system in its entirety.

Our experiments use BGP routing data from ROUTEVIEWS servers. For every BGP route, we first filter out and report the immediately apparent bogus routes, such as bogon prefixes [76]. Further, even though a prefix is in the allocated address space, if its prefix length is shorter than 8 or equal to 32 [209], we also consider it a bogon. Note that the list of bogon prefixes might be different for different ASs. For example, some stub customer ASs might allow their provider ASs to announce a default route 0.0.0.0/0 plus portions of the BGP routing table instead of a full BGP routing table. In such a case, 0.0.0.0/0 is not a bogon prefix. Next, we remove the private and unassigned AS numbers [5] from the AS-path and the continuously duplicate AS numbers. For example, an AS-path such as {2914, 19029, 26362, 65535, 65534, 65532, 65531, 26362} will be cleaned up as {2914, 19029, 26362}. These private AS numbers are typically used within ASs, but are not trimmed off when being exported outside. Finally, if the AS-path still contains loops, the route will be filtered.

16.7.1 Determining thresholds for legitimate objects

We build the routing information object sets $\mathbb{A}$ and $\mathbb{L}$ over a period of time and investigate the settings for inferred legitimate routing information objects.

Thresholds for legitimate objects

Figure 16.5 shows the CDFs of the uptime and the lifespan of prefix-originAS associations, directed AS-links and prefix-enrouteAS associations in the observation window January 1, 2006, to July 1, 2006, based on the data from EQIX server. From the full view of the distributions in Figures 16.5 (a), (b) and (c), we find that the majority of the routing objects are either extremely short-lived or persistent for the entire duration of the window.

The distributions suggest the proper values for the thresholds θ_u, θ_l and θ_e. If we zoom into the first eight days, as shown in Figures 16.5(d), (e) and (f), and if we set the thresholds to a value longer than one day, then the difference in percentage of objects classified as legitimate is marginal. Thus, for simplicity, we apply a one-day legitimate threshold for all the three objects, i.e. $\theta_u = \theta_l = \theta_e =$ one day. We also investigate the distributions for uptime and lifespan in several other observation windows with varying length. It turns out that the distributions are similar and that the one-day threshold is always a good choice.

It is worth noting that the gap between the curves of uptime and lifespan distributions for the three objects characterizes their differences in visibility. As mentioned before, the prefix-originAS associations are more visible to an observer and hence show up more persistently, i.e. their uptime is almost equal to their lifespan in most cases. In

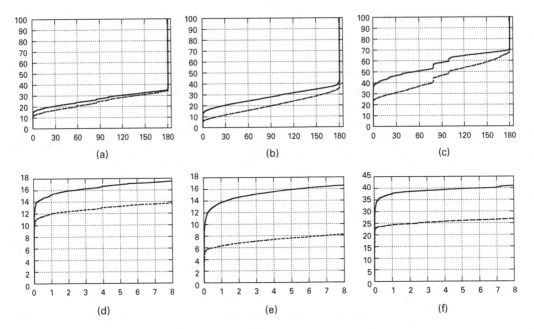

Fig. 16.5.   Distribution of uptime (solid line) and lifespan (dashed line) (from January 1, 2006, to July 1, 2006). The x-axis shows uptime/lifespan in days and the y-axis shows the CDFs; (a), (b) and (c) show the full view and (d), (e) and (f) show the zoom in view when $x \leq 8$ days.

contrast, the directed AS-links and prefix-enrouteAS associations show a larger gap between the two distributions, which can be attributed to their intermittent visibility, in that their uptime is less than their lifespan in most cases.

Observation-window size

Given the current time t, we try to find an observation window $[t-T, t)$ with proper size T that can account for as many legitimate objects as possible when we compare route announcements occurring in the "future" after t. Based on the routing data from EQIX, Figure 16.6 shows the percentage of legitimate objects in July 2006 that can be found legitimate in various observation windows that, immediately preceded July and lasted from one day to six months. It shows that, for the three objects, the longer the observation window, the more legitimate objects can be found. However, the growth becomes marginal when the window size is longer than $30 \sim 60$ days. Since the window size should be small to save storage space, we set the observation-window size $T = 30$ days for all three objects.

In addition, compared with the prefix-enrouteAS associations, the prefix-originAS associations and the directed AS-links are relatively stable over time. This can be explained by the fact that these two objects represent the stable structure of the routing infrastructure while the prefix-enrouteAS associations do not.

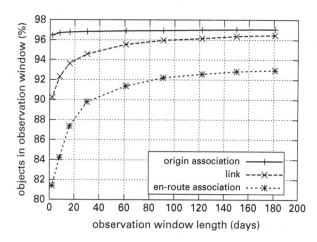

Fig. 16.6. Percentage of objects during July 2006 found in the observation windows of size T.

16.7.2 Validation of common-practice heuristics

In this section, we validate the common-practice heuristics by examining whether they can help identify the potentially legitimate routing objects while rendering few real malicious objects undetectable. We are not going to validate the attacker behavior heuristics since we believe that, as long as the attackers are rational, they will not use those sorts of routes to perform hijacking or spoofing.

Validation metrics
We validate each heuristic with the following two metrics.

Hit rate
The heuristics can help identify which are the potentially legitimate routing objects and which are the bogus ones. Consider an object which has a time point t such that it is not legitimate based on its past history during $[t - T, t)$ but will be legitimate based on the future history during $[t, t + T)$. If a heuristic, when applied against the past window, can justify the legitimacy of such a "will-be" legitimate object, the heuristic is said to *hit* the object. The percentage of the hit objects in the set of "will-be" legitimate objects is called the *hit rate* of the heuristic. Thus, the hit rate quantifies the power of a heuristic to predict the "will-be" legitimate objects.

Undetectable rate
The potentially legitimate routing information objects inferred with a heuristic might actually be bogus. For instance, if we employ the neighboring heuristic while an AS hijacks the prefixes of its neighbors, the hijacking would not be detected. The extent to which a heuristic renders bogus routes undetectable depends on how the attack is performed. We assume a *random attack model*, in which the malicious AS hijacks prefixes randomly in the entire IPv4 address space or spoofs AS-links to randomly chosen ASs in the Internet. Accordingly, we define the term *undetectable rate (under random*

attack) for an AS as the probability that a random attack launched by this AS becomes undetectable under a heuristic. Further, the *average undetectable rate* is the average over all ASs in the Internet.

Our experiments on BGP data from EQIX in March 2006 show that by effectively choosing the parameters of a heuristic, we can increase the hit rates to higher values, while not significantly compromising on the undetectable rates. Moreover, in Section 16.7.3, we will present detection strategies that combine these heuristics in a way that the detection performance improves even more.

Heuristics for inferring prefix-originAS associations

Next, we examine the address-expansion heuristics, neighboring heuristics and address-sharing heuristics, all of which infer potentially legitimate prefix-originAS associations. The heuristics actually expand the address space that an AS can legally claim. Suppose that, with the heuristics, the size of address space that an AS can claim is expanded from x to x'. Assume that the AS randomly chooses an address block in the IPv4 address space to hijack. Then the probability that the hijacking goes undetected is given by $(x' - x)/(2^{32} - x)$.

Address Expansion (AE) heuristic

Figure 16.7(a) shows the relevant change trends of hit rate and the average undetectable rate with the growth of the expansion factor δ. The points along the curve in Figure 16.7(a) from left to right correspond to values of the expansion factor δ from 0 to 10. It shows that when the expansion factor is between 0 and 5, the hit rate grows faster than the false negative rate and that it reaches around 35%, while the average undetectable rate is bounded by 0.01%. However, after that, the increase in the false negatives dominates. Therefore, we choose $\delta = 5$.

Neighboring (NB) and Address-sharing (SH) heuristics

In March 2006, we find 12 814 pairs of ASs that are not only neighbors, but also share address space. As a result, 59% of 21 661 pairs of ASs sharing address space were

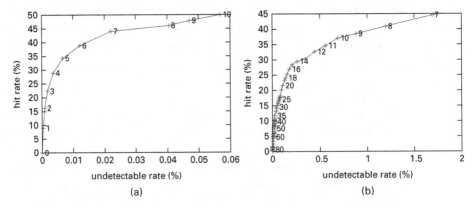

Fig. 16.7.   Hit rate versus average undetectable rate when applying heuristics: (a) AE; (b) backbone AS.

neighbors; meanwhile, 26% of 48 520 pairs of neighboring ASs shared address space. Here, two ASs are said to be neighbors if there is a legitimate directed AS-link from one to another; a prefix is said to be owned by an AS if the corresponding prefix-originAS association is legitimate. The observations show that the address boundary between ASs is not well defined. It is common that two neighboring ASs share address space, implying that they are very likely neighbors.

Heuristics for inferring directed AS-links

We examine the address-sharing and backbone AS heuristics that infer directed AS-links. Suppose that a heuristic increases the number of AS-links in the Internet from x to x' and that there are N ASs in the Internet. Then, the attacker ASs forge directed AS-links, which can be as many as $N(N-1)$ $- x$ if the AS-links are directional or $(N(N-1)/2) - x$ when we do not consider the direction of AS-links, with equal probability. The attacker ASs can forge $N(N-1)-x$ directed AS-links with equal probability. The chance that a fake link is undetectable is given by $(x' - x)/[N(N-1) - x]$.

Address-sharing heuristic

As mentioned before, with the address-sharing heuristic, the undetectable rate is about 0.003%. At the same time, the hit rate of the heuristic is 7.8%.

Backbone AS heuristic

Figure 16.7(b) shows the relative growth trend between the undetectable rate and the hit rate when the backbone AS in-degree threshold G is decreasing. The lower the threshold, the more ASs are qualified as backbone ASs, and the more AS-links are inferred. Thus, the hit rate becomes higher, but the undetectable rate is also growing. It shows that, when G is no more than 14, the growth of the undetectable rate becomes faster than that of the hit rate. Therefore, we set $G = 14$, which corresponds to a hit rate of 29% and an average undetectable rate of around 0.3%.

16.7.3   Evaluation of detection algorithm

After exploring the parameter settings for the detection algorithm, we go on to evaluate their performance. We first specify the metrics for evaluation and then apply our detection algorithm to detect bogus routes under various detection strategies.

Evaluation metrics

By applying the system to the BGP routing updates in a given period, we use the term *false-positive rate* to represent the percentage of legitimate routes that are misidentified as bogus among all legitimate routes and the term *false-negative rate* to indicate the percentage of bogus routes that are not identified as bogus among all bogus routes. Meanwhile, although the detection system reports every bogus route, the network operators actually investigate the corresponding suspicious objects since they are the "root-causes." So, each new suspicious object can be seen as an alarm. The number of alarms indicates the workload that the network operators require to deal with verification and mitigation. Accordingly, we define the *number of false alarms* as the number of legitimate objects that are misidentified as bogus and the *number of missed alarms* as the number of bogus objects that are not identified.

Detection strategies

By applying heuristics, we are able to infer potentially legitimate routing objects to reduce false positives but increase the false negatives. In order to achieve the best tradeoff, we explore the following four detection strategies in which different sets of heuristics are chosen and different levels of sensitivities are achieved. (1) *Basic strategy (B)* in which only the legitimate objects determined by the pseudo-code in Figure 16.3 are used for detection; this yields the highest number of false positives but the lowest number of false negatives. (2) *Rational attack strategy (R)*, where, in addition to the previous strategy, the potentially legitimate objects inferred by the attacker behavior heuristics are also used for detection. (3) *Common-practice strategy (C)*, where, in addition to the previous strategy, the potentially legitimate objects inferred by the common-practice heuristics are used for detection. (4) *Partial protection strategy (P)* based on the observation that the reported incidents in the Internet are usually duplicate-prefix hijacking and redistribution path-spoofing, so, while using the previous strategy, we detect duplicate-prefix hijacking and redistribution path-spoofing routes only, which yield the lowest false positives but highest false negatives. Finally, to each of the four strategies we also apply the Event Calibration (EC) to adjust the detection performance. As a result, each strategy has two substrategies: with or without EC.

Baseline evaluation

We first apply the algorithm onto the BGP updates during a period in which no prefix-hijacking or path-spoofing incidents are reported. With this assumption, we can estimate the baseline false positives that our system can produce. In this experiment, we choose the BGP updates from December 23 to December 29, 2006, from ROUTE-VIEWS. The routing information objects are inferred from the updates from six Tier-1 ISPs: AS701(Alternet), AS1239 (Sprint), AS7018 (AT&T), AS2914 (NTT-America), AS3356 (Level3) and AS3549 (Global Crossing), while only the routes from AS1239 (Sprint) are inspected for bogus route detection. Figure 16.8 shows the value of the four metrics achieved under different detection strategies. It shows that when the most stringent basic strategy, B, is applied, around 0.5% and 1.4% of the total routes are identified as prefix-hijacking and path-spoofing routes and around 1092 suspicious prefix-originAS associations and 427 directed AS-links raise alarms. Thus, on average, the network operators need to verify $156 + 61 = 217$ false alarms per day. If heuristics are incorporated, the false positives are reduced. For example, if the least stringent (partial protection strategy, P) is employed, the false-positive rate is reduced to 0.04% and 0.16% for prefix hijacking and path spoofing, respectively, for a 90% reduction when compared with the basic strategy. Moreover, the number of false alarms is reduced to about 19 per day.

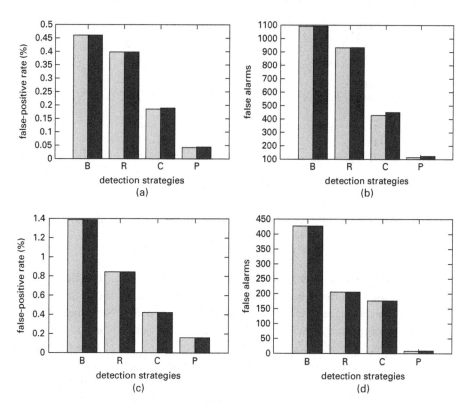

Fig. 16.8.   Baseline detection performance during December 23 to December 29, 2006. (a), (b) Prefix-hijacking detection; (c), (d) path-spoofing detection. Shown without EC (shaded gray) and with EC (black).

Contrary to our assumption, some of the alarms raised by our system did look suspicious, which implies that our system can achieve even lower false positives. For instance, after manual inspection of the suspicious objects, we find a very likely redistribution hijacking attack on December 27, 2006, in which an Indian ISP AS9498 (BBIL-AP) redistributed routes of 2703 prefixes all over the world from AS5511 (France Telecom) to AS1239 (Sprint) for almost an hour. If we consider these routes as bogus, then the false-positive rate for path spoofing can be further reduced to 0.01%. In addition, among the reported prefix-originAS associations, we find ten very suspicious ones. For instance, a New Jersey ISP (CYBERNET, AS6073) hijacked prefixes 204.117.112.0/24 of a Pennsylvania site (Big Brothers & Big Sisters of America, AS31906). ARIN WHOIS shows that this prefix is a "non-portable" prefix of Sprint.

To summarize, our baseline evaluation establishes the efficacy of our detection system in not misclassifying routes as bogus, since we were able to reduce the false positives for both prefix hijacking and path spoofing to values close to zero.

Impact of number of views on detection performance

We repeat the previous experiments by performing the detection based on the routing information objects inferred from the routing updates of one view (Sprint) and three

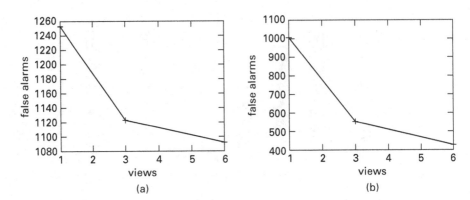

Fig. 16.9.   Performance of the detection algorithm as a function of the number of views.
(a) Prefix-hijacking detection; (b) path-spoofing detection.

views (Sprint, AT&T and Alternet), respectively. Figure 16.9 shows the number of false alarms reported when the basic strategy, B, is used. It shows that when the number of views increases, the number of false alarms decreases. Nevertheless, gradually the gain becomes marginal. Thus, we believe that having around three to six views yields a good enough performance at a moderate cost. Also, we can observe that after the number of views changes from one to three, the number of false alarms for path spoofing is reduced by almost one-half, while that for prefix hijacking is reduced by only about one-tenth. This further confirms that, unlike prefix-originAS associations, the visibility of directed AS-links can be drastically increased by increasing the number of views.

Evaluation with documented incidents

In this section, we apply our detection system to several documented incidents in 2004–2006 to examine the detection performance. The detection is based on the routing updates from EQIX during the three-day period around the incidents. Table 16.1 shows the date and the attacker ASs of four prefix-hijacking incidents and two path-spoofing incidents. The detection performance is depicted by the metrics under the most stringent basic strategy (B) and the least stringent partial protection + event calibration strategies (P+EC). Comparing the two strategies, it shows that the heuristics can help reduce the false positives by almost 90% while keeping the false negatives close to zero. In particular, Figure 16.10(a) shows the detection performance for the incident. Unlike other prefix-hijacking incidents, in which the attackers hijacked fractions of prefixes in the Internet and could be modeled as random attacks, the attacker in ConEd mainly hijacked prefixes of its customers. Thus, the common-practice heuristics do not identify more than 80% hijacked routes. Fortunately, the EC heuristics corrects most of the misidentification and adjusts the false-negative rate to 3.2%. In the two path-spoofing incidents, both ISPs in Hong Kong and Pakistan leaked routes of google.com and other sites to the Internet backbone and caused a large portion of the Internet to experience huge latencies for half an hour. Figure 16.10(b) shows that the detection system was able to detect the spoofing routes under all detection strategies.

Table 16.1. Detection performance for documented incidents

Name	Date	ASN	Strategy	+(%)	−(%)	#F	#M
Prefix hijacking							
TTNet	Dec. 24, 2004	9121	B	0.81	0	633	0
[4]			P+EC	0.11	0.04	101	4
ConEd	Jan. 22, 2006	27 506	B	0.39	0	614	0
[7]			P+EC	0.02	3.2	57	10
TTNet2	Feb. 26, 2006	9121	B	0.53	0	470	0
[11]			P+EC	0.07	0	48	0
NWNet	Jun. 7, 2006	23 520	B	0.37	0	822	0
[107]			P+EC	0.02	0	61	0
Path spoofing							
IS-AP-HK	Nov. 11, 2004	9729	P	0.85	0	335	0
[10]			P+EC	0.04	0	11	0
PK-TEL	Dec. 12, 2005	17 557	B	1.00	0	532	0
[17]			P+EC	0.01	0	15	0

#F = number of false alarms; #M = number of misses; +(%) and −(%) percentage of false positives and false negatives.

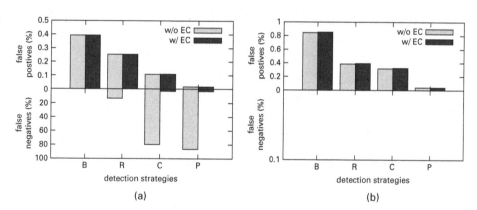

Fig. 16.10. Prefix-hijacking detection performance for different incidents. (a) ConED, (b) IS-AP-HK. Shown without EC (shaded gray) and with EC (black).

16.8 Lessons learned

In this chapter we presented a bogus route detection system that adopts a persistence-inference method of route information objects. We observed that, even though the BGP routes are highly dynamic, the routing system has a relatively stable structure. We capture the routing objects that represent this stable structure, i.e. the association between prefixes and origin ASs and the AS-links, and use the knowledge accumulated over time to detect the bogus BGP routes. In particular, the system is able to detect path spoofing with the aid of directed AS-links without knowing the hidden AS relationships. Further, in order to address the inherited shortcoming of the history-based approach, we take several measures to improve the detection knowledge base. First, we filter out transient objects likely to be associated with bogus routes that may affect the robustness of our base knowledge. Second, we explore several heuristics to infer the potentially legitimate routing objects. Further, we calibrate the detection results based on the intuition that the routes triggered by the same events share the same characteristics. We apply our detection system to real BGP data and show that we can achieve false-positive rates as low as 0.02% and raise no more than 20 alarms per day. Further, we show that our system can detect bogus routes of several documented incidents with almost 100% detection rate and about 0.02% false-positive rates.

Appendix A
How to link original and measured flow characteristics when packet sampling is used: bytes, packets and flows

In order to compile flow statistics, each router maintains a table of records indexed by flow key, e.g. 5-tuple of the flow. A flow is said to be *active* at a given time if there exists a record for its key. When a packet arrives at the router, the router determines if a flow is active for that key. If not, it instantiates a new record for that key. The statistics for the flow are updated for the packet, typically including counters for packets and bytes and arrival times of the first and most recent packet of the flow. Due to the fact that the router does not have knowledge of application-level flow structure, it must terminate the flow according to some criteria. The most commonly used criteria are the following: (i) inter-packet timeout, e.g. the time since the last packet observed for the flow exceeds some threshold; (ii) protocol syntax, e.g. observation of a FIN or RST packet of the TCP flow; (iii) aging, e.g. flows are terminated after a given elapsed time since the arrival of the first packet of the flow; (iv) memory management, e.g. flows might be terminated at any point in time to release memory. When a flow is terminated, its statistics are flushed for export and the associated memory is released for use by new flows.

The main resource constraint for the formation of flow statistics is at the router flow cache. Performing a lookup of packet keys and a counter increment at line rate would require the flow statistics to be stored in fast memory. However, core routers will carry increasingly large numbers of concurrent flows, necessitating large amounts of fast memory; this would be expensive. By sampling the packet stream in advance of the construction of flow statistics, the time window available for flow cache lookup is prolonged, enabling storage to be carried out in slower, less expensive, memory.

A number of different implementations are possible within the operational requirement of sampling packets at a given rate. Implementations include independent sampling of packets with probability $p = 1/N$ and periodic selection of every Nth packet from the full packet stream. In both cases, N is called the *sampling period*, i.e. the reciprocal of the average sampling rate. In the rest of this appendix, we consider period packet sampling due to its use in the current implementation of all versions of Cisco Netflow.

Independent of the specific implementation being used, packet sampling definitely enables operators to manage the router resources better. However, there are two factors that may hinder the effectiveness of the statistics exported. One is that new applications may generate packets in patterns that are not well captured by the sampling mechanism deployed. As a consequence, for a fixed sampling period Netflow is likely to under-report the number of flows associated to applications used to transfer packets in very short flows, and over-report the number of flows of applications that produce sparse flows (like peer-to-peer and streaming), e.g. packets over extended periods leading to flows that are likely to exceed the fixed aging threshold. The second factor is that packet sampling removes key characteristics for flow delineation from the packet stream. For example, termination of a TCP flow based on the observation of a FIN packet is hindered if the packet is not present in the sampled stream. Thus, inter-packet timeout is expected to become the dominant method of termination for TCP flows when the sampling rate is low.

We indicate to the reader that this section is not intended to provide the latest and greatest results achieved in statistical sampling, but it is intended to present a few estimators that might help ISPs in their day-by-day operation. For a more rigorous and detailed explanation of what is presented in this section, we direct the reader to refs. [70] and [69].

First, we present estimators to infer the sample packet stream statistics from the original flow. We report such estimators for (i) the total number of measured flows and (ii) the duration of an active flow. Second, we present estimators to infer the original flow statistics from sampled packet streams. We report such estimators for (i) bytes and packet counts of the original packet stream and (ii) the number of original TCP flows and their average size, e.g. number of packets per flow.

A.1   From original to measured data streams: average number of total flows and their duration

Consider an original flow comprising n packets distributed over a period of length t. Suppose packets are sampled from this flow at a deterministic rate $1/N$ and that measured flows formed with timeout T from the samples. We report two estimators to determine two quantities: (i) $f(n, t; N, T)$, the resulting number of measured flows and (ii) $a(n, t; N, T)$, the total time the flow was active, i.e. the time elapsed between the first and last sampled packet, plus the timeout T. Clearly, f and a are not determined without further assumptions concerning the spacing of packets within the original flow. In the following we assume packets of the original flow to be equally spaced in time. We denote the expected value of f and a by f_m and a_m, respectively. Thus, assuming a flow with random initial phase, we have the following:

$$f_m(n, t; N, T) = \begin{cases} 1 & \text{if } Nt \leq (n-1)T \text{ and } N < n \\ n/N & \text{otherwise} \end{cases} ; \qquad (A.1)$$

$$a_m(n, t; N, T) = \begin{cases} t\left(\frac{n-N}{n-1}\right) + T & \text{if } Nt \leq (n-1)T \text{ and } N < n. \\ nT/N & \text{otherwise} \end{cases} \quad (A.2)$$

The justification behind the forms of these estimators can be summarized as follows. Assume $n \leq N$. At most, one packet is selected with probability n/N. If the measured flow exists, it comprises one packet, and hence has duration T, e.g. the timeout. Thus $f_m = n/N$ and $a_m = Tn/N$. Otherwise, for $n > N$, the separation between original packets is given by $t/(n-1)$, and hence the separation between any selected packets is given by $Nt/(n-1)$. Suppose first that $Nt/(n-1) \leq T$. The separation of selected packets does not exceed the timeout, and selected packets form one flow: $f_m = 1$. On average this flow has n/N packets, so it is active for $a_m = t(n-N)/(n-1) + T$. On the other hand, if $Nt/(n-1) > T$, then n/N packets are selected on average, each packet giving rise to one flow. The expected number of active flows is then given by $a_m = nT/N$.

Now, given a set of original flows $1 = 1, 2, \ldots, S$ of duration t_i and comprising n_i packets present during an interval of duration D, we can estimate the total number of exported flows $\hat{F}$ and the mean number of active flows $\hat{A}$ as follows:

$$\hat{F} = \sum_{i=1}^{S} f_m(n_i, t_i; N, T) \quad (A.3)$$

and

$$\hat{A} = D^{-1} \sum_{i=1}^{S} a_m(n_i, t_i; N, T). \quad (A.4)$$

A.2 From measured to original data streams: total number of original bytes and packets

Given a *measured* flow f comprising n packets and b bytes obtained using a period packet sampling at rate $1/N$, the *original* number of packets P and bytes B can be estimated by multiplying the measured values n and p by N, e.g. $\hat{P} = Np$ and $\hat{B} = Nb$, where $\hat{}$ indicates an unbiased estimator of the original quantity. Furthermore, it can be easily proved that the standard error of $\hat{P}$, defined as $\sqrt{(Var(\hat{P}))}/P$, is bounded above by $\sqrt{(N/P)}$, with $Var(\hat{P}) = PN(1 - N^{-1})$. Similarly, it can be found that the standard error of $\hat{B}$, defined as $\sqrt{(Var(\hat{B}))}/B$, is bounded by $\sqrt{(N/P)}(b_{max}/b_{avg})$, where b_{max} and b_{avg} represent, respectively, the maximum and average packet sizes measured.

A.3 From measured to original data streams: total number of original TCP flows and their average size

In this section we report estimators presented in ref. [70] for the total number of original TCP flows and the associated average size, e.g. number of packets per flow. The estimator uses the additional information present in the flow details of exported flows to increase the accuracy of the overall estimate. Furthermore, note that TCP traffic is known to contribute to the majority of Internet traffic, and thus such a simple estimator ends up being very powerful on a daily basis. For more sophisticated estimators for overall traffic, we direct the reader to refs. [70] and [69].

The TCP signals the start and end of connections with packets that are distinguished by flags in the code bits of the TCP header. The first packet of a connection has a SYN flag set, whereas the last has the FIN flag set. Netflow traces include the cumulative OR of the code bits. Thus, by inspecting the code bits of the flow, we may determine whether or not the SYN and FIN flags were set on any packet detected in the flow. In the following, we refer to a packet with a SYN flag set as a SYN packet.

We now show that when packet sampling is used, the statistics of the flow code bits allow us to infer the number of original flows (let's call them M) that were present during the collection period. We will construct an estimator of M as follows. We assume that packets are selected with probability $1/N$. If the SYN packet is selected, then trivially the flow is sampled. Thus the number $\hat{m}$ of measured flows that contain a SYN packet has an expected value of M/N. Consequently, $\hat{M} = N\hat{m}$ is an unbiased estimator of M. One potential disadvantage of $\hat{M}$ is that it uses only those flows containing a SYN, i.e. only a subset of the measured flows as its data. An alternative that addresses this issue is the following. We assume that no splitting of flows takes place, e.g. we are assuming an infinite timeout T. We divide the original flows into three classes: S_1 comprises those flows for which the measured flow features *exactly one* SYN packet; S_2 comprises those flows for which the measured flow has *at least one* non-SYN packet. (Note that if any packet of a given original flow is sampled, then it must be either in S_1 or S_2.) Finally, S_3 denotes the set of original flows from which no packet was sampled. Thus, we can refine the estimator $\hat{M}$ as $\hat{M} = N\hat{s_1} + \hat{s_2}$, where $\hat{s_1}$ and $\hat{s_2}$ denote the measured numbers of flows stemming from S_1 and S_2.

Appendix B
Application-specific payload bit strings

This appendix lists the payload bit-strings used by a payload classifier.

Numbers in parentheses denote the beginning byte in the payload where each string is found. If there is no number the string is found at the beginning of the payload. Note that: "plen" denotes the size of the payload; \x denotes hex; && denotes AND; || denotes OR; plen - 2 = (1) denotes that the payload length minus 2 is given by the first byte in the payload

B.1   UDP bit strings

- eDonkey
  \xe3, \xc5 (all UDP flow packets must begin with one of these bytes)
- Gnutella
  \x47\x4e\x44,\x00\x01\x00\x00\x00\x00\x00(16,plen=23),"LIME"(23), \x01
  \x01\x00\x1f\x00\x00\x00 (16)
- FastTrack
  \x27\x00\x00\x00\x29\x80, \x27\x00\x00\x00\xa9\x80,
  \x28\x00\x00\x00\x29\x00, \x28\x00\x00\x00\xa9\x00,
  \x29\x00\x00\x00\x29, \x29\x00\x00\x00\xa9, \xc0\x28, \xc1 (plen=7),
  \x2a (plen=5)
- Direct Connect
  "$SR", "$Pin"
- Soribada/Goboogy
  \x10 (plen=2), \x51\x3a\x2b (3), "<peerplat>"
- Mp2p
  \x00\x00\x00 (2) && (\x00 (5) || \x01 (5) || \x02 (5) || \x03 (5) || \x04 (5))
- irc
  "USERNAME"
- DNS
  (\x00\x01\x00 (4) && \x00 (8) && \x00" (10)) && (\x00 (7)|| \x01 (7)),
  \x01\x02\x00\x07\xd1\x86\x3f\xc3
- Gaming

This appendix reflects work by Dr Thomas Karagiannis, http://www.cs.ucr.edu/ tkarag/papers/strings.txt

\xff\xff\xff\xff && ("details" (4) || "players" (4) || "ping" (4) || \x6d (4) ||
"getstatus" (4) || "getinfo" (4) || "status" (4) || "infoResponse" (4)),
\x00\x00 (2) && \x00 (6) && (\xc0 (7) || \x40 (7) || \x80),
"hostname" (5), \xfe\xfd\x00\xd6, \x00\xd6\x2b\x7d
\x5c && ("status" (1) || "game" (1) || "death" (1) || "kill" (1) || "keyhash" (1) ||
"info" (1)),
\x00\x00\x00\x0 && \x02\x00\x02\x00\x37 (12)

- streaming
  "applicat" (8)
- PeerEnabler
  "CL" && \x00 (4)
- SpamAssasin
  \x00 && \x04\x02\x00\x00\x00\x01 (2)
- SNMP
  \x30 && (plen - 2 = (1) || plen - 4 = (2)(3)

B.2   TCP bit strings

- eDonkey
  \xe3\x19 && \x00\x00 (3) \xe3\x19 && \x28\x00 (3) \xe3\x01 && \x00\x00 (3)
  \xe3\x11 && \x00\x00 (3) \xe3\x14 && \x00\x00 (3) \xe3\x15 && \x00\x00 (3)
  \xc5\x3f && \x00\x00 (3) \xe3\x42 && \x00\x00 (3) \xe3\x48 && \x00\x00 (3)
  \xe3\x29 && \x00\x00 (3) \xc5\x0d && \x00\x00 (3) \xe3\x41 && \x00\x00 (3)
  \xc5\x32 && \x00\x00 (3) \xc5\x5e && \x00\x00 (3) \xe3\x45 && \x00\x00 (3)
  \xe3\x62 && \x00\x00 (3) \xe3\x63 && \x00\x00 (3)
- Gnutella
  "GNUTELLA", "X-Query", "X-Guess", "X-Ultrap", "X-Ext-", "X-Try-", "X-
  Degree", "X-Lo", "X-Max-", "X-Version", "X-Dynami", "Server: Mor", "Server:
  Lim", "User-Agent: Lime", "Vendor-Message:", "GET /uri", "Busy Queued" (33),
  "HTTP/1.1 503 Que", "HTTP/1.1 503 Ful", "HTTP/1.1 503 Not"
- FastTrack
  "GIVE", "GET /.hash", "Retry-After:" (17), "Content-Range:" (14)
- Direct Connect
  "$Send", "$Get", "$Dir", "$ConnectT", "$Supports", "$Hello", "$MyINFO",
  "$Search" "$MyNick", "$Quit", "$Key", "$RevConn", "$Version", "$Lock", "$Hub-
  Name"
- BitTorrent
  \x13\x42\x69\x74 , \x00\x00\x00\x05\x04\x00\x00, \x00\x00\x00\x0d
  \x06\x00\x00, \x00\x00\x40\x09\x07\x00\x00 "GET /announce?", "GET/
  torrents", "GET /scrape", "info_hash" (in the url of an HTTP request)
- soulseek
  plen-4 = (0) &&

(\x01\x00\x00\x00 || \x03\x00\x00\x00 || \x07\x00\x00\x00 || \x05\x00\x00
\x00 || \x12\x00\x00\x00 || x1a\x00\x00\x00 || x09\x00\x00\x00 || \x28\x00\x00
\x00 || \x29\x00\x00\x00 || \x2a\x00\x00\x00 || \x41\x00\x00\x00||\x20\x00
\x00\x00)\x00\x00\x03\x31\x00\x00 (2)

- winmx
  \x31 (plen=1), "SEND" (plen=4), "GET" (plen=3) (within the first 6 packets of the flow only) \x50\x4e\x41\x00\x68\x56
- Ares
  "GET hash:", "PUSH", "GET sha1:", "HTTP/1.1 503 Bus"
- MP2P
  "MD5", "GO!!", "SIZ", "STR"
- GoBoogy/Soribada
  "GOT\x0d\x0aPro", "goboogy", "boogy", "GET /gethashinfo", "GET /getup-downin", "GET /peer", "GET /queue", "GET /?p2pmethod="
- PeerEnabler
  "GET /.file", "GET /.sig", "CDN0/0", "CL" && \x00 (4)
- GoToMyPc
  "GET /jedi?request", "GET /1?" && "=" (14), \x4d\x01\x00\x00 (plen=4)
- SSH
  "SSH"
- WEB
  "GET /", "POST", "HEAD", "HTTP/1.", "SEARCH", "PROPFIND", "HTTP", "GET"
- FTP
  "CWD", "PASV", "PORT", "200 PORT", "PWD\x0d\x0a", "250 OK. Current", "221" && bye\x2e\x0d\x0a" (end of payload), "220" && \x2e\x0d\x0a (end of payload)
- SMB
  "\xffSMB" (4), \x81\x00\x00\x44\x20\x43\x4b\x46 , \x82\x00\x00\x00 (plen=4), "No listen", "no tcp", \x2a (plen=1), "rctcpo", \x83\x00\x00\x01 (plen=5)
- nntp
  "CHECK ¡", "TAKETHIS ¡", "check ¡", "takethis ¡", "LISTGROUP", "ARTICLE", "\x0d\x0a=ybegin", ("MODE" || "mode") && ("READER" (5) || "STREAM" (5) || "reader" (5) || "stream" (5))
- mail
  "354 Enter mail", "250" && ("OK" (4) || "Ok" (4) || "ok" (4) || "sender" (4) || "recip-ient" (4)), "MAIL", "DATA\x0d\x0a", "RSET\x0d\x0a", "EHLO", "Received:", "+OK", "RCPT TO", "STAT\x0d\x0a" "* OK", "DONE\x0d\x0a", "* STATUS", "* FLAGS", "INBOX\x0d\x0a", "completed\x0d\x0a" (end of payload)
- ssh/ssl
  \x80 && plen-2 = (1) (\x03\x00 (1) || \x03\x01 (1)) && (\x14 || \x15 || \x16 || \x17) (within the first 5 packets of the flow only)
- chat

"PNG\x0d\x0a" (plen¡10), "USR", "CVR" && (4) is digit, "QNG" && (4) is digit, "CHG" && (4) is digit, "NLN idL", "NLN NLN", "YMSG" && \x00 (6), \x2a\x05 && x00\x00 (4) && plen =6, \x50\x4e\x41\x00\x01 "MSG", "JOI", \x2a\x02 && ("UPDATE_BUD" (6) || ("\x00" (4) && "\x00" (6) "\x00" (8) "\x00\x00" (10)))

- irc
  "PONG", "JOIN", "NICK", ":irc", "PING", "PRIVMSG", "WHO", "WATCH", "USERHOST"

- MySQL
  \x03 (4) && ("SELECT" (5)|| "select" (5), "INSERT" (5), "show"(5), update" (5), "UPDATE"(5), "SHOW" (5), "insert" (5))

- streaming
  "MMS" (12), "RTSP/1", "_PARAME" (3), "PLAY rtsp" , "ICY", "connected\x0d\0a \x0d\0a", "\x00\x00ML20" (10)

- SpamAssasin
  "a=" && "\x26" (3) && ("g" (2) || "c" (2)) "a=" && \x0d\x0a (3) && plen=5 "sn=" && "\x26srl (4), "-nsl", "cn=razor"

- Gaming
  "\xc2\x00" (plen=5) \x02\x00 (8) && \x00\x00 (4) && \x00 (1) && ("create" (11), "play" (11), "users_" (10) "\x00\x06app_so\x00\x00\x00"

Appendix C
BLINC implementation details

The *BLINC* implementation relies on three data structures; see the discussion in Section 13.6. Structure (C.1) captures the behavior of the *graphlet*s in Figures 13.5(d)–(g), (k) and (l), is populated by the majority of the traffic. Structure (C.2) focuses on failed connections, i.e. connections without any user data, to model attack traffic using the *graphlet*s in Figures 13.5 (a)–(c). Lastly, the complicated "criss-cross" interactions in the *graphlet*s of Figures 13.5 (h)–(j) are captured using structure (C.3).

All three structures consist of dictionaries (maps) of sets. The first level of each structure is a dictionary of all IPs in the trace, capturing the behavior of a source or a destination host (if the flow statistics are collected for bidirectional traffic, we can simply look at source IPs). For each flow a specific IP participates in, the structure is updated by following the appropriate path through the protocol (TCP/UDP) and source port dictionaries as indicated by the 5-tuple of the flow. At the last level (source port), we insert values for the sets corresponding to the destination IP, the destination port, the number of packets in the flow and the average packet size (i.e. bytes/packets). Structure (C.1) is *not* updated for failed flows, which populate structure (C.2).

$$IP \begin{bmatrix} TCP \begin{bmatrix} srcPort_1 & \begin{bmatrix} \{dstIPs\}, \{dstPorts\}, \\ \{\#pkts\}, \{avg_pktsize\} \end{bmatrix} \\ \vdots \\ srcPort_S \end{bmatrix} \\ \\ UDP \end{bmatrix} \qquad (C.1)$$

Structure (C.2) is adjusted to focus on the various attack *graphlet*s. In order to capture attacks, we are more interested in the destination fields. Thus, the main difference between structures (C.1) and (C.2) lies in the second level, where, instead of the source port, we now have a dictionary of destination ports. Also, failed flows necessitate the use of TCP as a transport protocol, and thus the protocol dictionary is omitted. Address-space attacks at particular ports are defined by a large number of failed flows to different destination IPs at the same destination port. On the other hand, port scans are identified by a large dictionary of destination ports, all of which have one (or more in case of multiple scans at the same time) and the same entry at the destination IP set. Finally, we use counter *#not_failed* to count the number of "payload" flows that may satisfy the criteria imposed by structure (C.1); if this number is above a specific threshold (4 in our experiments) the flow is deemed as non-attack traffic. Note that the structure is more efficient in detecting address space scans than port scans, since address scans appear more often in the analyzed traces:

$$IP \begin{bmatrix} dstPort_1 \\ \vdots \\ dstPort_P \end{bmatrix} \begin{bmatrix} \{dstIPs\}, \ int \ \#not\_failed \end{bmatrix} \tag{C.2}$$

Structure (C.3) captures *graphlets* with "criss-cross" behavior (*mail, ftp, streaming*). The structure stores interactions between source and destination IPs that communicate with more than one flow at different source and different destination ports. More specifically, for a flow to update the structure, the following must be true: the source port *does not* exist in the srcPorts set *and* the destination port *does not* exist in the dst-Ports set for the specific source–destination IP pair. As a result, *web* interactions will be excluded from this structure since one of the ports (source or destination) will constantly be unique (and possibly equal to 80). Note that this structure is an approximation of the "criss-cross" *graphlets*:

$$IP \begin{bmatrix} dstIP_1 \\ \vdots \\ dstIP_D \end{bmatrix} \begin{bmatrix} \{srcPorts\}, \{dstPorts\}, \{Proto\} \end{bmatrix} \tag{C.3}$$

In order to find the specific application form structure (C.3), we examine, for each source IP, the histograms of two lists that result from the union of (a) the *srcPorts* sets across all destination IPs (*srcPortsList*) and (b) the *dstPorts* sets across all destination IPs (*dstPortList*). These histograms will reveal source or destination ports that are commonly used for the specific IP when communicating with more than one flow with the same destination IPs at different source and destination ports. If ports are used in a random fashion, the histograms will have no peaks. Then, the following are true according to our *graphlets*:

- if there exist one or more peaks at the (*srcPortsList*) and one peak at the (*dstPortsList*) and only TCP is used for the peak ports, the IP is a *mail* server;
- if there exist one peak at the (*srcPortsList*) and one peak at the (*dstPortsList*) and both TCP and UDP are used for the peak ports, the IP is a *streaming* server;
- if there exist two peaks at the (*srcPortsList*) and no peak at the (*dstPortsList*) and only TCP is used for the peak ports, the IP is an *ftp* server.

Once the structures are populated after a first pass through the flow table, we simply traverse through all the rules and heuristics, starting from the least to the most specific. Note that, while the methodology does not incorporate timing in a direct way, it is incorporated indirectly by the time granularity at which the flow table is formed. All the structures may store information acquired during several time intervals. However, all entries of the structures are coupled with a timer value, which indicates the last time they were active, i.e. the last time the corresponding fields were seen in a flow. Entries inactive for large time intervals (sufficiently larger than the time corresponding to an iteration) are deleted from the structures. Deleting inactive entries both prevents memory saturation (note that we are dealing with millions of {IP, port} pairs) and speeds up processing.

Box C.1. *BLINC* pseudo-code

(1) **procedure** *BLINC*_FLOW_MAPPING
(2)   FT ← *Flow Table*
(3)   **for** all flows in FT **do**
(4)   check_attack   ▷ structure 2
(5)   **if** found **then** get next flow;
(6)   check_multiple_flows   ▷ struct3, ftp/mail/streaming
(7)   **if** found **then** get next flow;
(8)   **for** all entries in structure 1: **do**
(9)   **if** IP is server **then**   ▷ (uses one port)
(10)   **if** Is protocol TCP **then**
(11)   check_fanout_heuristic.
(12)   **if** dstPorts.size <= dstIps.size **then**
(13)   return chat; get next flow;
(14)   **else if** dstPorts.size > dstIps.size **then**
(15)   check_packet_size_heuristic
(16)   **if** pkts across flows constant AND
                                    dstPorts.size>>dstIps.size **then**
(17)   return SpamAs; get next flow;
(18)   **else**
(19)   return web; get next flow;
(20)   **else**
(21)   return NM; get next flow;
(22)   **if** IP uses same source port for TCP and UDP and not port 53 **then**
(23)   return P2P; get next flow;
(24)   **for** each srcport **do**
(25)   check_community_heuristic.
(26)   **if** found **then** get next flow;
(27)   check_cardinality_heuristic.
(28)   **if** dstPorts.size == dstIps.size **then**
(29)   return P2P; get next flow;
(30)   **else if** dstPorts.size > dstIps.size **then**
(31)   **if** Is protocol TCP **then**
(32)   return web; get next flow;
(33)   **else**
(34)   check_packet_size_heuristic
(35)   **if** pkt size across flows constant AND dstPorts.size >>
                                    dstIps.size **then**
(36)   return game; get next flow;
(37)   **else**
(38)   return NM; get next flow;
(39)   **End For**   ▷ src port
(40)   **End For**   ▷ entry in struct 1
(41)   **End For**   ▷ FT

To avoid processing of already classified servers or known {IP, port} pairs, we perform, at the end of each iteration, two different actions. First, if the classification agrees with the observed port number (for known services), we store the specific {IP, port} pair in a list with known pairs. Hence, flows containing known pairs in successive intervals will be automatically classified. Second, we apply the *recursive detection* heuristic. The heuristic moves into the known list: (a) {IP, port} pairs that talk to SpamAssasin servers (*mail*); (b) the destination {IP, port} pair of a *mail* server, when this *mail* server is the source IP and its service port appears as the destination port (a *mail* server that connected to another *mail* server); (c) similarly, the destination {IP, port} pair of a *DNS* (or NM) server when its service port appears as the destination port; (d) {IP, port} pairs communicating with known *gaming*-classified {IP, port} pairs.

For completeness, we include the pseudo-code for the mapping stage of *BLINC* in Box C.1.

Appendix D
Validation of direction-conforming rule

Due to the challenges of obtaining an AS topology annotated with AS relationships, it is infeasible to use the valley-free rule to identify redistribution path spoofing in the work. Alternatively, we apply the direction-conforming rule to the AS topology annotated with directed AS-links to carry out the detection. The following theorems show that the direction-conforming rule actually shows roughly equivalent efficiency.

THEOREM D.1
For an observer AS, a valley-free path in the AS topology annotated with AS relationships must be "direction-conforming" in the corresponding AS topology annotated with inferred directed AS-links.

THEOREM D.2
(1) For a Tier-1 AS, the direction-conforming paths in the AS topology annotated with inferred directed AS-links must be valley-free in the real AS topology annotated with AS relationships.
(2) For a non-Tier-1 AS, except the redistribution path-spoofing paths launched by the provider ASs, the direction-conforming paths must be valley-free.

In order to prove these theorems, we first investigate the mapping between the real AS topology annotated with AS relationships and the inferred AS topology annotated with directed AS-links.

Note that, similar to the analysis in the text, we assume that the inferred topology is "ideally" complete, namely it contains all legitimate directed AS-links that the observer AS should see. In order to infer a complete AS topology comprising of directed AS-links based on the route announcements from the observer AS, we assume an ideal inference scenario, in which the AS connections and relationships do not change over the inference period and every AS tries all possible valid routes.

As shown in Figure D.1, if we assume that the real AS topology annotated with AS relationships is shown in part (a) and the observer AS is AS A, then the detection system can capture the corresponding topology annotated with directed AS-links shown in part (b). The rules of mapping between a link annotated with AS relationships and a directed AS-link are as follows.

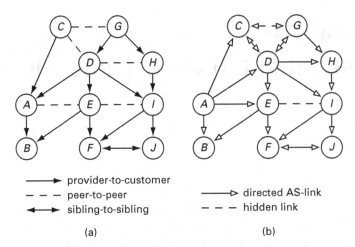

provider-to-customer
- - - peer-to-peer
sibling-to-sibling

directed AS-link
- - - hidden link

(a)   (b)

Fig. D.1.   (a) Real AS topology annotated with AS relationships. (b) Inferred AS topology annotated with AS-link from the perspective of observer AS A.

Adjacent links

All links adjacent to the observer AS, e.g. AB, AD and AE, irrespective of their AS relationships with A, are inferred as directed AS-links with the observer AS as upstream AS because the observer AS cannot see AS-paths going from its neighbors back to itself due to AS-path loop prevention.

Provider AS links

The *provider ASs* of the observer AS refer to either the direct or indirect providers of the observer AS. If any two of them are neighbors, they share a bidirectional AS-link irrespective of the type of AS relationship they have. The reason is that the topology is complete and the observer AS can see the links between its provider ASs in either directions. The example links are CD, DG and CG.

Other links

All links other than the ones belonging to the two families above can be either peer-to-peer, provider-to-customer or sibling-to-sibling links.

Peer-to-peer links

The peer-to-peer links between the provider ASs and non-provider ASs of the observer AS are directed AS-links in which the provider ASs are the upstream ASs. The reason is that, although these peer-to-peer links are visible to the observer AS, they are visible in the direction from provider ASs to non-provider ASs only. The deployed AS cannot see the AS-paths going in the other direction. Otherwise, the relevant non-provider ASs would be the provider AS of the deployed AS, since a peer-to-peer link has to follow a customer-to-provider link. DH is an example link.

Meanwhile, the peer-to-peer links between non-provider ASs of the observer AS are absent from the topology. The reason is that the routes that traverse these links should not be visible to the observer AS since they violate the "valley-free" rule. For example, the peer-to-peer link between E and I is invisible in the directed AS-link topology.

Provider-to-customer links

The provider-to-customer links are unidirectional links with the provider as the upstream AS and the customer as the downstream AS. Example links are HI and IJ.

Sibling links

All the sibling-to-sibling links are bidirectional; link FJ is an example. Given the above mapping between the links annotated with AS relationships and that with directed AS-links, we can prove Theorem D.1.

There are mainly three types of "valley-free" paths [82]. (1) Downhill paths: all the AS-links from the observer AS until the origin AS are provider-to-customer links. (2) Uphill–downhill paths: the path begins at the observer AS with customer-to-provider links then follows the provider-to-customer links and ends at the origin AS. (3) Uphill–plateau–downhill paths: the path is similar to on uphill–downhill path except that there is one, and only one, peer-to-peer link, that between the uphill and downhill parts.

Given the above mapping of links, it is trivial to show that all three types of valid paths must follow the direction-conforming rule. However, the direction-conforming path in an AS topology annotated with directed AS-links might not be valley-free. The exceptions are for the providers of the observer AS, because the directed AS-links between the provider ASs are bidirectional. If one of them redistributes the path between its providers or peers that are also the provider ASs of the observer AS, such routes are direction-conforming but not valley-free. For example, for observer AS A in Figure D.1, path $\{A, C, D, G, H, I, J\}$ is direction-conforming but not valley-free. Formally, we have the following lemma.

Lemma *Given a AS topology annotated with inferred directed AS-links of an observer AS, among all the direction-conforming paths of the observer ASs, except those caused by the redistribution path spoofing caused by the provider ASs of the observer ASs, other direction-conforming paths must be valley-free.*

We prove the lemma by contradiction. Assume a path is not valley-free, namely it is a redistribution path-spoofing path launched by a non-provider AS of the observer AS, say H in Figure D.1. Assume that H redistributes a path from a provider or a peer, say X, to another provider or peer. Then the path contains a link $H \rightarrow X$. According to the mapping rules, were X a peer of H, either directed AS-link $H \rightarrow X$ is absent or only $X \rightarrow H$ is present, e.g. $D \rightarrow H$. Were X a provider of H, only $X \rightarrow H$ would be present, e.g. $G \rightarrow H$. Accordingly, the path is not direction-conforming.

Because a Tier-1 AS has no provider AS, all its direction-conforming paths must be valid, which proves Theorem D.2.

References

1 http://www.ilog.com/products/cplex/.

2 http://www.planet-lab.org/.

3 http://www.packetizer.com/1pmc/h323/standards.html.

4 Anatomy of a leak: AS9121 (or How we learned to start worrying and hate maximum prefix limits), http://nanog.org/mtg-0505/underwood.html.

5 Autonomous system numbers, http://www.iana.org/assignments/as-numbers.

6 Bro, http://bro-ids.org/.

7 Con-Ed steals the 'Net, http://www.renesys.com/blog/2006/01/ coned_steals_the_net.shtml.

8 Dynamic proramming, http://www.ics.uci.edu/ dan/class/161/notes/6/Dynamic.html.

9 FBI busts alleged DDoS mafia, http://www.securityfocus.com/news/9411.

10 Googfle/Sprint having problems?, http://www.cctec.com/maillists/nanog/current/msg05514.html.

11 Pretty Good BGP and the Internet Alert Registry, http://www.nanog.org/mtg-0606/karlin.html.

12 Razor, http://razor.sourceforge.net/.

13 Ripe route information service, http://www.ripe.net/ris/index.html.

14 Route views project page, http://www.routeviews.org/

15 SNORT, http:http://www.snort.org/.

16 Wow, as7007!, http://www.merit.edu/mail.archives/nanog/1997-04/msg00340.html.

17 www.google.com latency/packet loss/very slow thru savvis, http://www.merit.edu/mail.archives/nanog/2005-12/msg00219.html.

18 APNIC, *Policies for IPv4 Address Space Management in the Asia Pacific Region*, Technical Report APNIC-086, December, 2005.

19 R. Addie, P. Mannersalo and I. Norros, Performance formulae for queues with Gaussian input. *Proceedings of 16th International Teletraffic Congress*, Edinburgh, UK, June, 1999, pp. 1169–1178.

20 B. Aiello, C. Kalmanek, P. McDaniel, S. Sen, O. Spatscheck and J. Van der Merwe, Analysis of communities of interest in data networks. *Passive and Active Network Measurement Workshop*, Boston, MA, March, 2005.

21 G. Almes, S. Kalidindi and M. Zekauskas, A one-way delay metric for IPPM. RFC 2679, September, 1999.

22 G. Almes, S. Kalidindi and M. Zekauskas, A one-way packet loss metric for IPPM. RFC 2680, September, 1999.

23 G. Almes, S. Kalidindi and M. Zekauskas, A round-trip delay metric for IPPM. RFC 2681, September, 1999.

24 G. Appenzeller, I. Keslassy and N. McKeown, Sizing router buffers. *ACM Sigcomm*, Portland, OR, August/September, 2004.

25 J. Apsidorf, K.C. Claffy, K. Thompson and R. Wilder, OC3MON: flexible, affordable, high performance statistics collection. *10th USENIX Conference on System Administration*, Chicago, IL, September/October, 2004.

26 J. Armitage, O. Crochat and J. Y. Le Boudec, Design of a survivable WDM photonic network. *IEEE Infocom*, Kobe, Japan, April, 1997.

27 V. Auletta, I. Caragiannis and P. Persiano, Randomized path coloring on binary trees. *3rd International Workshop on Approximation Algorihms for Combinatorial Optimization Problems (APPROX)*, Saarbrucken, Germany, September, 2000.

28 A. Aussem and F. Murtagh, Web traffic demand forecasting using wavelet-based multiscale decomposition. *International Journal of Intelligent Systems*, **16** (2001), 215–236.

29 M. Bailey, E. Cooke, F. Jahanian, D. Watson and J. Nazario, The blaster worm: then and now. *IEEE Security and Privacy* **3**:4 (2005), 26–31.

30 F. Baker and R. Coltun, OSPF Version 2 Management Information Base. RFC 1850, November, 1995.

31 P. Barford, J. Kline, D. Plonka and A. Ron, A signal analysis of network traffic anomalies. *ACM Sigcomm Workshop on Internet Measurement*, Marseille, France, November, 2002.

32 S. Basu and A. Mukherjee, Time series models for Internet traffic. *24th Conference on Local Computer Networks*, Lowell, MA, October, 1999.

33 T. Bates, R. Chandra and E. Chen, BGP route reflection: an alternative to full mesh IBGP. RFC 2796, April, 2000.

34 G. di Battista, M. Patrignani and M. Pizzonia, Computing the types of the relationships between autonomous systems. *IEEE Infocom*, San Francisco, CA, 2003.

35 B. Beauquier, J-C. Bermond, L. Gargano, P. Hell, S. Perennes and U. Vaccar, Graph problems arising from wavelength-routing in all-optical networks. *2nd IEEE Workshop on Optics and Computer Science (IPP'97)*, Genova, Switzerland, April, 1997.

36 L. Bernaille, R. Teixeira and K. Salamatian, Early application identification ACM. *CoNEXT*, Lisbon, Portugal, December, 2006.

37 S. Bhattacharyya, C. Diot, J. Jetcheva and N. Taft, POP-level and access-link-level traffic dynamics in a Tier-1 POP. *ACM Internet Measurement Workshop*, San Francisco, CA, November, 2001.

38 D. Bienstock and O. Gueluek, Computational experience with a difficult mixed-integer multicommodity flow problem. *Mathematical Programming*, **68** (1995), 213–237.

39 J. Bolot and P. Hoschka, Performance engineering of the World Wide Web: application to dimensioning and cache design. *5th International World Wide Web Conference*, Paris, France, May, 1996.

40 P. Brockwell and R. Davis, *Introduction to Time Series and Forecasting* (New York: Springer, 1996).

41 N. Brownlee, C. Mills and G. Ruth, Traffic flow measurement: architecture. RFC 2722, October, 1999.

42 J. Cao, D. Davis, S. Vander Weil and B. Yu, Time-varying network tomography. *Journal of the American Statistical Association*, **95**:452 (2000), 1063–1075.

43 J. Case, J. Davin, M. Fedor and M. Schoffstall, Simple Gateway Monitoring Protocol. RFC 1028, November, 1987.

44 J. Case, K. McCloghrie, M. Rose and S. Waldbusser, Structure of management information for version 2 of the Simple Network Management Protocol (SNMPv2). RFC 1902, January, 1996.

45 J. Case, R. Mundy, D. Partain and B. Stewart, Introduction to version 3 of the Internet-standard Network Management Framework. RFC 2570, April, 1999.

46 F. Castaneda, E.C. Sezer and J. Xu, Worm vs. worm: preliminary study of an active counter-attack mechanism. *Workshop on Rapid Malcode*, Fairfax, VA, October, 2004.

47 CERT Advisory CA-1999-04, *Melissa macro virus*, http://www.cert.org/advisories/ca-1999-04.html.

48 CERT Advisory CA-2000-04, *Love letter worm*, http://www.cert.org/advisories/ca-2000-04.html.

49 CERT Advisory CA-2002-25, *Integer overflow in XDR library*, http://www.cert.org/advisories/ca-2002-25.html.

50 D. Chakrabarti, S. Papadimitriou, D. Modha and C. Faloutsos, Fully automatic cross-associations. *ACM SIGKDD Conference on Knowledge Discovery and Data Mining*, Seattle, WA, August, 2004.

51 H. Chan, D. Dash, A. Perrig and H. Zhang, Modeling adoptability of secure BGP protocols. *ACM Sigcomm*, Pisa, Italy, September, 2006.

52 K. Chandra, C. You, G. Olowoyeye and C. Thompson, *Non-Linear Time-Series Models of Ethernet Traffic*. Technical report, CACT, June, 1998.

53 D.-F. Chang, R. Govindan and J.S. Heidemann, The temporal and topological characteristics of BGP path changes. *IEEE ICNP*, Atlanta, GA, November, 2003.

54 H. Chang, R. Govindan, S. Jamin, S. J. Shenker and W. Willinger, Towards capturing representative AS-level internet topologies. *Computer Networks*, **44**:6 (2004), 737–755.

55 I. Chlamtac, A. Ganz and G. Karmi, Lightpath communications: an approach to high bandwidth optical WANs. *IEEE/ACM Transactions on Communications*, **40**:7 (1992), 1171–1182.

56 K. Claffy, H.-W. Braun and G. Polyzos, A parametrizable methodology for Internet traffic flow profiling. *IEEE Journal of Selected Areas in Communications*, **13**:8 (1995), 1481–1494.

57 D. Clarke, The design philosophy of the DARPA Internet protocols. *ACM Sigcomm*, Stanford, CA, August, 1988.

58 Cisco Systems, Cisco IOS Netflow Version 9 Flow-Record Format, June, 2003.

59 Cisco Systems, NetFlow services and applications. http://www.cisco.com/warp/public/cc/pd/iosw/ioft/neflct/tech/napps_wp.htm. White paper.

60 Cisco Systems, Optimizing for network intra-pop interconnections with very short reach interfaces. White paper.

61 O. Crochat and J. Y. Le Boudec, Design protection for WDM optical networks. *IEEE Journal on Selected Areas in Communications*, **16**:7 (1998), 1158–1165.

62 O. Crochat, J. Y. Le Boudec and O. Gerstel, Protection interoperability for WDM optical networks. *IEEE Transactions on Networking*, **8**:3 (2000), 384–395.

63 M. Crovella and B. Krishnamurthy, *Internet Measurement: Infrastructure, Traffic and Applications*. (John Wiley & Sons Inc., 2006).

64 Dag 4 SONET network interface, http://dag.cs.waikato.ac.nz/dag/dag4-arch.html.

65 I. Daubechies, *Ten Lectures on Wavelets* (*CBMS–NSF Regional Conference Series in Applied Mathematics*) (Philadelphia, PA: SIAM, 1992).

66 C. Dewes, A. Wichmann and A. Feldmann, An analysis of Internet chat systems. *ACM Internet Measurement Conference*, Miami, FL, October, 2003.

67 S. Donnelly, *High precision timing in passive measurements of data networks*. Unpublished Ph.D. thesis, University of Waikato (2002).

68 F. Drake. Documenting Python Release 2.2.1, http://www.python.org/doc/current/download.html, April, 2002.

69 N. Duffield, C. Lund and M. Thorup, Properties and prediction of flow statistics from sampled packet streams. *ACM Internet Measurement Workshop*, Marseilles, France, November, 2002.

70 N. Duffield, C. Lund and M. Thorup, Estimating flow distributions from sampled flow statistics. *IEEE/ACM Transactions on Networking*, **13**:5 (2005), 933–946.

71 M. Elf, C. Gutwenger, M. Junger and G. Rinaldi, *Branch-and-Cut Algorithms for Combinatorial Optimization and Their Implementation in ABACUS*. Lecture Notes in Computer Science Vol. 2241 (London: Springer).

72 M. Enachescu, Y. Ganjali, A. Goel, T. Roughgarden and N. McKeown, Part III: Routers with very small buffers. *ACM Sigcomm Computer Communication Review*, **35**:3 (2005), 83–90.

73 Eric J. Harder, Sang-Kyu Lee and Hyeong-Ah Choi, On wavelength assignment in WDM optical networks. *4th International Conference on Massively Parallel Processing Using Optical Interconnections (MPPOI)*, Montreal, Canada, June, 1997.

74 T. Erlebach, K. Jansen, C. Kakalamanis, M. Mihail and P. Persiano, Optimal wavelength routing on directed fiber trees. *Theoretical Computer Science*, **221**:1–2.

75 C. Estan, S. Savage and G. Varghese, Automatically inferring patterns of resource consumption in network traffic. *ACM Sigcomm*, Karlsruhe, Germany, August, 2003.

76 N. Feamster, J. Jung and H. Balakrishnan, An empirical study of "Bogon" route advertisements. *ACM Sigcomm Computer Communication Review*, **35**:1 (2005), 63–70.

77 A. Feldmann, A. Greenberg, C. Lund, N. Reingold, J. Rexford and F. True, Deriving traffic demands for operational IP networks: methodology and experience. *IEEE/ACM Transactions on Networking*, **9**:3 (2001), 265–280.

78 Security Focus, Macos x software update arbitrary package installation vulnerability, http://online.securityfocus.com/bid/5176.

79 B. Fortz and M. Thorup, Internet traffic engineering by optimizing OSPF weights. *IEEE Infocom*, Tel Aviv, Israel, March, 2000.

80 B. Fortz and M. Thorup, Optimizing OSPF/IS-IS weights in a changing world, *IEEE Journal on Selected Areas in Communications. Special Issue on Advances in Fundamentals of Network Engineering*, **20**:4 (2002), 756–767.

81 C. Fraleigh, C. Diot, B. Lyles, S. Moon, P. Owezarski, K. Papagiannaki and F. Tobagi, Design and deployment of a passive monitoring infrastructure. *Passive and Active Measurement Workshop*, Amsterdam, The Netherlands, April, 2001.

82 L. Gao, On inferring autonomous system relationships in the Internet. *IEEE/ACM Transactions on Networking*, **9**:6 (2001), 733–745.

83 L. Gao, On inferring automonous system relationships in the Internet. *IEEE Infocom*, Tel Aviv, Israel, November, 2000.

84 M. Geral W. Charles, M. Fabian, Hmm profiles for network traffic classification (extended abstract). In *In Proceedings of the 2004 ACM workshop on Visualization and data mining for computer security*, Washington, DC, USA.

85 F. Giroire, A. Nucci, N. Taft and C. Diot, Increasing the robustness of IP backbones in the absence of optical level protection. *IEEE Infocom*, San Francisco, CA, March, 2003.

86 F. Glover and M. Laguna, *Tabu Search* (Dordrecht: Kluwer Academic Publishers).

87 F. Glover, E. Taillard and D. De Werra, A user's guide to Tabu Search. *Annals of Operations Research*, **41**:1–4 (1993), 3–28.

88 R. A. Golding, *End-to-End Performance Prediction for the Internet*, Technical Report UCSC-CRL-92-96, CISB, University of California, Santa Cruz, June, 1992.

89 N. K. Groschwitz and G. C. Polyzos, A time series model of long-term NSFNET backbone traffic. *IEEE International Conference on Communications*, New Orleans, LA, May, 1994.

90 A. Gunnar, M. Johansson and T. Telkamp, Traffic matrix estimation on a large IP backbone – a comparison on real data. *ACM Internet Measurement Conference*, Taormina, Italy, October, 2004.

91 S. Handelman, S. Stibler, N. Brownlee and G. Ruth, RTFM: new attributes for traffic flow measurement. RFC 2724, October, 1999.

92 J. Harmatos, A heuristic algorithm for solving the static weight assignment optimisation problem in OSPF networks. *Global Internet Conference*, Dallas, TX, November, 2001.

93 F. Hernandez-Campos, A. B. Nobel, F. D. Smith and K. Jeffay, *Statistical Clustering of Internet Communication Patterns Symposium on the Interface of Computing Science and Statistics*, Salt Lake City, UT, July, 2003.

94 D. S. Hochbaum. Approximation Algorithms for NP-Hard Problems (Boston, MA: PWS Publishers, 1996).

95 N. Hohn, D. Veitch and P. Abry, Cluster processes, a natural language for network traffic. *IEEE Transactions on Signal Processing, Special Issue on Signal Processing in Networking*, **51**:8 (2003) 2229–2244.

96 W. Hong, Hybrid models for representation of imagery data. Unpublished Ph.D. thesis, University of Urbana, IL (2006).

97 C. Huitema, *Routing in the Internet* (Englewood Cliffs, NJ: Prentice Hall PTR), 2000.

98 G. Huston, *ISP Survival Guide: Strategies for Running a Competitive ISP* (New York: John Wiley & Sons Inc., 1998).

99 G. Huston, Analyzing the Internet's BGP routing table. *The Internet Protocol Journal*, **4**:1 (2001).

100 G. Iannaccone, C. Chuah, R. Mortier, S. Bhattacharyya and C. Diot, Analysis of link failures in a large IP backbone. *ACM Internet Measurement Workshop (IMW)*, Marseille, France, November, 2002.

101 R. Jain, *The Art of Computer Systems Performance Analysis: Techniques for Experimental Design, Measurement, Simulation, and Modeling* (New York: John Wiley, 1991).

102 T. Johnson, S. Muthukrishnan, V. Shkapenyuk and O. Spatscheck, A heartbeat mechanism and its application in gigascope. *Very Large Data Bases (VLDB)*, Trondheim, Norway, August, 2005.

103 T. Karagiannis, A. Broido, N. Brownlee, K. Claffy and M. Faloutsos, Is P2P dying or just hiding? *IEEE Globecom*, Dallas, TX, November, 2004.

104 T. Karagiannis, A. Broido, M. Faloutsos and K. Claffy, Transport layer identification of P2P traffic. *ACM Internet Measurement Conference*, Taormina, Italy, October, 2004.

105 T. Karagiannis, K. Papagiannaki and M. Faloutsos, BLINC: multilevel traffic classification in the dark. *ACM Sigcomm*, Philadelphia, PA, August, 2005.

106 V. Karamcheti, D. Geiger and Z. Kedem, Detecting malicious network traffic using inverse distributions of packet contents. *ACM Sigcomm Workshop on Mining Network Data (MiNet)*, Philadelphia, PA, 2005.

107 J. Karlin, A fun hijack: 1/8, 2/8, 3/8, 4/8, 5/8, 7/8, 8/8, 12/8 briefly announced by AS 23520 (today), http://www.merit.edu/mail.archives/nanog/2006-06/msg00082.html.

108 J. Karlin, S. Forrest and J. Rexford, Pretty good BGP: improving BGP by cautiously adopting routes. *IEEE ICNP*, Santa Barbara, CA, November, 2006.

109 S. Kent, C. Lynn and K. Seo, Secure Border Gateway Protocol (Secure-BGP). *IEEE Journal of Selected Areas in Communications*, **18**:4 (2000), 582–592.

110 B.W. Kernighan and D.M. Ritchie, *The C programming language,* 2nd edn (Englewood Cliffs, NJ: Prentice Hall, 1988).

111 S. Keshav and R. Sharma, Issues and trends in router design. *IEEE Communications Magazine*, **36**:5 (1998), 144–151.

112 H.-A. Kim and B. Karp, Autograph: toward automated, distributed worm signature detection. *Proceedings of the 13th USENIX Symposium*, San Diego, CA, 2004.

113 B. Krishnamurthy, S. Sen, Y. Zhang and Y. Chen, Sketch-based change detection: methods, evaluation, and applications. *ACM Sigcomm Internet Measurement Conference*, Miami, FL, October, 2003.

114 C. Krügel, D. Mutz, W. K. Robertson and F. Valeur, Topology-based detection of anomalous BGP messages. *Proceedings of the 6th International on Recent Advances in Intrusion Detection*, Pittsburgh, PA, September, 2003.

115 V. Kumar, Approximating arc circular colouring and bandwidth allocation in all-optical ring networks. *First International Workshop on Approximation Algorithms for Combinatorial Optimization Problems (APPROX)*, Aalborg, Denmark, July, 1998.

116 M. Lad, D. Pei, D. Massey, Y. Wu, B. Zhang and L. Zhang, PHAS: a prefix hijack Alert System. *15th USENIX Security Symposium*, Vancouver, Canada, July 2006.

117 A. Lakhina, M. Crovella and C. Diot, Mining anomalies using traffic feature distributions. *ACM Sigcomm*, Pittsburgh, PA, August, 2005.

118 A. Lakhina, K. Papagiannaki, M. Crovella, C. Diot, E. Kolacyzk and N. Taft, Structural analysis of network traffic flows. *ACM Sigmetrics*, New York, June, 2004.

119 A. Lazarevic, L. Ertoz, V. Kumar, A. Ozgur and J. Srivastava, A comparative study of anomaly detection schemes in network intrusion detection. *Third SIAM International Conference on Data Mining*, San Francisco, CA, May, 2003.

120 H. Leijon. *Basic Forecasting Theories: A Brief Introduction.* Technical report, ITU, November, 1998.

121 E. Leonardi, M. Mellia and M. Ajmone Marsan, Algorithms for the topology design in WDM all-optical networks. *Optical Networks Magazine, Premiere Issue*, **1**:1 (2000), 35–46.

122 G. Liang and B. Yu, Pseudo likelihood estimation in network tomography. *IEEE Infocom*, San Francisco, CA, March, 2003.

123 L. Liv, X. Li, P. Wan and O. Frieder. Wavelength assignment in WDM rings to minimize sonet ADMs, *IEEE Infocom*, Tel Aviv, Israel, March, 2000.

124 O. Spatscheck, M. Roughan, S. Sen and N. Duffield. Class-of-service mapping for qos: A statistical signature-based approach to IP traffic classification). In *ACM Sigcomm Internet Measurement Conference*, Taormina, Italy, October, 2004.

125 D. MacKay, *Information Theory, Inference & Learning Algorithms* (Cambridge: Cambridge University Press, 2002).

126 K. McCloghrie and M. Rose. Management information base for network management of TCP/IP-based Internets: MIB-II. RFC 1213, March, 1991.

127 A.J. McGregor, J.P. Curtis, J.G. Cleary and M.W. Pearson. Measurement of Voice over IP Traffic. *Passive and Active Measurement*, Hamilton, New Zealand.

128 A. McGregor, M. Hall, P. Lorier and J. Brunskill, Low clustering using machine learning techniques. *Passive and Active Measurement*, Antibes Juan-les-Pins, France, March, 2004.

129 N. McKeown, A fast switched backplane for a gigabit switched router. *Business Communications Review*, **27**:12 (1997).

130 J. McQuilan, G. Falk and I. Richer, A review of the development and performance of the ARPANET routing algorithm. *IEEE/ACM Transactions on Communications*, **26**:12 (1978), 1802–1811.

131 B. McWilliams, Yaha worm takes out Pakistan Government's site. http://online. securityfocus.com/news/501.

132 J. Mahdavi and V. Paxson. IPPM Metrics for Measuring Connectivity. RFC 2678, September, 1999.

133 S. Mallat, A theory for multiresolution signal decomposition: the wavelet representation. *IEEE Transactions on Pattern Analysis and Machine Intelligence*, **11**, 674–693.

134 A. Markopoulou, G. Iannaccone, S. Bhattacharyya, C. N. Chuah and C. Diot, Characterization of failures in an IP backbone network. *IEEE Infocom*, Hong Kong, China, March, 2004.

135 M. Mathis and M. Allman, A framework for defining empirical bulk transfer capacity metrics. RFC 3148, July, 2001.

136 A. Medina, C. Fraleigh, N. Taft, S. Bhattacharyya and C. Diot, A taxonomy of IP traffic matrices. *SPIE ITCOM Conference: Workshop on Scalability and Traffic Control in IP Networks II* vol. 4868, Boston, MA, June, 2002.

137 A. Medina, N. Taft, K. Salamatian, S. Bhattacharyya and C. Diot, Traffic matrix estimation: existing techniques and new directions. *ACM Sigcomm*, Pittsburgh, PA, August, 2002.

138 M. Meo, D. Rossi, D. Bonfiglio, M. Mellia and P. Tofanelli, Revealing Skype Traffic: When Randomness Plays with You. *ACM Sigcomm*, Kyoto, Japan, Aug, 2007.

139 J. Mirkovic and P. Reiher, A taxonomy of DDoS attack and DDoS defense mechanisms. *ACM Sigcomm Computer Communication Review*, **34**:2 (2004), 39–53.

140 E. Modiano, Survivable routing of logical topologies in WDM networks. *IEEE Infocom*, Anchorage, AK, April, 2001.

141 A. Moore, J. Hall, C. Kreibich, E. Harris and I. Pratt, Architecture of a network monitor. *Passive and Active Measurement*, La Jolla, CA, April, 2003.

142 D. Moore, K. Keys, R. Koga, E. Lagache and K. Claffy, The CoralReef software suite as a tool for system and network administrators. *Usenix LISA*, San Diego, CA, December, 2001.

143 A. Moore and K. Papagiannaki, Toward the accurate identification of network applications. *Passive and Active Measurement*, Boston, MA, March, 2005.

144 D. Moore, V. Paxon, S. Savage, C. Shannon, S. Staniford and N. Weaver, The spread of the sapphire/slammer worm. (http://www.caida.org/publications/papers/2003/sapphire/sapphire.html)

145 D. Moore, C. Shannon and K. Claffy, Code-Red: a case study on the spread and victims of an internet worm, *ACM Internet Measurement Workshop*, Marseille, France, November, 2002.

146 A. Moore and D. Zuev, Internet traffic classification using Bayesian analysis techniques. *ACM Sigmetrics*, Banff, Canada, June, 2005.

147 J. Moy, OSPF Version 2. RFC 2328, April, 1998.

148 B. Mukherjee, D. Banerjee, S. Ramamurthy and A. Mukherjee, Some principles for designing a wide-area WDM optical network. *IEEE/ACM Transactions on Networking*, **4**:5 (1996), 684–695.

149 G. Nason and B. Silverman, The stationary wavelet transform and some statistical applications. In *Lecture Notes in Statistics: Wavelets and Statistics*, vol. 103 (New York: Springer, 1995), pp. 281–299.

150 J. Newsome, B. Karp and D. Song, Polygraph: automatically generating signatures for poly-morphic worms. *Proceedings of IEEE Security and Privacy Symposium*, Oakland, CA, May, 2005.

151 A. Nucci, R. Cruz, N. Taft and C. Diot, Design of IGP link weight changes for estimation of traffic matrices. *IEEE Infocom*, Hong Kong, China, March, 2004.

152 A. Nucci, B. Sansó, T. G. Crainic, E. Leonardi and M. A. Marsan, Design of fault-tolerant logical topologies in wavelength-routed optical IP networks. *IEEE Globecom*, San Antonio, TX, November, 2001.

153 A. Nucci, N. Taft, P. Thiran, H. Zang and C. Diot, Increasing the link utilization in IP over WDM networks. *Opticom*, Boston, MA, June, 2002.

154 D. Oran. OSI IS-IS Intra-domain routing protocol. RFC 1142, February, 1990.

155 K. Papagiannaki, R. Cruz and C. Diot, Network performance monitoring at small time scales. *ACM Internet Measurement Conference*, Miami, FL, October, 2003.

156 K. Papagiannaki, S. Moon, C. Fraleigh, P. Thiran, F. Tobagi and C. Diot, Analysis of mea-sured single-hop delay from an operational backbone network. *IEEE Infocom*, New York, June, 2002.

157 K. Papagiannaki, N. Taft, S. Bhattacharyya, P. Thiran, K. Salamatian and C. Diot, A prag-matic definition of elephants in Internet backbone traffic. *ACM Internet Measurement Workshop*, Marseille, France, November, 2002.

158 K. Papagiannaki, N. Taft and A. Lakhina, A distributed approach to measure IP traffic matrices. *ACM Internet Measurement Conference*, Taormina, Italy, October, 2004.

159 K. Papagiannaki, N. Taft, Z. Zhang and C. Diot, Long-term forecasting of Internet backbone traffic: observations and initial models. *IEEE Infocom*, San Francisco, CA, April, 2003.

160 K. Park, G. Kim and M. E. Crovella, On the effect of traffic self-similarity on network per-formance. *SPIE International Conference on Performance and Control of Network Systems*, Dallas, TX, November, 1997.

161 Pastry. The evolution of the US Internet peering ecosystem. http:http://research. microsoft.com/~antr/Pastry/.

162 V. Paxson, Measurements and analysis of end-to-end Internet dynamics. Unpublished Ph.D. thesis, University of California, Berkeley, (1997).

163 M. Pioro, A. Szentesi, J. Harmatos, A. Juttner, P. Gajowniczek and S. Kozdrowski, On OSPF related network optimisation problems. *IFIP ATM IP*, Ilkley, UK, July, 2000.

164 J. Quittek, T. Zseby, B. Claise, S. Zander, G. Carle and K. Norseth, Requirements for IP flow information export. RFC 3917, October, 2004.

165 M. O. Rabin, *Fingerprinting by Random Polynomials*. Technical Report TR-15-81, Center for Research in Computing Technology, Harvard University, 1981.

166 R. Ramaswami and K. N. Sivarajan, Design of logical topologies for wavelength-routed optical networks. *IEEE Journal of Selected Areas in Communications*, **14**:6 (1996), 840–851.

167 A. Ramachandran and N. Feamster, Understanding the network-level behavior of spammers. *ACM Sigcomm*, Pisa, Italy, September, 2006.

168 S. Ranjan, R. Swaminathan, M. Uysal and E. Knightly, DDoS-resilient scheduling to counter application layer attacks under imperfect detection. *IEEE Infocom*, Barcelona, Spain, April, 2006.

169 Y. Rekhter and T. Li, A Border Gateway Protocol 4 (BGP-4). RFC 1771, March, 1995.

170 Y. Rekhter, T. Li and S. H. Ed, A Border Gateway Protocol 4 (BGP-4). RFC 4271, January, 2006.

171 J. Rexford, J. Wang, Z. Xiao and Y. Zhang, BGP routing stability of popular destinations. *ACM Internet Measurement Workshop*, Marseilles, France, November, 2002.

172 A. Sang and S. Li, A predictability analysis of network traffic. *IEEE Infocom*, Tel Aviv, Israel, March, 2000.

173 B. Sansó, F. Soumis and M. Gendreau, On the evaluation of telecommunications network reliability using routing models. *IEEE/ACM Transactions on Communications*, **39**:39 (1991), 1494–1501.

174 B. Sansó, M. Gendreau and F. Soumis, An algorithm for network dimensioning under reliability considerations. *Annals of Operation Research*, **36**:1 (1992), 263–274.

175 J. S. Sauver, Route injection and spam. http://www.uoregon.edu/ joe/maawg8.

176 M. Schoffstall, M. Fedor, J. Davin and J. Case, A simple network management protocol (SNMP). RFC 1157, May, 1990.

177 S. Sen, O. Spatscheck and D. Wang, Accurate, scalable in-network identification of P2P traffic using application signatures. *IADIS International Conference WWW Internet*, Madrid, Spain, October, 2004.

178 M. Shensa, The discrete wavelet transform: wedding the à Trous and Mallat algorithms. *IEEE Transactions on Signal Processing*, **40** (1992), 2464–2482.

179 W. Simpson, PPP in HDLC-like framing. RFC 1662, July, 1994.

180 S. Singh, C. Estan, G. Varghese and S. Savage, Automated worm fingerprinting. *Proceedings of the 6th ACM/USENIX Symposium on Operating System Design and Implementation (OSDI)*, San Francisco, CA, December, 2004.

181 V. A. Siris and F. Papagalou, Application of anomaly detection algorithms for detecting SYN flooding attacks. *IEEE Globecom*, Dallas, TX, November, 2002.

182 A. Soule, A. Nucci, E. Leonardi, R. Cruz and N. Taft, How to identify and estimate the top largest traffic matrix elements in a changing environment. *ACM Sigmetrics*, New York, June, 2004.

183 A Soule, K. Salamatian, A. Nucci and N. Taft, Traffic matrix tracking using Kalman filters. *Performance Evaluation Review*, **33**:3 (2005), 24–31.

184 J.-L. Starck and F. Murtagh, Image restoration with noise suppression using the wavelet transform. *Astronomy and Astrophysics*, **288** (1994), 342–348.

185 J. W. Steward. *BGP4: Inter-Domain Routing in the Internet*. (Reading, MA: Addison Wesley, 1999).

186 J. Stewart. Sobig.e: evolution of the worm. http://www.lurhq.com/sobig-e.html.

187 B. Stroustrup. *The C++ Programming Language*, 3rd edn (Reading, MA: Addison Wesley, 2004).

188 L. Subramanian, S. Agarwal, J. Rexford and R. Katz, Characterizing the Internet hierarchy from multiple vantage points. *IEEE Infocom*, New York, June, 2002.

189 A. Sudduth. The what, why, and how of the 1988 internet worm. http://snowplow.org/ tom/worm/worm.html.

190 Symantec. W.32 benjamin worm. http://securityresponse.symantec.com/avcenter/venc/ data/w32.benjamin.worm.html.

191 A. Tannenbaum. *Computer Networks* (Englewood Cliffs, NJ: Prentice Hall, 1985).

192 R. Teixeira, T. Griffin, A. Shaikh and G.M. Voelker, Network sensitivity to hot-potato disruptions. *ACM Sigcomm*, Portland, OR, August, 2004.

193 K. Thompson, G. Miller and R. Wilder, Wide area internet traffic patterns and characteristics. *IEEE Network*, **11**:6 (1997), 10–23.

194 Y. Vardi, Estimating source-destination traffic intensities from link data. *Journal of the American Statistical Association*, **9**:433 (1996), 365–377.

195 D. Veitch, F. Baccelli, S. Machirajiu and J. Bolot, On optimal probing for delay and loss measurement, *ACM Internet Measurement Conference*, San Diego, CA, October, 2007, pp. 291–302.

196 W. N. Venables and B. D. Ripley, *Modern Applied Statistics with S-PLUS* (New York: Springer, 1999).

197 M. Vojnovic and A. Ganesh, Reactive patching: a viable worm defense strategy? http://research.microsoft.com/ milanv/PerfTut05.pps.

198 A. Wagner and B. Plattner, Entropy based worm and anomaly detection in fast IP networks. *IEEE 14th International Workshop on Enabling Technologies: Infrastructures for Collaborative Enterprises (WET ICE), STCA Security Workshop*, Linkoping, Sweden, June, 2005.

199 S. Waldbusser, Remote network monitoring management information base. RFC 1757, February, 1995.

200 J. Walker, *A Primer on Wavelets and Their Scientific Applications* (Boca Raton, FL: Chapman & Hall, 1999).

201 Y. Wang, Z. Wang and L. Zhang, Internet traffic engineering without full mesh overlaying. *IEEE Infocom*, Anchorage, AK, April, 2001.

202 H. Wang, D. Zhang and K.G. Shin, Change-point monitoring for detection of DoS attacks. *IEEE Transactions on Dependable and Secure Computing*, **1**:4 (2004), 193–208.

203 N. Weaver, V. Paxson, S. Staniford and R. Cunningham, A taxonomy of computer worms. *Workshop on Rapid Malcode (Archive)* Washington, D.C., October, 2003.

204 S. Wehner, *Analyzing Worms and Network Traffic using Compression. Journal of Computer Security*, **15**:3 (2007), 303–320.

205 R. White, Architecture and deployment considerations for secure origin bgp (sobgp). Internet draft draft-white-sobgp-architecture-01, May, 2005.

206 S. Willis, J. Burruss and J. Chu, Definitions of managed objects for the fourth version of the Border Gateway Protocol (BGP-4) using SMIv2. RFC 1657, July, 1994.

207 D. Wischik and N. McKeown, Part I: Buffer sizes for core routers. *ACM Sigcomm Computer Communication Review*, **35**:3 (2005), 75–78.

208 R. Wolski, Dynamically forecasting network performance using the network weather service. *Journal of Cluster Computing*, **1** (1998), 119–132.

209 B. Woodcock, Best practices in IPv4 anycast routing, 2002. http://www.pch.net/resources/papers/ipv4-anycast/ipv4-anycast.ppt.

210 J. Wu, Z. Morley Mao, J. Rexford and J. Wang, Finding a needle in a haystack: pinpointing significant BGP routing changes in an IP network. *Networked Systems Design and Implementation*, Boston, MA, May, 2005.

211 K. Xu and Z. M. Mao, Accurate real-time identification of IP prefix hijacking. *IEEE Symposium on Security and Privacy*, Oakland, CA (2007), pp. 3–17.

212 K. Xu, Z. Zhang and S. Bhattacharyya, Profiling Internet backbone traffic: behavior models and applications. *ACM Sigcomm*, Philadelphia, PA, August, 2005.

213 B. Yener and T. Boult, A study of upper and lower bounds on minimum congestion routing in lightwave networks. *IEEE Infocom*, Toronto, Canada, June, 1994.

214 P. Yu, A. Goldberg and Z. Bi, Time series forecasting using wavelets with predictor-corrector boundary treatment. *7th ACM SIGKDD International Conference on Knowledge Discovery and Data Mining*, San Francisco, CA, August, 2001.

215 Y. Zhang, M. Roughan, N. Duffield and A. Greenberg, Fast accurate computation of large-scale IP traffic matrices from link loads. *ACM Sigmetrics*, San Diego, CA, June, 2003.

216 Y. Zhang, M. Roughan, C. Lund and D. Donoho, An information theoretic approach to traffic matrix estimation. *ACM Sigcomm*, Karlsruhe, Germany, August, 2003.

217 Q. Zhao, Z. Ge, J. Wang and J. Xu, Robust traffic matrix estimation with imperfect information: making use of multiple data sources. *ACM Sigmetrics*, Saint Malo, France, 2006.

218 S. Ehlert, S. Petgang, T. Magedanz, and D. Sisalem, Analysis and Signature of Skype VoIP Session Traffic. *Communications, Internet, and Information Technology*, St. Thomas, USVI, USA.

Index

Printed in the United States
by Baker & Taylor Publisher Services